Bifurcation Theory, Mechanics and Physics

Mathematics and Its Applications

Bifurcation Theory, Mechanics and Physics

edited by

C. P. Bruter
Mathématiques, UER Sciences, Université de Paris 12, France

A. Aragnol
Université d'Aix Marseille, France

and

A. Lichnerowicz
Collège de France, Paris, France

D. Reidel Publishing Company

A MEMBER OF THE KLUWER ACADEMIC PUBLISHERS GROUP

Dordrecht / Boston / Lancaster

Library of Congress Cataloging in Publication Data

CIP

Main entry under title:

Bifurcation theory, mechanics and physics.

(Mathematics and its applications)
Includes index.
1. Bifurcation theory—Addresses, essays, lectures. 2. Differential equations, Partial—Addresses, essays, lectures. 3. Mechanics—Addresses, essays, lectures. 4. Physics—Addresses, essays, lectures. I. Bruter, Claude Paul. II. Aragnol, A., 1937– . III. Lichnerowicz, André, 1915– . IV. Series: Mathematics and its applications (D. Reidel Publishing Company)
QA374.B54 1983 515.3'53 83–11110
ISBN 90–277–1631–5

Published by D. Reidel Publishing Company
P.O. Box 17, 3300 AA Dordrecht, Holland

Sold and distributed in the U.S.A. and Canada
by Kluwer Academic Publishers,
190 Old Derby Street, Hingham, MA 02043, U.S.A.

In all other countries, sold and distributed
by Kluwer Academic Publishers Group,
P.O. Box 322, 3300 AH Dordrecht, Holland

Printed in The Netherlands

TABLE OF CONTENTS

FOREWORD

This volume presents the proceedings of a colloquium inspired by the former President of the French Mathematical Society, Michel Herve. The aim was to promote the development of mathematics through applications.

Since the ancient supports the new, it seemed appropriate to center the theoretical conferences on new subjects.

Since the world is movement and creation, the theoretical conferences were planned on mechanics (movement) and bifurcation theory (creation). Five aspects of mechanics were to be presented, but, unfortunately, it has not been possible to include the statistical mechanics aspect. So that only four aspects are presented:

- Classical mechanics (Hamiltonian, Lagrangian, Poisson) (W.M. Tulczyjew, J.E. White, C.M. Marle).
- Quantum mechanics (in particular the passage from the classical to the quantum approach and the problem of finding the explicit solution of Schrödinger's equation)(M. Cahen and S. Gutt, J. Leray).
- Fluid mechanics (meaning problems involving partial differential equations. One of the speakers we hoped would attend the conference was in Japan at the time, however his lecture is presented in these proceedings.) (J.F. Pommaret, H.W. Shi)
- Mathematical "information" theory (S. Guiasu)

Traditional physical arguments are characterized by their great homogeneity, and mathematically expressed by the compactness property. In such cases, there is a kind of duality between locality and globality, which allows the use of the infinitesimal in global considerations.

In the papers, infinitesimal methods appear through the use of infinitesimal operators (in particular differential forms and Lie groups), and through the use of Taylor's series expansion (jet bundles: the use of this language is the most convenient in the search for solutions of partial differential equations on Riemannian spaces). Global considerations appear through the use of global energy functions, and extremal or variational principles (see the

C. P. Bruter et al. (eds.), Bifurcation Theory, Mechanics and Physics, vii–x.

paper by L. Nirenberg). None of these principles is really well understood.

Three aspects of bifurcation theory are embraced:

- Bifurcation in ordinary differential equations (the lecturer centered his talk on his work on some special Hopf bifurcations. The results are included in the reviewed paper written by a colleague who could not attend the colloquium (W. Broer).
- Numerical methods in bifurcation theory tied to the Lyapunov-Schmidt procedure (J. Rappez).
- Bifurcation in partial differential equations involving the the Lyapunov-Schmidt procedure and singular theory (M. Golubitsky).

At this point, a feeling arises that group considerations play an increasing role in the study of bifurcation phenomena. These can be understood as the results of bifurcations of group (and pseudo-group) actions.

Monge's method, expanded by W. Shi, is used by W. H. Shi to treat a non-trivial example. The physical significance of this example is criticized by J. Leray; however the method remains strong despite this criticism. Besides I would like to point out that we are not always sure which mathematical formulation of a physical problem is the best.

This volume presents a large number of open problems concerning topological methods which are useful to show the influence of the topology of the space of solutions induced by the functional equations to be solved and the nature of the boundary conditions. The arising or vanishing of topological obstructions are obviously bound to shock and bifurcation phenomena. In any case, the Monge-Shi method has to be handled with care in order to take into account Borel's phenomenon (a well-chosen variational coefficient induces the non-analyticity of the unique solution of the partial differential equation with analytical data) or the turbulence phenomena. The problem already posed by small denominators in classical mechanics suggest that the role of number theory will increase in the study of refinements in bifurcation theory.

The last paper presented on bifurcation theory (M. Golubitsky) mainly concerns the Bénard problem. This paper is followed by an illustrative article (S. Fauve and A. Lichaber) on experiments showing turbulence phenomena and chaos. At that moment, the homogeneity of the physical state is somehow perturbed but through renormalization, a discrete modelization applies.

Numerical analysis can only use discrete models. Non-standard analysis can be seen as a convergence technique from the discrete to the non-discrete. The next paper (C. Lober and C. Reder) uses this recognition to solve some classical partial differential equations. Automata defined on finite sets (time excepted) can but admit periodic or quasi-periodic regimes. Extensions of such models through non-standard analysis might preserve some periodicities and be convenient models for some natural phenomena, including chemical systems.

The next paper (P. Dousson) is devoted to the mathematical study through quasi-autonomous "ordinary" differential or chemical systems satisfying Wei's axiomatic equations.

The volume concludes with papers (F.A. Grünbaum, M. Kleman, Y. Bouligand) on new mathematical applications to subjects which have recently been developed: tomography on the one hand, liquid crystals on the other hand. If tomography leads to the development of analysis, liquid crystals have given rise to the first useful applications of algebraic topology - through homotopy groups - to the study of physical structures. Liquid crystals appeal also to new mathematical studies in Euclidean, Riemannian, and particularly hyperbolic geometry.

The talk given on a use of catastrophe theory leading to a positive inhibition of hemophilia was not written because of health problems suffered by the author. This is regrettable since, for the first time, the treatment of a disease until now incurable, has been made possible through the use of mathematics. However, an audio-cassette, prepared by J.P. Duport, is available from him.

To end with, I would like to suggest the study of three physical problems, all related to morphogenesis.

(1) Study experimentally and mathematically the physical morphogenesis introduced by Leduc at the beginning of the century.

Note that in the study of biological morphology involving membranes, differential geometry based on surface metrics (Cartan's metrics) should have an advantage over differential geometry based on line metrics (Riemann's metrics).

(2) Study experimentally and mathematically the trajectories of air molecules in a real balloon inflated by blowing it up. (Of course, this problem can be refined by introducing different kinds of local constraints on the shape of the balloons, and by inflating the balloon in different ways.)

(3) Study experimentally and mathematically the evolution in a convex body of sound waves created by a tiny shock on the boundary or inside the body. (The problem interests not only acousticians, but also morphogenesists; think of the problem of fecondation and very early embryology.)

I would like to thank Professors André Aragnol and André Lichnerowicz for their warm help in the preparation of the Colloquium.

The Colloquium was financially supported by the Faculté des Sciences de Marseille-Luminy, the Université d'Aix-Marseille II, the Centre National de Recherche Scientifique, and the Direction de la Coopération et des Relations Internationales of the Ministère de l'Education Nationale. The writer and all the participants address their thanks to these organizations, and to the staff of the Centre International des Rencontres Mathématiques de Luminy, who did a magnificent job of organizing our meeting.

C. P. BRUTER

EDITOR'S PREFACE

Growing specialization and diversification have brought a host of monographs and textbooks on increasingly specialized topics. However, the "tree" of knowledge of mathematics and related fields does not grow only by putting forth new branches. It also happens, quite often in fact, that branches which were thought to be completely disparate are suddenly seen to be related.

Further, the kind and level of sophistication of mathematics applied in various sciences has changed drastically in recent years: measure theory is used (non-trivially) in regional and theoretical economics; algebraic geometry interacts with physics; the Minkowsky lemma, coding theory and the structure of water meet one another in packing and covering theory; quantum fields, crystal defects and mathematical programming profit from homotopy theory; Lie algebras are relevant to filtering; and prediction and electrical engineering can use Stein spaces. And in addition to this there are such new emerging disciplines as "completely integrable systems", "chaos, synergetics and large-scale order", which are almost impossible to fit into the existing classification schemes. They draw upon widely different sections of mathematics.

This program, Mathematics and Its Applications, is devoted to such (new) interrelations as exampla gratia:

- a central concept which plays an important role in several mathematical and/or scientific specialized areas;
- new applications of the results and ideas from one area of scientific endeavor into another;
- influences which the results, problems and concepts of one field of enquiry have and have had on the development of another.

The Mathematics and Its Applications programme tries to make available a careful selection of books which fit the philosophy outlined above. With such books, which are stimulating rather than definitive, intriguing rather than encyclopaedic, we hope to contribute something towards better communication among the practitioners in diversified fields.

C. P. Bruter et al. (eds.), Bifurcation Theory, Mechanics and Physics, xi–xii.

It used to be that physics (especially mechanics) and large parts of mathematics were inextricably intertwined. We have seen a period of separation and specialization in this respect. And now that some powerful new tools have been developed to a fine point (especially bifurcation theory, symplectic geometry and symmetry (group) ideas) they are again applied to mechanics - now enriched with problems of quantum mechanics. At the same time, these fields and a newcomer, experimental mathematics (or computer modelling), are raising fascinating mathematical questions and generating conjectures. The stated aim of the colloquium of which this collection of 19 survey papers constitutes the proceedings was "to promote the development of mathematics through applications" which is precisely one of the guiding principles of this book series.

The unreasonable effectiveness of mathematics in science ...

Eugene Wigner

Well, if you knows of a better 'ole, go to it.

Bruce Bairnsfather

What is now proved was once only imagined.

William Blake

As long as algebra and geometry proceeded along separate paths, their advance was slow and their applications limited. But when these sciences joined company they drew from each other fresh vitality anf thenceforward marched on at a rapid pace towards perfection.

Joseph Louis Lagrange

Amsterdam, April 1983 Michiel Hazewinkel

Wlodzimierz M. Tulczyjew

HAMILTONIAN, CANONICAL AND SYMPLECTIC FORMULATIONS OF DYNAMICS

1. INTRODUCTION

Dynamics of mechanical systems is traditionally formulated in terms of Hamiltonian vector fields [1]. Time dependent dynamics of a mechanical system is usually described by a time tependent Hamiltonian vector field or by a Poincaré-Cartan form. Recently two new formulations of dynamics have been proposed: the canonical formulation due to Lichnerowicz [3][5] and the symplectic formulation [2][6]. The canonical formulation emphasizes the importance of the Poisson structure of the phase space of a mechanical system and is related to a new approach to quantum mechanics proposed by Lichnerowicz [4]. The symplectic formulation based on the geometry of Lagrangian submanifolds of symplectic manifolds is important mainly because of its applications to relativistic dynamics [7] and because of the new interpretation of the Legendre transformation it provides [8].

The present lecture contains a review of the different formulations of time independent and time dependent dynamics of nonrelativistic mechanical systems.

2. HAMILTONIAN FORMULATION OF TIME INDEPENDENT DYNAMICS

Let (P,ω) be a symplectic manifold.

DEFINITION 2.1. A vector field

$$X:P \to TP \tag{2.1}$$

is said to be *Hamiltonian* if the form $X \lrcorner\, \omega$ is exact. A function

$$H:P \to \mathrm{R} \tag{2.2}$$

such that

$$X \lrcorner\, \omega = -dH \tag{2.3}$$

is called a *Hamiltonian* for X.

C. P. Bruter et al. (eds.), Bifurcation Theory, Mechanics and Physics, 1–11.

Nonrelativistic dynamics is usually formulated in terms of Hamiltonian vector fields. The symplectic manifold (P,ω) represents the phase space of a mechanical system and trajectories of the system are solution curves of a system of ordinary differential equations represented geometrically as a Hamiltonian vector field.

3. CANONICAL FORMULATION OF TIME INDEPENDENT DYNAMICS

Let (P,ω) be a symplectic manifold and let G be the unique 2-vector field on P satisfying

$$G \llcorner (u \lrcorner \omega) = u \tag{3.1}$$

for each vector u in P.

It is known that G satisfies the *Schouten bracket* condition

$$[G,G] = 0. \tag{3.2}$$

Consequently (P,G) is a *Poisson manifold* [3][5]. The Poisson bracket $\{f,g\}$ of two functions f and g on P is defined by

$$\{f,g\} = \langle G, df \wedge dg \rangle. \tag{3.3}$$

The *Jacobi identity*

$$\{f,\{g,h\}\} + \{g,\{h,f\}\} + \{h,\{f,g\}\} = 0 \tag{3.4}$$

is a consequence of (3.2).

Time independent dynamics can be formulated in terms of the Poisson structure. The Hamiltonian vector field X characterized by (2.3) is defined explicitely by

$$X = -G \llcorner dH. \tag{3.5}$$

This formulation of dynamics in terms of the Poisson structure is the canonical formulation prefered by Lichnerowicz. Although equivalent to the usual Hamiltonian formulation the canonical formulation suggests a different generalization to the time dependent case.

4. LAGRANGIAN SUBMANIFOLDS

DEFINITION 4.1. A *Lagrangian submanifold* of a symplectic

manifold (P,ω) is a submanifold $N \subset P$ such that $\omega|N = 0$ and $\dim N = \frac{1}{2}\dim P$.

Let M be a manifold of dimension m. Let θ_M denote the canonical 1-form defined on the cotangent bundle T^*M by

$$\langle u,\theta_M\rangle = \langle T\pi_M(u),\tau_{T^*M}(u)\rangle, \tag{4.1}$$

where u is an element of the tangent bundle TT^*M,

$$T\pi_M : TT^*M \to TM \tag{4.2}$$

is the tangent mapping of the cotangent bundle projection

$$\pi_M : T^*M \to M \tag{4.3}$$

and

$$\tau_{T^*M} : TT^*M \to T^*M \tag{4.4}$$

is the tangent bundle projection.

PROPOSITION 4.1. *The relation*

$$\mu^*\theta_M = \mu \tag{4.5}$$

holds for each 1-*form* $\mu : M \to T^*M$.

Proof. For each element u of TM we have

$$\begin{aligned} \langle u,\mu^*\theta_M\rangle &= \langle T\mu(u),\theta_M\rangle \\ &= \langle T\pi_M(T\mu(u)),\tau_{T^*M}(T\mu(u))\rangle \\ &= \langle u,\mu(\tau_M(u))\rangle \\ &= \langle u,\mu\rangle . \end{aligned}$$

Hence, $\mu^*\theta_M = \mu$.

As is well known $(T^*M, d\theta_M)$ is a symplectic manifold. We denote the symplectic form $d\theta_M$ by ω_M.

PROPOSITION 4.2. *Let* $F : M \to \mathrm{R}$ *be a differentiable function. The image* $N = \mathrm{im}(dF)$ *of the differential* $dF : M \to T^*M$ *is a Lagrangian submanifold of* (T^*M,ω_M).

Proof. The condition $\dim N = \frac{1}{2}\dim T^*M$ is obviously satisfied and $\omega_M|N = 0$ is equivalent to $(dF)^*\omega_M = 0$. From Proposition 4.1 we deduce

$$(dF)^*\omega_M = (dF)^*d\theta_M = d(dF)^*\theta_M = ddF = 0.$$

Hence, N is a Lagrangian submanifold.

DEFINITION 4.2. The Lagrangian submanifold $N = \mathrm{im}\ (dF)$ is said to be *generated* by F and F is called a *generating function* of N.

The Lagrangian submanifold generated by a function F is the subset of T^*M on which the equality

$$\theta_M = d(F \circ \pi_M) \tag{4.6}$$

holds.

PROPOSITION 4.3. *Let C be a submanifold of M. The set*

$$N = \{p \in T^*M;\ x = \pi_M(p) \in C,\ \langle v,p\rangle = 0 \text{ for each } v \text{ in } T_xC\} \tag{4.7}$$

is a Lagrangian submanifold of (T^*M,ω_M).

Proof. The set N is obviously a submanifold of T^*M of dimension equal to dim M. If w is a vector tangent to N then

$$\langle w,\theta_M\rangle = \langle T\pi_M(w),\tau_{T^*M}(w)\rangle = 0$$

because $T\pi_M(w)$ is tangent to C and $\tau_{T^*M}(w)$ belongs to N. It follows that $\theta_M|N = 0$. Hence, $\omega_M|N = 0$.

5. SYMPLECTIC FORMULATION OF TIME INDEPENDENT DYNAMICS

Let $(P,\ )$ be a symplectic manifold. The mapping

$$\beta: TP \to T^*P: u \mapsto u \lrcorner\, \omega \tag{5.1}$$

is a vector bundle isomorphism. The cotangent bundle T^*P has a canonical symplectic structure independent of ω. We denote by θ_P the canonical 1-form on T^*P, and by ω_P the symplectic form $d\theta_P$. Let

$$\chi = \beta^*\theta_P \tag{5.2}$$

and

$$\rho = d\chi = \beta^*\omega_P. \tag{5.3}$$

The pair (TP,ρ) is a symplectic manifold and β is a symplectomorphism.

PROPOSITION 5.1. *The image* $\dot{D} = \mathrm{im}\, X$ *of a Hamiltonian vector field* $X:P \to TP$ *is a Lagrangian submanifold of* (TP,ρ).

Proof. Let $H:P \to \mathrm{R}$ be a Hamiltonian for X. Then $X = -G \llcorner dH = \beta^{-1}(-dH)$. Hence, $\dot{D}$ is a Lagrangian submanifold since it is the inverse image by the symplectomorphism β of the Lagrangian submanifold of (T^*P,ω_P) generated by $-H$.

Symplectic formulations of physical theories consist in representing the constitutive equations or the dynamical equations of physical systems as Lagrangian submanifolds of suitable symplectic manifolds.

Let (P,ω) be the phase space of a mechanical system.

DEFINITION 5.1. A *dynamical system* in (P,ω) is a Lagrangian submanifold $\dot{D}$ of the symplectic manifold (TP,ρ).

The image of a Hamiltonian vector field is a special case of a dynamical system. Other examples are encountered in relativistic dynamics and in time dependent dynamics.

6. HAMILTONIAN FORMULATION OF TIME DEPENDENT DYNAMICS [1]

Let (P,ω) be a symplectic manifold.

DEFINITION 6.1. A *time dependent vector field* on P is a mapping

$$X:P \times \mathrm{R} \to TP \tag{6.1}$$

such that for each t in R the mapping

$$X_t:P \to TP:p \mapsto X(p,t) \tag{6.2}$$

is a vector field.

DEFINITION 6.2. A time dependent vector field X is said to be *Hamiltonian* if for each t the vector field X_t is Hamiltonian. A function

$$H:P \times \mathrm{R} \to \mathrm{R} \tag{6.3}$$

such that for each t the function

$$H_t:P \to \mathrm{R}:p \mapsto H(p,t) \tag{6.4}$$

is a Hamiltonian for X_t is called a *time dependent Hamiltonian* for X.

We denote by

$$pr_1 : P \times \mathrm{R} \to P \tag{6.5}$$

and

$$t : P \times \mathrm{R} \to \mathrm{R} \tag{6.6}$$

the canonical projections. A vector field $\frac{\partial}{\partial t}$ is defined on $P \times \mathrm{R}$ by

$$\langle \frac{\partial}{\partial t}, d(f \circ pr_1) \rangle = 0 \tag{6.7}$$

for each function f on P, and

$$\langle \frac{\partial}{\partial t}, dt \rangle = 1. \tag{6.8}$$

Given a time dependent vector field X we define a field $\overline{X}$ on $P \times \mathrm{R}$ tangent to fibres of the projection pr_1 by requiring that the restriction $\overline{X}_t$ of $\overline{X}$ to the fibre over t be equal to X_t.

DEFINITION 6.3. The vector field

$$\tilde{X} : P \times \mathrm{R} \to T(P \times \mathrm{R}) \tag{6.9}$$

defined by

$$\tilde{X} = \overline{X} + \frac{\partial}{\partial t} \tag{6.10}$$

is called the *suspension* of the time dependent vector field X.

Time dependent dynamics is formulated in terms of time dependent Hamiltonian vector fields and their suspensions.

Let (P, ω) represent the phase space of a mechanical system. Let H be a time dependent Hamiltonian and X the associated time dependent Hamiltonian vector field. Trajectories of the system in the phase-time space $P \times \mathrm{R}$ are integral curves of the suspension $\tilde{X}$ of X. These trajectories are parametrized by time. They can be projected to P without any loss of information. Alternately we can disregard the parametrization and consider integral manifolds of the distribution $\tilde{D}$ on $P \times \mathrm{R}$ spanned by $\tilde{X}$.

7. THE POINCARE-CARTAN FORM

Let (P, ω) be the phase space of a mechanical system. We denote by

$$pr_1 : P \times \mathrm{R} \to P \tag{7.1}$$

and

$$t : P \times \mathrm{R} \to \mathrm{R} \tag{7.2}$$

the canonical projections of $P \times \mathrm{R}$ onto P and R. Let H be a time dependent Hamiltonian. We define a 2-form

$$\Omega = pr_1{}^*\omega - dH \wedge dt \tag{7.3}$$

on $P \times \mathrm{R}$. This form is closed. If P is the cotangent bundle T^*Q of a configuration manifold Q and ω is the differential of the canonical 1-form θ_Q on T^*Q then Ω is exact:

$$\Omega = d(pr_1{}^*\theta_Q - Hdt). \tag{7.4}$$

DEFINITION 7.1. The 1-form

$$\Theta = pr_1{}^*\theta_Q - Hdt \tag{7.5}$$

is called the *Poincaré-Cartan form.*

The form is degenerate. Trajectories of the mechanical system in P R are the integral manifolds of the characteristic distribution

$$\tilde{D} = \{u \in T(P \times \mathrm{R});\ \langle u \wedge v, \Omega\rangle = 0 \text{ for each } v \text{ in } T(P \times \mathrm{R}) \text{ such that } \tau_{P\times\mathrm{R}}(v) = \tau_{P\times\mathrm{R}}(u)\}. \tag{7.6}$$

The distribution $\tilde{D}$ is of dimension 1. It is spanned by the vector field

$$\tilde{X} : P \times \mathrm{R} \to T(P \times \mathrm{R}) \tag{7.7}$$

characterized by

$$\mathrm{im}\ \tilde{X} \subset \tilde{D} \tag{7.8}$$

and

$$\langle \tilde{X}, dt\rangle = 1. \tag{7.9}$$

Trajectories of the system, when properly parametrized, become integral curves of $\tilde{X}$ and can be projected to P. The vector field $\tilde{X}$ is the suspension of the time dependent vector field

$$X = Tpr_1 \circ \tilde{X}. \tag{7.10}$$

8. CANONICAL FORMULATION OF TIME DEPENDENT DYNAMICS

Let (P,ω) be a symplectic manifold. As we have seen in Section 3, P has a canonical Poisson structure represented by the 2-vector field G defined in (3.1). As in the preceding section we denote by pr_1 and t the canonical projections of $P \times \mathrm{R}$ onto P and R respectively. The product manifold $P \times \mathrm{R}$ is not a symplectic manifold. There is however a canonical Poisson structure on $P \times \mathrm{R}$ represented by the 2-vector field $\tilde{G}$ defined by

$$<\tilde{G}, d(f \circ pr_1) \wedge d(g \circ pr_1)> = <G, df \wedge dg> \circ pr_1 \tag{8.1}$$

for arbitrary functions f and g on P, and

$$<\tilde{G}, dh \wedge dt> = 0 \tag{8.2}$$

for any function h on $P \times \mathrm{R}$.

The formulation of time dependent dynamics in terms of $\tilde{G}$ is specially simple. Let (P,ω) be the phase space of a system and let H be a time dependent Hamiltonian. Time parametrized trajectories of the system in $P \times \mathrm{R}$ are obtained as integral curves of the vector field

$$\tilde{X} = \bar{X} + \frac{\partial}{\partial t} , \tag{8.3}$$

where

$$\bar{X} = -\tilde{G} \llcorner dH. \tag{8.4}$$

9. SYMPLECTIC FORMULATION OF TIME DEPENDENT DYNAMICS

Let (P,ω) be a symplectic manifold. The product $P \times T^*\mathrm{R}$ is a symplectic manifold with a symplectic form μ defined by

$$\mu = \pi_1{}^*\omega - \pi_2{}^*\omega_{\mathrm{R}} , \tag{9.1}$$

where

$$\pi_1 : P \times T^*\mathrm{R} \to P \tag{9.2}$$

and

$$\pi_2 : P \times T^*\mathrm{R} \to T^*\mathrm{R} \tag{9.3}$$

are canonical projections, and ω_R is the canonical 2-form on $T^*\mathrm{R}$. The cotangent bundle $T^*\mathrm{R}$ is isomorphic to $\mathrm{R}\times\mathrm{R}$. If

$$\tau : T^*\mathrm{R} \to \mathrm{R} \tag{9.4}$$

is the cotangent bundle projection and

$$\varepsilon : T^*\mathrm{R} \to \mathrm{R} \tag{9.5}$$

is the second canonical projection then

$$\omega_\mathrm{R} = d\varepsilon \wedge d\tau. \tag{9.6}$$

Dynamics can be formulated in terms of a Lagrangian submanifold of the symplectic manifold $(T(P\times T^*\mathrm{R}),\sigma)$. The symplectic form σ is defined by

$$\sigma = \gamma^*\omega_{P\times T^*\mathrm{R}} \,, \tag{9.7}$$

where $\omega_{P\times T^*\mathrm{R}}$ is the canonical 2-form on $T^*(P\times T^*\mathrm{R})$ and γ is the mapping

$$\gamma : T(P\times T^*\mathrm{R}) \to T^*(P\times T^*\mathrm{R}) : u \mapsto u \,\lrcorner\, \mu. \tag{9.8}$$

Let $H : P\times\mathrm{R} \to \mathrm{R}$ be a time dependent Hamiltonian of a mechanical system. Identifying $T^*\mathrm{R}$ with $\mathrm{R}\times\mathrm{R}$ we define a section κ of the fibration

$$\eta = id_P \times \tau : P\times T^*\mathrm{R} \to P\times\mathrm{R} \tag{9.9}$$

by

$$\kappa : P\times\mathrm{R} \to P\times T^*\mathrm{R} : (p,t) \mapsto (p,t,H(p,t)). \tag{9.10}$$

The submanifold

$$K = \mathrm{im}\,\kappa \tag{9.11}$$

of the phase-time-energy space $P \times T^*\mathrm{R}$ is called the energy hypersurface. It follows from Proposition 4.3 that the set

$$Q = \{w \in T^*(P \times T^*\mathrm{R});\ r = \pi_{P \times T^*\mathrm{R}}(w) \in K,$$
$$\langle v, r \rangle = 0 \text{ for each } v \text{ in } T_r K\} \qquad (9.12)$$

is a Lagrangian submanifold of $(T^*(P \times T^*\mathrm{R}), \omega_{P \times T^*\mathrm{R}})$. Consequently

$$\dot{E} = \gamma^{-1}(Q) \qquad (9.13)$$

is a Lagrangian submanifold of $(T(P \times T^*\mathrm{R}), \ )$. Being a submanifold of the tangent bundle, $\dot{E}$ can be interpreted as a system of ordinary differential equations on $P \times T^*\mathrm{R}$. Solution curves of this system are trajectories of the mechanical system in the phase-time-energy space $P \times T^*\mathrm{R}$.

Department of Mathematics and Statistics
The University of Calgary

REFERENCES

[1] Abraham, R. and Marsden, J.E.: 1978, Foundations of Mechanics, Benjamin-Cummings.

[2] Kijowski, J. and Tulczyjew, W.M.: 1978, A symplectic framework for field theories, Lecture notes in Physics 107, Springer-Verlag, New York.

[3] Lichnerowicz, A.: 1976, 'Variétés symplectiques, variétés canoniques et systèmes dynamiques', in Topics in Differential Geometry, Academic Press, New York, pp. 57-85.

[4] Lichnerowicz, A.: 1982, 'Deformations and quantization', in Dynamical Systems and Microphysics, Academic Press, New York, pp. 27-60.

[5] Marle, C.-M.: 1982, 'Lie group actions on Poisson and canonical manifolds', in Dynamical Systems and Microphysics, Academic Press, New York, pp. 61-73.

[6] Tulczyjew, W.M.: 1974, 'Hamiltonian systems, Lagrangian systems and the Legendre transformation', Symposia Mathematica 16, 247-258.

[7] Tulczyjew, W.M.: 1977, 'A symplectic formulation of relativistic particle dynamics', Acta Phys. Polon B8, 431-477.

[8] Tulczyjew, W.M.: 1977, 'The Legendre transformation', Ann. Inst. H. Poincaré 27, 101-114.

J. Enrico White

ITERATED TANGENTS AND LAGRANGIAN DYNAMICS

INTRODUCTION

Classically, a mechanical system is determined by the data of a Lagrangian $L: T_1(M) \times R \to R$, or of its Legendre Transform a Hamiltonian $H: T^1(M) \times R \to R$. Here, M is a smooth manifold, $T_1(M)$ is the tangent space and $T^1(M)$ the cotangent space. M is referred to as the (finite-dimensional) manifold of configurations. In local coordinates on M the Euler-Lagrange equations: $\frac{d}{dt}\left[\frac{\partial L}{\partial v}\right] = \frac{\partial L}{\partial x}$ or respectively, Hamilton's equations: $\frac{dx}{dt} = \frac{\partial H}{\partial p}$ and $\frac{dp}{dt} = -\frac{\partial H}{\partial x}$ then determine the dynamics. These pre-relativistic formalisms are traditionally derived, and their invariance interpreted, from a variational principle such as Hamilton's Least Action Principle.

Here we develop a derivation of these formalisms from a more special principle called the Generalized Energy Gradient Principle. It is more special than the Least Action Principle in that it admits naturally Lagrangians of the form: (in local coordinates) $L(x,v,t) = T(x,v,t) + w(x,v,t) - V(x,t)$ with T homogeneous quadratic in velocity v, w linear in velocity v, and V a time-dependent potential, assumed independent of velocity. In exchange for the restriction in generality on the Lagrangian it may offer a certain advantage in explicating the classical formalism. It derives all of the dynamics directly from the data of a Riemannian metric on $M \times R$ giving an easy deduction of the "naturality" with respect to coordinate changes and "holonomic" constraints from the corresponding geometric naturality properties of the metric. In addition, it gives a clear picture of the relation of the Lagrangian dynamics with the dynamics of a special, standard example: the geodesic flow in the tangent space of a Riemannian manifold.

C. P. Bruter et al. (eds.), Bifurcation Theory, Mechanics and Physics, 13–46.

Before proceeding to the body of the paper we sketch for sake of illustration and motivation the application of a simpler principle, the Energy Gradient Principle to this last example, the geodesic flow. The relation between the Energy Gradient and the Generalized Energy Gradient principles will be explained later.

Here and in the sequel we use the word metric to refer to a symmetric, covariant 2-tensor field and refer to the nondegenerate case as a regular metric. This liberty with the language will simplify the form of several statements. We make no general assumptions about the signature of regular metrics.

EXAMPLE 1: The geodesic flow

Suppose that M is a smooth finite-dimensional manifold with a metric tensor G. There is a metric on the tangent space $T_1(M)$ called the promotion of the metric G and denoted δG. The association $[M,G] \rightarrow [T_1(M), \delta G]$ has the following properties:

a) δG is regular if G is regular
b) if regular, δG is never definite
c) if f: $M \rightarrow N$ is smooth then

$$[T_1(f)]^*(\delta G) = \delta[f^*(G)] \qquad [1]$$

where $T_1(f)$ is the tangent map, and "*" denotes pull-back. This promoted metric has been referred to as the "Metric II" by [Yano and Ishihara] and its properties as the promotion of a sectorform field are developed in [White].

Now associate with the Riemannian manifold [M,G] (so G is assumed regular) the kinetic energy function $T_G(V) = \frac{1}{2} G\langle V,V\rangle$ for V in $T_1(M)$. Then the Energy Gradient Principle is the following:

The geodesic flow in $T_1(M)$ associated with regular metric G is given by the following infinitesimal generator. It is the gradient of the kinetic energy T_G with respect to the promoted metric δG. That is, if V in $T_1(M)$ and $X|_V$ is the tangent vector at V determined by the infinitesimal generator of this flow then for $Y|_V$ an arbitrary tangent vector at V we have:

$$\underline{\delta G\langle X|_V , Y|_V \rangle = d(T_G)|_V \langle Y|_V \rangle} \qquad [2]$$

The "naturality properties" of the geodesic flow with respect to coordinate changes on M and with respect to smooth immersions (when they pull G back to a regular metric) then follow easily from equations [1] and [2].

#

In this example, the dynamics are entirely determined by the metric on M. And it is natural to raise two questions:

(a) If conservative "forces" are introduced via a potential on M, does a gradient principle still apply, and if so, can it be formulated in terms of the data of a single metric?

(b) If the metric itself, and the forces are allowed to depend on the time, what is the appropriate generalization of this principle?

The answer to question (a) is yes to both parts, and to question (b) the Generalized Energy Gradient Principle will give the appropriate generalization; it will lead to the symplectic formalism via the Cartan form on $T^1(M) \times R$. Our emphasis here will be on the Lagrangian rather than the Hamiltonian formulation of the dynamics, however.

The remainder of the paper will be devoted to three things. First, we sketch the machinery needed from the method of iterated tangents, particularly the notion of a sectorform field, that can justify equation [1] and give meaning to the idea of the promoted metric. This is developed in much greater detail in [White] along with various applications in differential geometry. Next, we state the Generalized Energy Gradient Principle which can be taken as a starting point for the deduction of the Euler-Lagrange equations (in the case that the Lagrangian has the special form mentioned above). From this we derive the E-L equations and give the answers to questions (a) and (b), and show in what sense the Energy Gradient Principle invoked above is a special case. Finally, we show how to pass from this viewpoint to the symplectic formulation as developed for example in [Souriau] or in [Abraham & Marsden], but we do this using the calculus of sectorforms and without the detour to the cotangent bundle.

PART 1: PROMOTING THE METRIC

The Canonical Simplicial Fiber Bundle Functor associates to each smooth finite-dimensional manifold M a sequence of "natural fiber bundles":

$$\begin{array}{ccccc} T_1(M) & T_2(M) & T_3(M) & \ldots\ldots & T_k(M) \\ \downarrow q_1 & \downarrow q_2 & \downarrow q_3 & & \downarrow q_k \ \ldots\ldots \\ M & M & M & & M \end{array}$$

where $T_k(M)$ has the interpretation of being $T_1(T_1(\ldots(T_1(M)\ldots))$ (k times).

[Many of the algebraic properties of this functor are developed (for the first time to the author's knowledge) by [Tulczyjew] in a somewhat more general setting than they are discussed here and in [White]. The author wishes to thank Claude-Paul Bruter for bringing this to his attention.]

We extend the definition to $T_0(M) = M$. For each $k \geq 1$ there is a set of k bundle maps $\{D^i(k) \mid 1 \leq i \leq k\}$ with $D^1(1)$ the tangent projection,

$$\begin{array}{ccc} T_k(M) & \xrightarrow{D^i(k)} & T_{k-1}(M) \\ \downarrow q_k & & \downarrow q_{k-1} \\ M & \xrightarrow{id} & M \end{array}$$

Such that each $D^i(k)$ is the projection of a vector bundle. These bundle maps are natural, that is the association which corresponds each M to a simplicial fiber bundle (ignoring degeneracy operators) is a functor. Therefore, the fiber of each $D^i(k)$ has a natural vector space structure over each point in M. That vector space structure is intrinsic giving an extension of the linear structure of the tangent space over each point. The elements of these fiber spaces $T_k(M)$ are called k-sectors or sectors. For the case $k = 1$ they are, of course, tangent vectors; and for larger k they have several interpretations. For example, at each point x in M, they can be thought of as the "elements" or the infinitesimal generators of a system of k commuting flows. We give a quick description of the structure of $T_k(M)$ showing that its elements can be thought of as generalized tensors, or higher-order geometric objects. Fore more details, see [White].

Some notational preliminaries:

Let E represent R^m with a fixed set of linear coordinate functions $(x^1, x^2, \ldots, x^m)$ together with its dual, a fixed basis $(e_1, e_2, \ldots, e_m)$ for R^m. Also let E' represent R^n with linear coordinate functions $(y^1, y^2, \ldots, y^n)$ and dual basis $(c_1, c_2, \ldots, c_n)$. Let M be a smooth m-dimensional manifold, and N a smooth n-dimensional manifold. M is covered with open sets U_a with for each a, an open set 0_a of E and a local homeomorphism $\Phi_a : 0_a \to U_a$. We call Φ_a a local frame and the corresponding Φ_a^{-1} a local chart. We assume the maps $\Phi_b^{-1} o \Phi_a$ are smooth (hence diffeomorphisms) on their domains of definition. Finally, a map f: M → N is smooth if Φ_b^{-1} of $o\Phi_a$ is smooth on its domain of definition.

Let M and N be smooth manifolds. If f:M → N is smooth and $x_o \in M$ with $f(x_o) = y_o$ write f: $(M, x_o) \to (N, y_o)$. The smooth germ of f at x_o is the equivalence class of smooth maps (M, x_o) to (N, y_o) which agree with f in some neighborhood of x_o. Denote the germ of f at x_o with the same symbol f.

Now suppose that $f:(M, x_o) \to (N, y_o)$ is a smooth germ at x_o. The k-jet of f, denoted $f_k : (M, x_o) \to (N, y_o)$ is the equivalence class of germs $(M, x_o) \to (N, y_o)$ with respect to the following equivalence relation:

say that f is k-equivalent to g (denoted $f \underset{k}{\sim} g$) if for θ: $(E', 0) \to (N, y_o)$ and Φ: $(E, 0) \to (M, x_o)$ all partial derivates $\left.\frac{\partial^{/a/}(\Theta^{-1} o\ f o \Phi)^j}{\partial x^a}\right|_0$ are equal to $\left.\frac{\partial^{/a/}(\Theta^{-1} o\ g o \Phi)^j}{\partial x^a}\right|_0$

where $a = (a_1, a_2, \ldots, a_m)$, a_i nonnegative integers, $\sum_{i=1}^{m} a_i = /a/ \leq k$, and $1 \leq j \leq n = \dim N$, and let a! be the integer $(a_1!)(a_2!)\ldots(a_m!)$.

It happens that this definition of k-equivalence is independent of the frame germs. Also, there is a well-defined composition of k-jets: $f_k o g_k = (fog)_k$ when f and g are composable germs. In particular, $f:(E,0) \to (E',0)$ has k-jet represented by the Taylor polynomial whose j^{th} component function is:

$$\sum_{/a/ \leq k} \left(\frac{1}{a!}\right) \frac{\partial^{/a/} f^j}{\partial x^a}\Big|_0 (x^1)^{a_1} (x^2)^{a_2} \ldots\ldots (x^m)^{a_m} =$$

$$\sum_{/a/ \leq k} \left(\frac{1}{a!}\right) \frac{\partial^{/a/} f^j}{\partial x^a}\Big|_0 x^a$$

Bundle of k-frames over M

Let $G_k(m)$ be the group of invertible k-jets of smooth mappings $(E,0) \to (E,0)$ and denote an element of $G_k(m)$ g_k. Multiplication is computed by truncated composition of representative polynomials. Note that a k-jet is invertible if and only if its 1-jet is. For each x_0 in M let $F_k^{x_o}$ be the set of invertible k-jets $(E,0) \to (M,x_0)$. Then $G_k(m)$ acts on the right on $F_k^{x_o}$ by the rule $\Phi_k \cdot g_k = (\Phi \circ g)_k$ and this action is simply transitive.

Now let $F_k(M)$ be the set $\bigcup_{x_0 \in M} F_k^{x_o}$. There is an obvious right action of $G_k(m)$ on $F_k(M)$ which takes fibers to themselves over the projection $F_k(M) \to M$. If $F_k(M)$ is locally trivialized using frames $\Phi: W \to U \subset M$ (W open in E) in the following way: for t_{z_o} the translation $z \to z_0 + z$ in E, map $G_k(M) \times U \to F_k(M)$ by $(g_k,x) \to (\Phi \circ t_{\Phi^{-1}(x)} \circ g)_k$ then $F_k(M)$ inherits a smooth manifold structure for which the projection to M is a bundle projection, in fact, the projection of a principle $G_k(m)$ bundle. This bundle is the bundle of k-frames over M denoted

$$\begin{array}{c} F_k(M) \\ \downarrow \\ M \end{array}$$

The tensor bundles are associated with the 1-frame bundles in the same way that the k-sector bundles are associated with the k-frame bundles. We make this idea precise below.

......

The sector bundles $T_k(M)$

Let V be a finite-dimensional vector space on which $G_k(m)$ acts smoothly on the left. Here, we do not assume the action to be linear. Form the product $F_k(M) \times V$, and let $G_k(m)$ act on $F_k(M) \times V$ on the left by the rule: $g_k.(\Phi_k, v) = (\Phi_k g_k^{-1}, g_k.v)$. Denote the set of orbits of this action

$F_k(M) \times_{G_k(m)} V$. This is a smooth manifold with the quotient topology and projects on M by the projection of a fiber bundle with fiber V. Denote the orbit (equivalence class) of (Φ_k, v) : $[\Phi_k, v]$.

With this setup, the following equation may be interpreted as a "transformation law" $[\Phi_k, g_k.v] = [\Phi_k.g_k, v]$. Thus an orbit such as $[\Phi_k, v]$ may be thought of as a generalized tensor at the target of Φ_k in M. The assignment to each local frame $\Phi : U \subset M$ of a smooth function $F: W \to V$ (satisfying the obvious compatibility conditions on overlaps) gives a cross-section of the bundle $F_k(M) \times_{G_k(m)} V$.

This defines a (local) field by the rule $x \to [(\Phi \circ t_{\Phi^{-1}(x)})_k, F(\Phi^{-1}(x))]$, for $x \in M$.

EXAMPLE 2: The tangent bundle

Here let V = E (as above) and represent an element of E as an mxl matrix (X^i), $1 \leq i \leq m$, of components with respect to the standard basis (e_i). Define left action of $G_1(m) = Gl(m)$ on E by:

$$g_1.(X^i) = \left(\frac{\partial g^j}{\partial x^i}\bigg|_0 X^i \right)$$

(summation convention).

This defines the tangent bundle on M, denoted $\begin{matrix} T_1(M). \\ \downarrow q_1 \\ M \end{matrix}$

Here each fiber has the structure of a vector space since the $G_1(m)$ action was linear on V = E.

The definition of the k-sector bundle:

$T_k(M)$ follows the pattern. Let the standard k-1 simplex
$\downarrow q_k$
M

be the collection of nonempty subsets of the set $\{1,2,\ldots,k\}$, An i face of the simplex (for $0 \leq i \leq k-1$) is a subset with i+1 elements. We denote the set of faces of the standard k-1 simplex: $\Gamma(k-1)$, thus $\Gamma(k-1)$ has $2^k - 1$ elements. In general, we shall designate an element of $\Gamma(k-1)$ as an ordered k-tuple $(a_1,a_2,\ldots,a_k)$ with each a_j equal to 0 or 1 ($a_j = 1$ if j is included). Note that the tuple $(0,0,\ldots,0)$ does not occur.

Now define E_k to be the $m(2^k-1)$ dimensional vector space of maps from $\Gamma(k-1) \to E$. Then $\gamma \in \Gamma(k-1)$ is a sequence $(\gamma_1, \gamma_2,\ldots,\gamma_k)$ as above and we may denote an element of E_k as a "matrix" (X^i_γ) where for each $\gamma \in \Gamma(k-1)$ $X^i_\gamma e_i$ is the image of γ.

Now letting $F = R^k$ with standard linear coordinates $(t^1,t^2,\ldots,t^k)$ we may define the following equivalence relation on k-jets of maps $(F,0) \to (M,x_o)$. Say that two jets f_k and h_k : $(F,0) \to (M,x_o)$ are F-related if for some frame-jet Φ_k: $(E,0) \to (M,x_o)$ the Taylor polynomial representatives of $\Phi_k^{-1} \circ f_k$ and $\Phi_k^{-1} \circ h_k$ have the same coefficients for the monomials $t^\gamma = (t^1)^{\gamma_1} \ldots . (t^k)^{\gamma_k}$ for all $\gamma \in \Gamma(k-1)$ (hence $\gamma_i = 0$ or 1, not all 0). It is a straightforward exercise in polynomial algebra to show that this definition is independent of frame-jet Φ (See also [Tulczyjew]).

The k-sectors will be the equivalence classes of k-jets $(F,0) \to M$ with respect to this equivalence relation. Note that for k = 1, 1-jets are F-related if and only if they are equal. This is not true, of course, for larger k. Now suppose that Φ_k is a k-frame at x_o in M. Then for each element (X^i_γ) in E_K we define an F-related class of jets from $(F,0) \to (M,x_o)$ in this way. Let f_k: $(F,0) \to (E,0)$ by

$$(t^1,t^2,\ldots t^k) \to \sum_{\gamma\in\Gamma(k-1)} X^i_\gamma t^\gamma.$$ We shall in fact

adapt the summation convention to represent this polynomial $X^i_\gamma\ t^\gamma$. Then the F-related class of $\Phi_k o f_k\ : (F,0) \to (M,x_o)$ is uniquely determined by the data of (X^i_γ) and Φ_k.

Now define the following left action of $G_k(m)$ on E_k. Each element (X^i_γ) in E_k may be thought of as an F-related class of k-jets $(F,0) \to (E,0)$ by the association above (that corresponded with (X^i_γ) the polynomial f_k). Then for g_k in $G_k(m)$ define $g_k.(X^i_\gamma)$ to be (Y^i_γ) where $Y^i_\gamma\ t^\gamma$ is in the F-related class of $(g\ o\ f)_k$. This clearly defines a left action of $G_k(m)$ on E_k. We may thus follow the procedure described above to construct the bundle $F_k(M) \times_{G_k(m)} E_k$ and this bundle we identify with $T_k(M)$.

$$F_k(M) \times_{G_k(m)} E_k \xrightarrow{} \qquad T_k(M)$$

$$\downarrow q_k \qquad\qquad \downarrow q_k$$

$$M \qquad\qquad M$$

Thus a k-sector at x_o in M is a class $[\Phi_k,\ X^i_\gamma]$ with x_o the target of frame-jet Φ_k. And if f_k is the polynomial mapping (F,0) to (E,0): $X^i_\gamma\ t^\gamma$ then it is the F-related class of $\Phi_k o\ f_k$.

This last statement makes sense since the "transformation law" $[\Phi_k g_k,\ (X^i_\gamma)] = [\Phi_k, g_k.(X^i_\gamma)]$ simple expresses the commutativity of the following diagram of mappings of F-related classes:

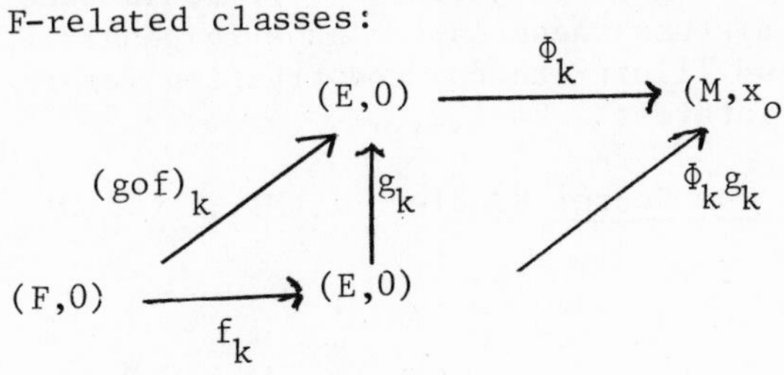

Finally, the association $M \to \begin{array}{c} T_k(M) \\ \downarrow q_k \\ M \end{array}$ is extended to a functor by the following consideration. If $h: M \to N$ is a smooth map of manifolds and if $[\Phi_k, (X^i_\gamma)]$ is a k-sector at x_o in M, then let f_k be the associated F-related class: $(F,0) \to (M,x_o)$. Then it is again not difficult to see that the F-related class of $(h\circ f)_k$ is well-defined (the reasoning is the same as that used to show that F-relatedness is a well-defined relation, that is, independent of frame). This gives a bundle map and the desired functoriality.

$$\begin{array}{ccc} T_k(M) & \xrightarrow{T_k(h)} & T_k(N) \\ \downarrow q_k & & \downarrow q_k \\ M & \xrightarrow[h]{} & N \end{array}$$

All of this construction pretty much generalizes that of the tangent bundle functor. But it is clear that for $k > 1$, the projections q_k are not the projections of vector bundles. It is shown in [White] that the spaces $T_k(M)$ can be naturally identified with $\underbrace{T_1(\ldots T_1(M)..)}_{k \text{ times}}$ and therefore that $T_k(M)$ can be identified with the tangent space $T_1[T_{k-1}(M)]$. There are k different such identifications and these give rise to the boundary operators $D^i(k) : T_k(M) \to T_{k-1}(M)$, $1 \leq i \leq k$. The $D^i(k)$ are essentially tangent projections. It would take us too far afield to discuss these latter maps in generality. Instead, we illustrate the construction for the cases of immediate interest: $k = 1,2,3$.

<u>EXAMPLE 3</u>: <u>The Sector Bundles</u> $\begin{array}{c} T_1(M) \\ \downarrow q_1 \\ M \end{array}$, $\begin{array}{c} T_2(M) \\ \downarrow q_2 \\ M \end{array}$, $\begin{array}{c} T_3(M) \\ \downarrow q_3 \\ M \end{array}$

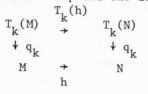

a) $\begin{array}{c} T_1(M) \\ \downarrow q_1 \\ M \end{array}$ is, as we saw in EXAMPLE 2, the tangent bundle.

We make the natural identifications using the fact that E_1 may be identified with E, and that for $F = R^1$, two 1-jets are F-related if and only if they are equal.

b) Suppose k = 2. Then $\gamma(1)$, the standard 1-simplex, is the set of nonempty subsets of $\{1,2\}$. Denote an element of E_2: X^i ___ A^r ___ Y^j, $1 \leq i,j,r \leq m$.

Thus, E_2 is isomorphic with E^3, and the "position" of the component corresponds to the "face" of the simplex. Suppose that $g_2 \in G_2(m)$, and as 2-jet from $(E,0) \to (E,0)$ it has Taylor polynomial representative:

$$\sum_{/a/ \leq 2} \left[\frac{1}{a!}\right] \frac{\partial^{/a/} g^i}{\partial x^a} x^a$$

Define the left action of $G_2(m)$ on E_2 then by the formula:

$$g_2 \cdot (X^i \_\_\_ A^r \_\_\_ Y^j) = \left(\frac{\partial g^i}{\partial x^\alpha} X^\alpha \quad \text{---} \quad \frac{\partial g^r}{\partial x^\gamma} A^\gamma + \frac{\partial^2 g^r}{\partial x^\alpha \partial x^\beta} X^\alpha Y^\beta \quad \text{---} \quad \frac{\partial g^j}{\partial x^\beta} Y^\beta \right) \qquad [3]$$

(Here and in the sequel, unsubscripted derivatives are taken at 0.)

It follows from earlier considerations, or can be easily shown directly, that this is a left action. The associated bundle $\begin{array}{c} F_2(M) \times_{G_2(m)} E_2 \\ \downarrow q_2 \\ M \end{array}$ is the 2-sector bundle $\begin{array}{c} T_2(M) \\ \downarrow q_2 \\ M \end{array}$ which was defined earlier. Generically, a 2-sector at x_o in M will be denoted $[\Phi_2, X^i$ ___ A^r ___ $Y^j]$

where Φ_2 is a 2-jet of a frame at x_o.

In the same spirit, a 1-sector, or a tangent vector at point x_0 in M will be denoted $[\Phi_1, X^i]$ for Φ_1 the 1-jet of a frame at x_o.

Next, we give an explicit description of the "boundary maps" $D^i(2)$, i = 1,2. Let Φ_2 be the 2-jet of a frame at $x_0 \varepsilon$ M, and let Φ_1 be its 1-jet. And let F_2 be the 2-sector at x_0 $[\Phi_2, X^i \_\_\_ A^r \_\_\_ Y^j]$. Then define

$$D^1(2)\ (F_2) = [\Phi_1, Y^j]\ , \text{ and }\ D^2(2)\ (F_2) = [\Phi_1, X^i]$$

These prescriptions define the bundle maps

$$\begin{array}{ccc} T_2(M) & \xrightarrow{D^i(2)} & T_1(M) \\ \downarrow q_2 & & \downarrow q_1 \\ M & \xrightarrow[id]{} & M \end{array}$$

mentioned earlier. They become vector bundle projections when linear structure is put on their fibers in the following way: Consider the fiber of $D^2(2)$ over $[\Phi_1, X^i]$. Let F_2 be as above, and let $\tilde{F}_2$ be the 2-sector in the same fiber $[\Phi_2, X^i \_\_\_ \tilde{A}^r \_\_\_ \tilde{Y}^j]$. Then define the sum $F_2 + \tilde{F}_2$ to be the 2-sector $[\Phi_2, X^i \_\_\_ A^r + \tilde{A}^r \_\_\_ Y^j + \tilde{Y}^j]$.

A check with the "transformation law," [3] shows the definition is good in the fiber over $[\phi_1, X^i]$. Scalar multiplication is defined in that fiber in the obvious way. As mentioned earlier, this linear structure is "really" the tangent space structure when $T_2(M)$ is interpreted as $T_1(T_1(M))$. The fibers of $D^1(2)$ are similarly endowed with intrinsic linear structure.

c) Suppose k = 3. $\Gamma(2)$, the standard 2-simplex, is the set of nonempty subsets of {1,2,3}. Denote an element of

$$E_3: \quad \begin{array}{ccccc} X^i & \_\_\_ & A^r & \_\_\_ & Y^j \\ & \diagdown & & \diagup & \\ B^s & & E^u & & C^t \\ & \diagdown & & \diagup & \\ & & Z^k & & \end{array}$$

all superscripts between 1 and m. Then E_3 is isomorphic as linear space with E^7 and again the position of the component corresponds with the face of the simplex. Now suppose that $g_3 \in G_3(m)$ has the form

$$\sum_{/a/ \leq 3} \left[\frac{1}{a!}\right] \frac{\partial^{/a/} g^i}{\partial x^a} x^a$$

then define the left action of $G_3(m)$ on E_3 by the following formula:

$$g_3 \cdot \left(\begin{array}{ccccc} X^i & \_\_ & A^r & \_\_ & Y^j \\ & \diagdown & & \diagup & \\ & B^s & E^u & C^t & \\ & & \diagdown \quad \diagup & & \\ & & Z^k & & \end{array} \right) =$$

$$\begin{array}{ccccc} \dfrac{\partial g^i}{\partial x^\alpha} X^\alpha & \text{——} & \left[\dfrac{\partial g^r}{\partial x^\gamma} A^\gamma + \dfrac{\partial^2 g^r}{\partial x^\alpha \partial x^\beta} X^\alpha Y^\beta \right] & \text{——} & \dfrac{\partial g^j}{\partial x^\beta} Y^\beta \\ \diagdown & & & & \diagup \\ \left[\dfrac{\partial g^s}{\partial x^\lambda} B^\lambda + \dfrac{\partial^2 g^s}{\partial x^\alpha \partial x^\theta} X^\alpha Z^\theta \right] & & \tilde{E}^u & & \left[\dfrac{\partial g^t}{\partial x^\nu} C^\nu + \dfrac{\partial^2 g^t}{\partial x^\beta \partial x^\theta} Y^\beta Z^\theta \right] \\ & \diagdown & & \diagup & \\ & & \dfrac{\partial g^k}{\partial x^\theta} Z^\theta & & \end{array} \qquad [4]$$

with

$$\tilde{E}^u = \frac{\partial^3 g^u}{\partial x^\alpha \partial x^\beta \partial x^\theta} X^\alpha Y^\beta Z^\theta + \frac{\partial^2 g^u}{\partial x^\alpha \partial x^\nu} X^\alpha C^\nu + \frac{\partial^2 g^u}{\partial x^\beta \partial x^\lambda} Y^\beta B^\lambda +$$

$$\frac{\partial^2 g^u}{\partial x^\theta \partial x^\gamma} Z^\theta A^\gamma + \frac{\partial g^u}{\partial x^\xi} E^\xi$$

Again, the interpretation of sectors as F-related classes of jets shows that this is indeed a left action (or it can be shown directly). The associated bundle $F_3(M) \times_{G_3(m)} E_3$ is the 3-sector bundle $T_3(M)$

$$\begin{array}{c} F_3(M) \times_{G_3(m)} E_3 \\ \downarrow q_3 \\ M \end{array} \qquad \qquad \begin{array}{c} T_3(M) \\ \downarrow q_3 \\ M \end{array}$$

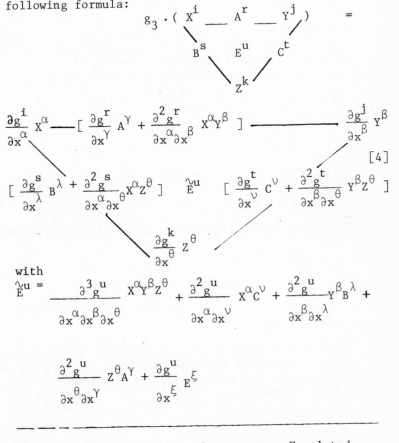

As with 2-sectors, we denote a 3-sector at x_o in M generically:

$$\left[\Phi_3,\ \begin{array}{ccccc} X^i & \_\_ & A^r & \_\_ & Y^j \\ & \diagdown & & \diagup & \\ & B^s & E^u & C^t & \\ & & \diagdown\ \diagup & & \\ & & Z^k & & \end{array}\right]$$

with Φ_3 the 3-jet of a frame at x_o.

Next we give a description of the boundary map $D^1(3)$: $T_3(M) \rightarrow T_2(M)$. The other two boundary maps will be defined by obvious extension. Let Φ_3 be a 3-jet of a frame at $x_o \in M$ and let F_3 be the 3-sector at x_o :

$$\left[\Phi_3,\ \begin{array}{ccccc} X^i & \_\_ & A^r & \_\_ & Y^j \\ & \diagdown & & \diagup & \\ & B^s & E^u & C^t & \\ & & \diagdown\ \diagup & & \\ & & Z^k & & \end{array}\right].$$

Then $D^1(3)\ (F_3) = [\Phi_2,\ Y^j \_\_ C^t \_\_ Z^k]$ for Φ_2 the 2-jet of Φ_3.

[Also, $D^2(3)\ (F_3) = [\Phi_2, Z^k \_\_ B^s \_\_ X^i]$, and $D^3(3)(F_3) = [\Phi_2,\ X^i \_\_ A^r \_\_ Y^j]$.]

Following this prescription, $D^1(3)$ will give a bundle map:

$$\begin{array}{ccccc} & T_3(M) & \rightarrow & T_2(M) & \\ q_3 & \downarrow & D^1(3) & \downarrow & q_2 \\ & M & \underset{id}{\rightarrow} & M & \end{array}$$

as will $D^2(3)$ and $D^3(3)$.

$D^1(3)$ is the projection of a <u>vector</u> bundle when the fiber is equipped with the following linear structure. Suppose that F_3 is as above, and $\tilde{F}_3$ is another 3-sector in the fiber,

$$\tilde{F}_3 = \left[\Phi_3,\ \begin{array}{ccccc} \tilde{X}^i & \_\_ & \tilde{A}^r & \_\_ & Y^j \\ & \diagdown & & \diagup & \\ & \tilde{B}^s & \tilde{E}^u & C^t & \\ & & \diagdown\ \diagup & & \\ & & Z^k & & \end{array}\right].$$

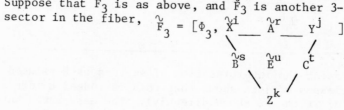

Then define the sum

$$F_3 + \tilde{F}_3 = \left[\Phi_3, \begin{array}{ccccc} (X^i + \tilde{X}^i) & \_\_\_\_ & (A^r + \tilde{A}^r) & \_\_\_ & Y^j \\ & (B^s + \tilde{B}^s) & (E^u + \tilde{E}^u) & C^t & \\ & & Z^k & & \end{array}\right].$$

Then it follows easily from equation [4] that this is a good definition, as is the corresponding obvious definition for scalar multiplication. This concludes the examples of low dimensional sector bundle structure. The pattern for the definition of the linear structure on the fibers of $D^i(k)$ should be clear from this.

We are now in a position to define sectorform fields. Given a smooth manifold M a k-sectorform field on M is a smooth map $A^k : T_k(M) \to R$ with the property that it is linear on each fiber of each $D^i(k): T_k(M) \to T_{k-1}(M)$ for $1 \leq i \leq k$.

A 1-sectorform at $x_0 \in M$ is just a 1-form at x_0. Next let Φ_2 be a frame jet at $x_0 \in M$. Then a 2-sectorform at x_0 may be represented (with reference to Φ_2) as a sum $a_{ij} + b_r$ where a_{ij} is an mxm matrix and b_r is a 1xm matrix, and where contraction of this sectorform on any 2-sector at x_0 $[\Phi_2, X^i \_\_\_ A^r \_\_\_ Y^j]$ is: $a_{ij}X^iY^j + b_rA^r$.

Of course, the components of the sectorform at x_0 depend on the frame-jet, but the contraction formula above can be used (in conjunction with equation [3]) to give the transformation law for sectorforms. In a similar way, if $\Phi: W \to U \subset M$ is a local frame, then a local 2-sectorform on U may be defined as a smooth map from W to the space of maps from E_2 to R of the form $a_{ij} + b_r$.

In a similar way one sees that a 3-sectorform at x_0 is specified by the data of a 3-jet of a frame at x_0, Φ_3, and by a sum $a_{ijk} + b_{it} + c_{js} + d_{kr} + e_u$ where the contraction with any 3-sector at x_0 $\left[\Phi_3, \begin{array}{ccccc} X^i & \_\_\_ & A^r & \_\_\_ & Y^j \\ & B^s & E^u & C^t & \\ & & Z^k & & \end{array}\right]$ is:

$a_{ijk}X^iY^jZ^k + b_{it}X^iC^t + c_{js}Y^jB^s + d_{kr}Z^kA^r + e_uE^u$. And local 3-sectorform fields are defined in the obvious way with reference to local frames.

We shall not give the general characterization here (see [White]) but we observe that every (smooth) covariant tensor field is a sectorform field; if the rank of the covariant tensor field is k, then it is a k-sectorform field. For example, suppose a metric on M has local coordinates with respect to some local frame $g_{ij}(z)$. Then its contraction with a 2-sector whose components with respect to that frame at x_0 are: X^i___ A^r___ Y^j is $g_{ij}(z_0)X^iY^j$. (Here, $\Phi(z_0) = x_0$)

Now the most important concept for the promotion of the metric is the following. If A^k is a smooth k-sectorform field on M, then there are (k+1) new sectorform fields called the differentials of A^k, and denoted $\{ d_jA^k | 1 \leq j \leq k+1\}$. Each differential is a smooth (k+1)-sectorform field and the construction satisfies this naturality property with respect to pullbacks.

Suppose f: N → M is a smooth map of manifolds, and that A^k is a k-sectorform field on M. Then the bundle map

$$\begin{array}{ccc} T_k(N) & \xrightarrow{T_k(f)} & T_k(M) \\ \downarrow q_k & & \downarrow q_k \\ N & \xrightarrow{f} & M \end{array}$$

maps the intrinsic spaces, the fibers of $D^i(k)$ to themselves, and is linear on them (straightforward verification). Therefore, there is a well-defined pullback of A^k under $T_k(f)$. For simplicity, denote the pullback, which is a k-sectorform field on N: $f^*(A^k)$. Then for each j, $1 \leq j \leq k+1$,

$$f^*[d_j(A^k)] = d_j[f^*(A^k)] . \qquad [5]$$

We shall not give here the general expression for the differentials of sectorform fields. But we show how to compute one of the differentials in the case that the initial sectorform field is a metric. We then interpret this differential as the promotion of that metric.

Thus suppose that M has a metric tensor G defined and that in some frame G has components $g_{ij}(z)$ for $z \varepsilon V$, $\Phi : V \to U \subset M$ the local frame. Then d_3G has the following local components in the local frame Φ: $\left.\dfrac{\partial g_{ij}}{\partial x^k}\right|_z + g_{sj}(z) + g_{it}(z)$. What this means is that the <u>contraction</u> of d_3G with any sector at $x_o \varepsilon M$ with components

$$\left[(\dot{\Phi} o t_{z_o})_3,\ \begin{array}{ccccc} X^i & \text{—} & A^r & \text{—} & Y^j \\ & B^s\diagdown & E^u & \diagup C^t & \\ & & Z^k & & \end{array} \right] \text{(where } \Phi(z_o) = x_o \text{, and}$$

t_{z_o} is the usual translation of E) is

$$\left.\frac{\partial g_{ij}}{\partial x^k}\right|_{z_o} X^iY^jZ^k + g_{sj}(z_o)B^sY^j + g_{it}(z_o)\ X^iC^t \qquad [6]$$

Evidently, d_3G is not a tensor field on M. However when interpreted as 2-sectorform field on $\overline{T_1(M)}$ in a certain way, it exhibits itself as a metric tensor on $\overline{T_1(M)}$. This will yield the promotion of G, a metric δG on $\overline{T_1(M)}$.

The two other differentials of the metric G are obtained from d_3 essentially through the action of the symmetric group on sectorforms. We write down the component descriptions here by giving their contractions with the 3-sector used in equation [6].

$$d_1G < F_3 > \ = \left.\frac{\partial g_{jk}}{\partial x^i}\right|_{z_o} X^iY^jZ^k + g_{js}(z_o)\ B^sY^j + g_{rk}(z_o)\ A^rZ^k \qquad [6']$$

$$d_2G < F_3 > \ = \left.\frac{\partial g_{ki}}{\partial x^j}\right|_{z_o} X^iY^jZ^k + g_{ti}(z_o)\ C^tX^i + g_{kr}(z_o)\ Z^kA^r \qquad [6'']$$

We shall actually define the promotion of G, δG using $d_1 G$. To do this, suppose that $V = [\Psi_1, X^i] \varepsilon\, T_1(M)$. Then an element of $T_V[T_1(M)]$ i.e., a tangent vector to $T_1(M)$ <u>at V</u> may be represented as an element of $T_2(M)$ in the following way:

$X|_V = [\Psi_2,\ X^i \underline{\quad} A^r \underline{\quad} Y^j]$. Then the linear structure of the tangent space $T_V[T_1(M)]$ is structure defined above (EXAMPLE 2(b)) for this set as the $D^2(2)$ fiber over V. Thus if another tangent vector at V has components $[\Psi_2,\ X^i \underline{\quad} B^s \underline{\quad} Z^k]$ and we call it $Y|_V$, then

$$X|_V + Y|_V = [\Psi_2,\ X^i \underline{\quad} (A^r + B^r) \underline{\quad} (Y^j + Z^j)].$$

With this notation, we define the <u>promotion of G</u> by the formula:

$$\delta G|_V < X|_V, Y|_V > \ = \ d_1(g) < F_3 > \ = \ \frac{\partial g_{jk}}{\partial x^i}\bigg|_{z_o} X^i Y^j Z^k \ + \ g_{js}(z_o) B^s Y^j \ + \ g_{rk}(z_o) A^r Z^k$$

In terms of a local frame Φ: $W \to U \subset M$ with $\Phi(z) = x$, if the matrix for G at x_o is written $g_{ij}(z_o)$ then the matrix for δG at V (in terms of the induced frame on $T_1(M)$) can be written:

$$\begin{vmatrix} \dfrac{\partial g_{jk}}{\partial x^i}\bigg|_{z_o} X^i & g_{js}(z_o) \\ g_{rk}(z_o) & 0 \end{vmatrix}$$

each block being an mxm matrix.

PART 2: THE GENERALIZED ENERGY GRADIENT PRINCIPLE

Having explained what the promotion of the metric G is in the previous section, let us observe how it is used to derive the Energy Gradient Principle of EXAMPLE 1. In local coordinates with Ψ, V, $X_{|V}$, $Y_{|V}$, and G as in the last section, equation [2] determines the infinitesimal generator of the geodesic flow $X|_V = [\Psi_2, X^i \_\_\_ A^r \_\_\_ Y^j]$ by the equation:

$$[\, Y^j \,,\, A^r \,] \begin{bmatrix} \left.\dfrac{\partial g_{jk}}{\partial x^i}\right|_{z_o} X^i & g_{sj}(z_o) \\ g_{rk}(z_o) & 0 \end{bmatrix} \begin{bmatrix} Z^k \\ B^s \end{bmatrix} =$$

$$d(T_G)|_V \langle Y \rangle|_V = \frac{1}{2} \left.\frac{\partial g_{ij}}{\partial x^k}\right|_{z_o} X^i X^j Z^k + g_{sj}(z_o) B^s X^j$$

for all Z^k and B^s, for X^i fixed.

Setting Z^k to zero implies that $X^i = Y^i$ for all i, using regularity of the metric, and then setting $B^s = 0$, we derive the equation:

$$\left.\frac{\partial g_{jk}}{\partial x^i}\right|_{z_o} X^i X^j Z^k + g_{rk}(z_o) A^r Z^k = \frac{1}{2} \left.\frac{\partial g_{ij}}{\partial x^k}\right|_{z_o} X^i X^j Z^k \text{ (for all } Z^k) \qquad [7]$$

---

easily recognized as the (time independent) Euler-Lagrange equation

$$\frac{d}{dt}\left[\frac{\partial T_G}{\partial v}\right] = \frac{\partial T_G}{\partial x}.$$

Notice also the appearance of the Riemann-Christoffel symbols in equation [7].

We now formulate the physical hypothesis from which

the Lagrangian and Hamiltonian formalisms can be derived. This hypothesis makes use, as do all prerelativistic approximations to relativistic dynamics of a canonical, and fictitious, separation of "space" and "time" variables.

Let M be the m-dimensional manifold of "configurations" and let (x,t) represent a general point of MxR. It will be assumed that all physically meaningful coordinate changes map fibers of the canonical projection $\begin{matrix} M \times R \\ \downarrow \Pi \\ R \end{matrix}$ to themselves.

Now let $\pi = T_1(\Pi)$ be the tangent map of the "time" projection and write $\pi = \pi_1 \times \pi_2$ where

$\pi(x,t\ ;\ X^i,\tau) = (t\ ,\tau)$ in $T_1(R)$.

Then for each pair (t,τ) in $R \times R = T_1(R)$ we may define:

$$\pi_1^{-1}(t) = H_t \subset T_1(MxR),$$

$$\pi_2^{-1}(\tau) = H^\tau \subset T_1(MxR), \text{ and}$$

$H_t \cap H^\tau = H_t^\tau \subset T_1(MxR)$. The H_t and H^τ give codimension-1 foliations of $T_1(MxR)$ and it is easy to see that each leaf of each foliation is diffeomorphic in a natural way (given our restriction on coordinate changes) to $T_1(M)xR$. In the development to follow, we shall be particularly interested in H^1. Notice that the tangent maps of all special coordinate changes in MxR map H^1 and all H_t to themselves, and they map H_t^1 which, for each t, is isomorphic with $T_1(M)$ to itself.

Next, introduce a regular metric A on MxR and say that A satisfies Hypothesis 1 if:

HYPOTHESIS 1 For all t, for each $V \varepsilon H_t^1$, the promoted metric δA satisfies the condition that it give an isomorphism $T_V(H^1) \rightarrow T_V(H_t)^*$ #

Finally, define an "energy" function $L: T_1(MxR) \rightarrow R$ by the rule $L_A(V) = \frac{1}{2} A(x,t) \langle V,V \rangle$ for $V \varepsilon T_1(MxR)$

Under these hypotheses, define the Generalized Energy Gradient Vector Field on H^1 in this way.

At $V \varepsilon H^1_t$ define the tangent sector $X|_V$ in $T_V[H^1]$:

for each $Z|_V \varepsilon T_V(H_t)$ $\qquad \delta A|_V \langle X|_V, Z|_V \rangle = dL|_V \langle Z|_V \rangle$.

[8]

This gives an invariant definition of a global vector field on H^1 according to Hypothesis 1. Any admissible change of coordinates will preserve the leaves of the foliation π, and will define with respect to the pullback of A under the coordinate change precisely the same vector field owing to the naturality property (Equation [1]) of the promotion operation. This is the Generalized Energy Gradient Principle. As mentioned earlier, it can be taken on equal footing with a variational principle such as Hamilton's Least Action Principle as the starting point for classical analytical mechanics.

Now for the sake of this discussion, we assume that the metric A satisfies the following:

HYPOTHESIS 2: A is a regular metric on MxR, and its restriction to each fiber of the canonical projection MxR gives a regular metric on that fiber. $\downarrow \Pi$ R

It is not difficult to see that if A satisfies Hypothesis 2, then it satisfies Hypothesis 1. We show how this goes, and at the same time establish some notation. In local coordinates about $(x,t) \varepsilon$ MxR, we write tangent vectors U and V in the form

$$V = [\Phi_1, \binom{X^i}{\tau}] \text{ and } = [\Phi_1, \binom{Y^i}{\sigma}]$$ and the τ and σ

"time" components of the velocity. Then in those local coordinates $A(x,t) \langle V, U \rangle$ may be written

$$a_{ij}(x,t)X^iY^j + w_i(x,t)X^i\sigma + w_j(x,t)Y^j\tau + K(x,t)\tau\sigma \qquad [9]$$

In this case, we represent the matrix of $A(x,t)$ as the square matrix:

$$\begin{array}{c|cc|} & m & 1 \\ \hline m & a_{ij}(x,t) & w_i(x,t)^T \\ & & \\ 1 & w_j(x,t) & K(x,t) \end{array}$$

Now a general tangent vector $X|_V$ in $T_V[T_1(MxR)]$, that is, vector tangent to V in $T_2(MxR)$ can be written.

$X|_V = [\Phi_2, (X^i_\tau) \_\_ (A^r_\rho) \_\_ (Y^j_\sigma)]$. Then to say that V belongs to H^1 is to say that $\tau = 1$. To say that $X|_V \; \varepsilon \; T_V[H^1]$ is to say that $\tau = 1$ <u>and</u> $\rho = 0$.

Finally, to say that $X|_V \; \varepsilon \; T_V[H_t]$ is to say that $\sigma = 0$.

Now in these local coordinates, the matrix for the <u>promotion</u> of A, δA, at $V\varepsilon\, T_1(MxR)$ may be written:

$$\begin{array}{c|c|c||c|c|} & -m- & -1- & -m- & -1- \\ m & \left(\dfrac{\partial a_{jk}}{\partial x^i} X^i + \dfrac{\partial a_{jk}}{\partial t}\tau\right) & \left(\dfrac{\partial w_j^T}{\partial x^i} X^i + \dfrac{\partial w_j^T}{\partial t}\tau\right) & a_{js} & w_j^T \\ \hline 1 & \left(\dfrac{\partial w_k}{\partial x^i} X^i + \dfrac{\partial w_k}{\partial t}\tau\right) & \left(\dfrac{\partial K}{\partial x^i} X^i + \dfrac{\partial K}{\partial t}\tau\right) & w_s & K \\ \hline\hline m & a_{rk} & w_r^T & 0 & 0 \\ \hline 1 & w_k & K & 0 & 0 \end{array}$$

(everything evaluated at (x,t)) [10]

The regularity hypothesis guarantees that the $(m+1)x(m+1)$ blocks on the skew-diagonal are invertible, and Hypothesis 2 guarantees that the mxm blocks "a_{rk}" and "a_{js}" are invertible. This, with the characterization just given of the vectors tangent to H^1 and H_t gives a simple proof that Hypothesis 1 is satisfied if Hypothesis 2 is.

Letting V be as above in these local coordinates we have the "lagrangian":

$$L(V) = \frac{1}{2} a_{ij}(x,t)X^iX^j + w_i(x,t)X^i\tau + \frac{1}{2} K(x,t)\tau^2 \quad [11]$$

We show that the vector field on H^1 determined by this data via equation [8] is the same as the vector field on $T_1(M) \times R \equiv H^1$ determined by the Euler-Lagrange equations for the Lagrangian (on $T_1(M) \times R$ in local coordinates):

$$\tilde{L}(x,X^i,t) = \frac{1}{2} a_{ij}(x,t)\; X^iX^j + w_i(x,t)\; X^i + \frac{1}{2}\; K(x,t)$$

---

[12]

Recall that on H^1, $\tau = 1$.

We thus substitute in equation [8] letting $X|_V \in T_V[H^1]$ and $Z|_V \in T_V[H_t]$ have the local components with respect to some fixed frame-germ at (x,t):

$$X|_V: \binom{X^i}{1} \text{---} \binom{A^r}{0} \text{---} \binom{Y^j}{\sigma}$$

$$Z|_V: \binom{X^i}{1} \text{---} \binom{B^s}{\gamma} \text{---} \binom{Z^k}{0}$$

[13]

Computing the contraction $\delta A|_V < X|_V, Z|_V >$ we obtain:

$$\frac{\partial a_{jk}}{\partial x^i}X^iY^jZ^k + \frac{\partial a_{jk}}{\partial t}Y^jZ^k + \frac{\partial w_k}{\partial x^i}X^iZ^k\sigma + \frac{\partial w_k}{\partial t}Z^k\sigma + a_{rk}A^rZ^k + a_{sj}B^sY^j +$$

$$K\sigma\gamma \;+\; w_s B^s \sigma \;+\; w_j Y^j \gamma \qquad [14]$$

(everything evaluated at (x,t))

And then, computing $dL|_V < Z|_V >$ we get:

$$\frac{1}{2}\frac{\partial a_{ij}}{\partial x^k} X^i X^j Z^k + \frac{1}{2}\frac{\partial K}{\partial x^k} Z^k + a_{sj} B^s X^j + K\gamma +$$

$$\frac{\partial w_i}{\partial x^k} Z^k X^i + w_s B^s + w_i X^i \gamma \qquad [15]$$

(everything evaluated at (x,t))

Now since equation [8] must hold for all $Z|_V$ of the above form we see that setting Z^k and B^s to 0 we have:

$$(w_j Y^j + K\sigma)\gamma = (w_i X^i + K)\gamma$$

And setting Z^k and γ to 0 we have:

$$(a_{sj} Y^j + w_s \sigma)\, B^s = (a_{sj} X^j + w_s) B^s$$

Writing these as the matrix equation:

$$\begin{vmatrix} a_{sj} & w_s^T \\ w_j & K \end{vmatrix} \begin{vmatrix} (Y^j - X^j) \\ (\sigma - 1) \end{vmatrix} = \begin{vmatrix} 0 \\ 0 \end{vmatrix}$$

shows, from regularity of A, that $X^j = Y^j$ and $\sigma = 1$.

From this, it is clear that equation [8] reduces to:

$$\frac{\partial a_{jk}}{\partial x^i} X^i X^j + \frac{\partial a_{jk}}{\partial t} X^j + a_{rk} A^r + \frac{\partial w_k}{\partial x^i} X^i + \frac{\partial w_k}{\partial t} =$$

$$\frac{1}{2}\frac{\partial a_{ij}}{\partial x^k} X^i X^j + \frac{\partial w_i}{\partial x^k} X^i + \frac{1}{2}\frac{\partial K}{\partial x^k} \qquad [16]$$

(everything evaluated at (x,t))

---

and recognizing that $\frac{\partial \tilde{L}}{\partial v}$ at (x,v,t) is $a_{ij}X^i + w_j$ we conclude that $X|_V$ is the tangent vector at V for the the Euler-Lagrange flow associated with Lagrangian $\tilde{L}$ (equation [12])on $T_1(M)xR$. Thus:

<u>THEOREM 1</u> The Euler-Lagrange flow on $T_1(M)xR$ associated with $\tilde{L}: T_1(M)xR \to R$ where $\tilde{L}$ is derived according to equation [12] from metric A satisfying Hypothesis 2 is, modulo the identification of H^1 with $T_1(M)xR$, the Generalized Energy Gradient Vector Field on H^1 whose definition is given in equation [8] . #

This flow is <u>entirely</u> determined by A. It is <u>invariantly</u> defined for coordinate changes in MxR which preserve the leaves of time. In some ways, this result is unsatisfactory because the regularity restriction on A puts constraints on $\tilde{L}$ which can, in some cases, be relaxed. Thus, we introduce:

<u>HYPOTHESIS 3</u>: A metric A on MxR satisfies hypothesis 3 if its restriction to each fiber of the canonical projection $\begin{matrix} MxR \\ \downarrow \Pi \\ R \end{matrix}$ gives a regular metric on that fiber.
A, itself, need not be regular. This guarantees that the Lagrangian $\tilde{L}$ is "regular."

Now suppose A satisfies Hypothesis 3. Letting $X|_V$ and $Y|_V$ have local components of [13] for $V \in H^1_t$ and letting A be as in [9] and the associated L_A as in [11] it is not difficult to deduce from expressions [14] and [15] that:

<u>THEOREM 2</u> If metric A on MxR satisfies Hypothesis 3, with the notation as above, then for $V \varepsilon H^1_t$ there is a <u>unique</u> $X|_V \ \varepsilon \ T_V [H^1]$ satisying the conditions:

a) $\sigma = 1$, and

b) $\delta A|_V < X|_V , Z|_V > = dL|_V < Z|_V >$

for <u>all</u> $Z|_V \ \varepsilon \ T|_V [H_t]$

Further the vector field on H^1 determined by these conditions is the Euler-Lagrange vector field on $T_1(M)xR$ (modulo identification of $T_1(M)xR$ with H^1) with respect to Lagrangian $\tilde{L}$ of equation [12].

<u>Proof</u>: The condition $\sigma = 1$ guarantees that $X^i = Y^i$ for all i. Then equating expressions [14] and [15], substituting and cancelling determines the A^r in the resulting equation ([16]) which is the Euler-Lagrange equation for $\tilde{L}$. #

While condition (b) may be satisfied by other vectors in the degenerate case above, we shall always chose the $X|_V$ determined by the <u>two</u> conditions above, and shall continue to call the resulting vector field on H^1 the Generalized Energy Gradient Vector Field. We consider now the relationship of the <u>Generalized</u> E.G.P. with the Energy Gradient Principle discussed in Example 1. For this, suppose that <u>the metric A is independent of the time</u>. Suppose that $\bar{A}$ is the induced metric on M, and that $\bar{L}$ is the induced Lagrangian. Writing in local coordinates the function $\bar{L} : T_1(M) \to R$ has the form

$$\bar{L}(x,X^i) = \frac{1}{2} a_{ij}(x) X^i X^j + w_i(x) X^i + \frac{1}{2} K(x) \qquad [17]$$

where we continue to assume that A satisfies Hypothesis 3, and so the matrix $a_{ij}(x)$ is invertible. We would like to express the Generalized Energy Gradient Vector Field in terms of the metric $\bar{A}$ and the Lagrangian $\bar{L}$ as a vector field on $T_1(M)xR$ which is the "suspension" [Abraham & Marsden] of a time independent vector field on $T_1(M)$. To do this, we first observe that $\bar{L}$ has a decomposition $\bar{L} = \bar{L}_1 + \bar{L}_2$ where we write in local coordinates:

$$\bar{L}_1(x,X^i) = \frac{1}{2} a_{ij}(x)X^iX^j + \frac{1}{2} K(x) \quad \underline{\text{and}}$$

$$\bar{L}_2(x,X^i) = w_i(x)X^i$$

This decomposition does not depend on the choice of coordinates in M, but is invariantly defined (ultimately) by the metric A. In particular, as a function from $T_1(M)$ to R $\bar{L}_2$ is simply a differential 1-form on M, invariantly determined with respect to coordinate changes in M by the metric A.

Now if at $V \varepsilon H^1_t$, $X|_V$ is the tangent vector in $T_V[H^1]$ given in THEOREM 2 for the metric A (which is assumed to satisfy Hypothesis 3) we see that in local coordinates the components of $X|_V$ satisy the conditions $\sigma = 1$, $X^i = Y^i$ for all i, and the equation derived from [16]:

$$\frac{\partial a_{jk}}{\partial x^i} X^iX^j + a_{rk}A^r = \frac{1}{2}\frac{\partial a_{ij}}{\partial x^k} X^iX^j + \frac{1}{2}\frac{\partial K}{\partial x^k} +$$

$$\left[\frac{\partial w_i}{\partial x^k} - \frac{\partial w_k}{\partial x^i}\right] X^i$$

(everything evaluated at x) [18]

Now the tangent vector $Z|_V$ determines tangent vectors at $x \in M$ whose local components may be written in the obvious charts: X^i & Z^k. Say the components of U are X^i and the components of W are Z^k at x. Then we recognize the expression

$$\left[\frac{\partial w_i}{\partial x^k} - \frac{\partial w_k}{\partial x^i}\right] X^iZ^k \quad \text{as the } \underline{\text{contraction}}\ d\bar{L}_2\langle W,U\rangle$$

where "d" represents <u>exterior derivative</u>. Of course, $d\bar{L}_2$ is also an invariantly defined differential 2-form on M determined by A. Finally, suppose we represent in these local coordinates at x in M, at U in $T_x(M)$ the tangent vectors

$\bar{X}|_U$ with components: X^i ___ A^r ___ Y^j , and

$\bar{Z}|_U$ with components: X^i ___ B^s ___ Z^k [19]

Then we may write $d\bar{L}_2 < W,U >$ as the contraction $-d\bar{L}_2 < \bar{Z}|_U >$ here thinking of $d\bar{L}_2$ as a 2-sectorform field, and $\bar{Z}|_U$ as a 2-sector at x.

With all of these notational choices, we may now observe: for each $U \varepsilon T_x(M)$ for x in M there is a unique tangent vector $\bar{X}|_U \ \varepsilon \ T_U[T_1(M)]$ which satisfies the equation:

$$\delta\ \bar{A}|_U < \bar{X}|_U ,\ \bar{Z}|_U > \ = d\bar{L}_1|_U < \bar{Z}|_U > \ - d\bar{L}_2|_U < \bar{Z}|_U >$$

$$\text{all } \bar{Z}|_U \ \varepsilon\ T_U\ [T_1(M)] \qquad [20]$$

The existence and uniqueness of $\bar{X}|_U$ are guaranteed by the regularity of $\bar{A}$ as metric on M, and by the fact that the right-hand side is linear in $\bar{Z}|_U$ considered as element of $T_U\ [T_1(M)]$.

Now the "suspension" of the vectorfield $\bar{X}$ on $T_1(M)$ is the vectorfield on $T_1(M)$xR which associates to (U,t) the tangent vector "$\bar{X}|_U + \frac{\partial}{\partial t}$". We are finally in a position to state the theorem that relates the two principles:

THEOREM 3 Suppose the metric A on MxR is time independent, and that $\bar{A}$ and $\bar{L} = \bar{L}_1 + \bar{L}_2$ are defined as above. Suppose also that A satisfies Hypothesis 3.

Then the Generalized Energy Gradient Vector field for A (as defined in Theorem 2) is time-independent. It is the <u>suspension</u> of the vector field on $T_1(M)$ defined in equation [20] (modulo the usual identification of H^1 with $T_1(M)xR$).

<u>Proof</u>: This follows immediately from equation [18] together with a short calculation of the left-hand side of [20] like the one immediately preceding equation [17].

We observe in particular that if $d\bar{L}_2 = 0$ for the metric A then the term $\frac{1}{2}K(x)$ simply contributed the classical negative potential to the kinetic energy, and in that case, the dynamic on $T_1(M)$ is a gradient dynamic again.

PART 3: THE EQUATIONS OF MOTION FROM THE SYMPLECTIC VIEWPOINT

Continuing with the notation and language of Part 2, suppose given a metric A on MxR that satisfies Hypothesis 3. Define the following 1-form on $H^1 \subset T_1(MxR)$.
For $V \varepsilon H^1$ and $X|_V \ \varepsilon \ T_V [H^1]$ let

$$\omega \langle X|_V \rangle = A \langle V,W \rangle$$ (where if $X|_V$ has local components of [13] then V has components (X^i_1) and W has components (Y^j_σ)) [21]

The contraction which defines ω above is just the contraction of the metric as 2-sectorform with $X|_V$ as 2-sector at (x,t).

Next, let θ be the 1-form on H^1: $\theta = \omega - \tilde{L}dt$ where $\tilde{L}: H^1 \to R$ is essentially defined from A in equation[12]. The symplectic structure that we want is the closed 2-form on H^1: dθ, where "d" again denotes exterior derivative. We shall prove the following theorem.

THEOREM 4 At $V \varepsilon H^1_t$ we have for any $X|_V \ \varepsilon \ T_V [H^1]$ the following two statements equivalent (keeping the local components and local coordinates of [13]):

a) $\sigma = 1$ and $\delta A|_V \langle X|_V, Z|_V \rangle = dL|_V \langle Z|_V \rangle$

for all $Z|_V$ in $T_V [H_t]$

and

b) $\sigma = 1$ and $X|_V \lrcorner \, d\theta = 0$.

Proof: Let $X|_V$ and $Y|_V$ be tangent vectors in $T_V [H^1]$ and suppose that with respect to the (overworked) frame of [13] they have components:

$$X|_V : \binom{X^i}{1} \_\_ \binom{A^r}{0} \_\_ \binom{Y^j}{\sigma}$$

$$Y|_V : \binom{X^i}{1} \_\_ \binom{B^s}{0} \_\_ \binom{Z^k}{\tau}$$

Then $d\omega|_V < X|_V, Y|_V > = (d_2A - d_3A)$

$$< \binom{X^i}{1} \_\_ \binom{A^r}{0} \_\_ \binom{Y^j}{\sigma} > \quad \binom{B^s}{0} \quad * \quad \binom{Z^k}{\tau} \quad *$$

where the starred entries are arbitrary in the 3-sector in this contraction and where the differentials d_2A and d_3A of the 2-sectorform field A have been defined essentially in equations [6] and [6"]. Notice that the difference $d_2A - d_3A$ does not depend on the starred entries while each of the sectorforms d_2A and d_3A does. This is a common occurrence in such constructions. In particular, the sectorform above called the "balance" of the metric gives, in this manner, an antisymmetric bilinear pairing of elements of $T_V[T_1(M)]$. For concreteness, we compute the contraction:

$$d\omega|_V < X|_V, Y|_V > = \left[\frac{\partial a_{ik}}{\partial x^j} - \frac{\partial a_{ij}}{\partial x^k} \right] X^i Y^j Z^k$$

$$+ \quad a_{rk} A^r Z^k - a_{sj} B^s Y^j$$

$$+ \quad \frac{\partial a_{ik}}{\partial t} X^i Z^k \sigma - \frac{\partial a_{ij}}{\partial t} X^i Y^j \tau$$

$$+ \quad \frac{\partial K}{\partial x^j} Y^j \tau - \frac{\partial K}{\partial x^k} Z^k \sigma \quad +$$

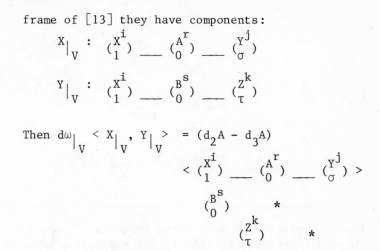

$$\left[\frac{\partial w_k}{\partial x^j} - \frac{\partial w_j}{\partial x^k}\right]Y^j Z^k + \frac{\partial w_k}{\partial t}[\sigma Z^k - \tau Y^k]$$

$$+ w_r A^r \tau - w_s B^s \sigma$$

$$+ \frac{\partial w_i}{\partial x^j} X^i Y^j \tau - \frac{\partial w_i}{\partial x^k} X^i Z^k \sigma \qquad [22]$$

This calculation is made using formulas [6] and [6"] with the modifications implicit in the expression used to represent the promoted metric [10].

$$\text{Next } d\tilde{L}|_V \wedge dt \, \langle X|_V, Y|_V\rangle = \frac{1}{2}\frac{\partial a_{ij}}{\partial x^k} X^i X^j [Y^k \tau - Z^k \sigma]$$

$$+ a_{ij} X^i [A^j \tau - B^j \sigma] + \frac{\partial w_i}{\partial x^k} X^i [Y^k \tau - Z^k \sigma]$$

$$+ \frac{1}{2}\frac{\partial K}{\partial x^j}[Y^j \tau - Z^j \sigma] + w_i [A^i \tau - B^i \sigma] \qquad [23]$$

Putting these together we find that:

$$d\Theta|_V \langle X|_V, Y|_V\rangle = \left[\frac{1}{2}\frac{\partial K}{\partial x^j} Y^j - \frac{1}{2}\frac{\partial a_{ij}}{\partial x^k} X^i X^j Y^k - \frac{\partial w_k}{\partial t} Y^k\right.$$

$$\left. - \frac{\partial a_{ij}}{\partial t} X^i Y^j - a_{ir} X^i A^r \right] \tau$$

$$+ \left[\frac{\partial a_{ik}}{\partial x^j} X^i Y^j - \frac{\partial a_{ij}}{\partial x^k} X^i Y^j + \frac{\partial w_k}{\partial x^j} Y^j - \frac{\partial w_j}{\partial x^k} Y^j + a_{rk} A^r\right.$$

$$- \frac{1}{2}\frac{\partial K}{\partial x^k}\sigma + \frac{\partial w_k}{\partial t}\sigma + \frac{\partial a_{ik}}{\partial t}X^i\sigma + \frac{1}{2}\frac{\partial a_{ij}}{\partial x^k}X^iX^j\sigma\ ]\ Z^k$$

$$+ [\ a_{is}X^i\sigma - a_{sj}Y^j\ ]\ B^s$$

[24]

Now assume that condition (b) is satisfied. Then setting τ and Z^k_s to 0 gives $X^j = Y^j$ for all j. Then setting τ and B^s to 0, remembering that $\sigma = 1$ and $X^j = Y^j$ we see that we obtain equation [16], therefore condition (a) is satisfied.

On the other hand, suppose condition (a) is satisfied. Then $\sigma = 1$ and $X^j = Y^j$ for all j. Now let $Z|_V = \sigma Y|_V - \tau X|_V$. Then $Z|_V \in T_V[H_t]$ and in fact has local components:

$$Z|_V : \binom{X^i}{1} \_\_ \binom{\sigma B^s - \tau A^s}{0} \_\_ \binom{\sigma Z^k - \tau Y^k}{0}.$$

Thus writing the equation $\delta A|_V \langle X|_V, Z|_V \rangle - dL|_V \langle Z|_V \rangle = 0$,

for <u>this</u> choice of $Z|_V$ and substituting as in [14] and [15] (remembering that $\sigma = 1$, and $X^j = Y^j$) we conclude that for such an $X|_V$ equation [24] is satisfied for <u>all</u> $Y|_V$ in $T_V[H^1]$. This completes the proof. #

J. Enrico White
Bates College
Lewiston, Maine 04240

REFERENCES:

Abraham, R. and Marsden, J. Foundations of Mechanics, Rev. ed. 1978, Benjamin/Cummings

Berger, M. Lectures on Geodesics in Riemannian Geometry, Tata Institute Publications, 1965

Hertz, H. The Principles of Mechanics, Dover, 1956

Souriau, J.M. Structure des Systemes Dynamiques, Dunod, 1970

Tulczyjew, W. Les jets generalises, C.R. Acad. Sc. Paris t. 281

A Symplectic Formulation of Particle Dynamics, Lec. Notes in Math, Vol. 570

White, J.E. The Method of Iterated Tangents with Applications in Local Riemannian Geometry Pitman, 1982

Yano, K. and Ishihara, S. Tangent and Cotangent Bundles, Marcel Dekker, 1973

Charles-Michel Marle

POISSON MANIFOLDS IN MECHANICS

1. INTRODUCTION

Poisson structures were defined and studied by Lichnerowicz (1976, 1977), who recognized their importance in mechanics and mathematical physics. Under the name of Hamiltonian structures, several other authors gave various definitions of Poisson structures, equivalent to the definition used by Lichnerowicz: among others, we refer to Iacob and Sternberg (1979), Kuperschmidt and Manin (1977), Symes (1980 a and b). Poisson structures are in fact a particular case of local Lie algebras, studied by Kirillov (1974, 1976). In this introduction, we will indicate some of the reasons which account for the growing importance of Poisson structures in mechanics.

1.1. Poisson manifolds as reduced phase spaces of Hamiltonian systems

Classically (Abraham and Marsden, 1978, Arnold, 1974), a Hamiltonian mechanical system is mathematically described by a symplectic manifold (M,Ω), called the phase space of the system, and a differentiable function $H : M \to \mathbb{R}$, called the Hamiltonian of the system. The time evolution of the system is described by integral curves of the Hamiltonian vector field $\#dH$, defined by the property :

$$i(\#dH)\Omega = - dH \quad .$$

Let G be a Lie group of symmetries of the system, that means, a Lie group which acts on the manifold M, by an action which preserves the symplectic 2-form Ω and the Hamiltonian H; for all $g \in G$, we have :

$$g^{\star}\Omega = \Omega \quad ; \quad g^{\star}H = H \circ g = H \quad .$$

Let us assume that the set P of orbits of the G-action on M has a differentiable manifold structure, such that the

C. P. Bruter et al. (eds.), Bifurcation Theory, Mechanics and Physics, 47–76.

canonical projection $\pi: M \to P$ is a submersion. The manifold P is called the reduced phase space of the system. It can be shown that the Hamiltonian vector field #dH projects onto M. For studying motions of the system, one may first look at integral curves of the projected vector field, on the reduced phase space P. But in general P is no more a symplectic manifold: it is a Poisson manifold.

A classical example of such a situation is the Euler-Poinsot motion of a rigid body with a fixed point (Arnold, 1974): the phase space is the cotangent bundle $T^*(SO(3))$ to the rotation group SO(3); the Hamiltonian H is a left-invariant Riemannian metric H on the group SO(3), which may be defined by a symmetric map $I: \mathfrak{g} \to \mathfrak{g}^*$, called the inerty operator; here $\mathfrak{g}$ is the Lie algebra of the rotation group, and $\mathfrak{g}^*$ its dual space. The reduced phase space is $\mathfrak{g}^*$, and the projection on $\mathfrak{g}^*$ of the Hamiltonian vector field #dH, leads to the Euler differential equation:

$$\frac{d\mu}{dt} = - \operatorname{ad}^*_{I^{-1}(\mu)}\mu \quad , \quad (\mu \in \mathfrak{g}^*) \quad ,$$

where ad^* stands for the coadjoint representation of $\mathfrak{g}$.

Poisson structures on reduced phase spaces may also be encountered in more general situations, for instance when such a reduced phase space is the set of leaves of a foliation of a symplectic manifold, instead of the space of orbits of a Lie group action; see Proposition 2.10 and Example 2.11 in the following.

1.2. Poisson manifolds a spaces of values of momentum maps

As in 1.1, let (M,Ω,H) be a classical Hamiltonian system, and G a Lie group which acts on M by a Hamiltonian action; for each $g \in G$, we have:

$$g^*\Omega = \Omega \quad ,$$

and moreover, there exists a differentiable map $J: M \to \mathfrak{g}^*$, such that for all $X \in \mathfrak{g}$:

$$i(X_M)\Omega = - d\langle J , X\rangle \quad .$$

Here $\mathfrak{g}$ is the Lie algebra of G, $\mathfrak{g}^*$ its dual space, and X_M the vector field on M defined by:

$$X_M(x) = \frac{d}{dt} \exp(-tX).x\Big|_{t=0} \quad , \qquad (x \in M).$$

The map J was first defined by Souriau (1969), and is called the momentum of the G-action. Souriau has shown that there exists an affine action of G on $\mathfrak{G}^\star$:

$$a : G \times \mathfrak{G}^\star \to \mathfrak{G}^\star$$

$$a(g,\xi) = Ad^\star_g \xi + \theta(g)$$

(where $\theta: G \to \mathfrak{G}^\star$ is a symplectic 1-cocycle of G), for which the map J is equivariant. Moreover, there exists on $\mathfrak{G}^\star$ a Poisson structure, called the modified Kirillov-Kostant-Souriau structure associated with the symplectic cocycle θ, for which J is a Poisson morphism (see example 2.3, 3°, and definition 2.8 below).

The Hamiltonian H is no more assumed G-invariant, but we assume that it may be written as:

$$H = \hat{H} \circ J \quad ,$$

where $\hat{H} : \mathfrak{G}^\star \to \mathbb{R}$ is some differentiable function. Then it can be shown that the Hamiltonian vector field #dH on the symplectic manifold M, and the Hamiltonian vector field $\#d\hat{H}$ on the Poisson manifold $\mathfrak{G}^\star$, (def. 2.6 below), are J-related.

An example of such a situation is the Euler-Lagrange motion of a rigid body in a gravity field, with a fixed point on its revolution axis. The phase space is $T^\star(SO(3))$, just as for the Euler-Poinsot motion. But the Lie group G used now is the group of displacements of a three-dimensional Euclidean space, $SO(3) \times \mathbb{R}^3$ (semi-direct product). See for instance Iacob and Sternberg (1979).

1.3. Remark

The two ways by which Poisson structures appear, described in 1.1 and 1.2, are related by the following fact: under suitable assumptions, the set of values of the momentum map J appears as the set of leaves of the generalized foliation of the manifold M defined by the symplectic orthogonal of the set of subspaces tangent to orbits of the G-action. This property seems closely related to example 2.11 given below.

1.4. Poisson and canonical manifolds

For the mathematical description of Hamiltonian mechanical systems with time-dependent Hamiltonians and constraints, Lichnerowicz (1976, 1979) has defined and studied canonical manifolds, which are Poisson manifolds with an additionnal structure. For more details on this subject, the reader is referred to the papers of Lichnerowicz quoted above, and to Marle (1982).

1.5. Poisson manifolds and completely integrable systems

A considerable interest was raised up recently by completely integrable Hamiltonian systems: see the works of Lax (1968), Adler (1979), Adler and Van Moerbeke (1980, a and b), Iacob and Sternberg (1979), Kazhdan, Kostant and Sternberg (1978), Kostant (1979), Mischenko and Fomenko (1978), Moser (1975), Olshanetsky and Perelomov (1976, 1979), Ratiu (1980), Reyman and Semenov-Tian-Shansky (1979, 1981), Symes (1980, a and b); see also the conference of Verdier (1980) at the Séminaire Bourbaki. In these works, Poisson structures appear; they are mainly of Kirillov-Kostant-Souriau type, and defined on dual spaces of Lie algebras. As will be seen in paragraph 5 below, some of the involution theorems obtained in these works may be put under a simpler and more general form, by the use of general Poisson structures.

2. POISSON MANIFOLDS: ELEMENTARY PROPERTIES AND EXAMPLES

2.1. Definition

A Poisson structure on a differential manifold M, is defined by a bilinear map :

$$C^\infty(M,\mathbb{R}) \times C^\infty(M,\mathbb{R}) \to C^\infty(M,\mathbb{R}) \quad ,$$

called Poisson bracket, and noted : $(f,g) \mapsto \{f,g\}$, satisfying the following properties:

i) the Poisson bracket is skew-symmetric:

$$\{g,f\} = - \{f,g\} \quad ;$$

ii) it is a derivation in each of its arguments :

$$\{f_1 f_2, g\} = \{f_1, g\} f_2 + f_1 \{f_2, g\} \quad ;$$

$$\{f, g_1 g_2\} = \{f, g_1\} g_2 + g_1 \{f, g_2\} \quad ;$$

iii) it satisfies the Jacobi identity :

$$\{f,\{g,h\}\} + \{g,\{h,f\}\} + \{h,\{f,g\}\} = 0 \quad .$$

With such a structure, the manifold M is called a Poisson manifold.

2.2. Remark

The space $C^\infty(M,\mathbb{R})$ of differentiable functions on a Poisson manifold M, is endowed with two algebraic structures:

an associative algebra structure, defined by the ordinary product $(f,g) \mapsto fg$;

a Lie algebra structure, defined by the Poisson bracket $(f,g) \mapsto \{f,g\}$.

These two structures are related by the property ii) of definition 2.1, which may be put under the following form : for any $f \in C^\infty(M,\mathbb{R})$, let ad_f be the linear endomorphism of $C^\infty(M,\mathbb{R})$:

$$g \mapsto ad_f(g) = \{f,g\} \quad ;$$

then, ad_f is a derivation of the associative algebra structure defined by the ordinary product.

A real vector space with an associative algebra structure and a Lie algebra structure related in such a way, will be called a Poisson algebra. Many properties of Poisson manifolds are in fact properties of the corresponding Poisson algebra, and remain valid for any Poisson algebra. This idea is developed by Ouzilou (1981).

2.3. Examples

1°) Let (M,Ω) be a symplectic manifold. The 2-form Ω defines an isomorphism :

$$(\#) : T^\star M \to TM \quad ;$$

by definition, for all $x \in M$, $\alpha \in T^\star_x M$, $\#\alpha$ is the unique vector of $T_x M$ such that :

$$i(\#\alpha)\,\Omega = -\alpha \quad .$$

The Poisson bracket associated with the symplectic structure on M, is the bilinear map from $C^\infty(M,\mathbb{R}) \times C^\infty(M,\mathbb{R})$ into $C^\infty(M,\mathbb{R})$:

$$(f,g) \mapsto \{f,g\} = \#df.g = -\#dg.f = \Omega(\#df \, , \#dg) \quad .$$

One may check that this Poisson bracket satisfies the properties of Definition 2.1. This shows that any symplectic manifold has an underlying Poisson structure.

2°) Let $\mathcal{G}$ be a real, finite dimensional Lie algebra; the bracket of two elements X and Y of $\mathcal{G}$ will be noted [X,Y]. Let $\mathcal{G}^\star$ be the dual space of $\mathcal{G}$. For all f and g $\in C^\infty(\mathcal{G}^\star,\mathbb{R})$, and all x $\in$ M, we set :

$$\{f,g\}(x) = \langle x \, , [df(x) \, , dg(x)]\rangle \quad .$$

One may check that this Poisson bracket satisfies the properties of Definition 2.1. This proves that the dual space $\mathcal{G}^\star$ of a real, finite dimensional Lie algebra $\mathcal{G}$, has a natural Poisson structure. This structure was defined by A. Kirillov (1974), B. Kostant (1970), and J.-M. Souriau (1969).

3°) With the same hypotheses and notations as in the last example, let $\Theta : \mathcal{G} \times \mathcal{G} \to \mathbb{R}$ be a skew-symmetric bilinear form on $\mathcal{G}$, such that, for all X, Y and Z $\in \mathcal{G}$:

$$(\star) \quad \Theta(X \, , [Y,Z]) + \Theta(Y \, , [Z,X]) + \Theta(Z \, , [X,Y]) = 0 \quad .$$

We will say that Θ is a symplectic 2-cocycle of $\mathcal{G}$ (with values in $\mathbb{R}$), or, when Θ is looked at as a linear map from $\mathcal{G}$ into its dual space $\mathcal{G}^\star$, a symplectic 1-cocycle of $\mathcal{G}$ with values in $\mathcal{G}^\star$.

For all f and g $\in C^\infty(\mathcal{G} \, , \mathbb{R})$ and x $\in$ M, we set :

$$\{f,g\}_\Theta(x) = \langle x \, , [df(x) \, , dg(x)]\rangle - \Theta(df(x) \, , dg(x)) \quad .$$

One may check again that this Poisson bracket satisfies the properties of Definition 2.1. The corresponding Poisson structure on $\mathcal{G}^\star$ will be called the modified Kirillov-Kostant-

Souriau structure associated with the symplectic cocycle Θ. When $\Theta = 0$, it reduces to the Kirillov-Kostant-Souriau structure defined in the preceding example.

2.4. <u>Proposition</u>

Let M be a Poisson manifold. There exists on M a unique two times contravariant, skew-symmetric tensor field Λ such that, for all f and g $\in C^\infty(M , \mathbb{R})$:

$$(\star) \qquad \{f,g\} = \Lambda(df , dg) \quad .$$

The tensor field Λ is called the Poisson tensor field of M.

<u>Proof</u>. The defining property $(\star)$ shows that Λ is unique. For proving its existence, it is sufficient to check that for all f and g $\in C^\infty(M , \mathbb{R})$ and all $x \in M$, $\{f,g\}(x)$ depends only on df(x) and dg(x). But this results from property ii) in Definition 2.1.

2.5. <u>Remark</u>

Let Λ be a two times contravariant, skew-symmetric tensor field on a manifold M. The formula $(\star)$ in Proposition 2.4 defines a bilinear map $(f,g) \mapsto \{f,g\}$, from $C^\infty(M,\mathbb{R}) \times C^\infty(M,\mathbb{R})$ into $C^\infty(M,\mathbb{R})$. This map satisfies properties i) and ii) of Definition 2.1, but in general it does not satisfy property iii). Lichnerowicz (1977) has shown that this map satisfies property iii), if and only if the tensor field Λ is such that:

$$(\star\star) \qquad [\Lambda , \Lambda] = 0 \quad ,$$

the bracket in this formula being the Schouten bracket, (Schouten 1954).

This shows that a Poisson structure on a manifold may be defined by a two times contravariant, skew-symmetric tensor field Λ, which satisfies property $(\star\star)$ above. This is the definition of Poisson structures initially introduced by Lichnerowicz (1977); it is equivalent to Definition 2.1.

2.6 <u>Definition</u>

Let M be a Poisson manifold, and Λ its Poisson tensor field. We will note $(\#)$ the morphism, from the cotangent bundle

$T^{\star}M$ into the tangent bundle TM, which associates with any $x \in M$ and $\alpha \in T^{\star}_{x}M$, the unique vector $\#\alpha \in T_{x}M$ such that, for all $\beta \in T^{\star}_{x}M$:

$$<\beta\ ,\ \#\alpha> = \Lambda_{x}(\alpha\ ,\ \beta) \quad .$$

Let $f \in C^{\infty}(M\ ,\ \mathbb{R})$. The vector field $\#df$ will be called the Hamiltonian vector field with Hamiltonian function f. It is characterized by the fact that, for any $g \in C^{\infty}(M\ ,\ \mathbb{R})$:

$$\#df\ .\ g = \{f,g\} \quad .$$

2.7. Proposition

Let M be a Poisson manifold. The map :

$$f \mapsto \#df$$

is a Lie algebra homomorphism of $C^{\infty}(M\ ,\ \mathbb{R})$ (with the Lie algebra structure defined by the Poisson bracket), into the space $\mathscr{C}(M)$ of C^{∞} vector fields on M (with the Lie algebra structure defined by the usual bracket).

Proof. We must check that, for all f and $g \in C^{\infty}(M\ ,\ \mathbb{R})$:

$$\#d\{f,g\} = [\#df\ ,\ \#dg] \quad .$$

But for any third element h of $C^{\infty}(M\ ,\ \mathbb{R})$, we have :

$$\#d\{f,g\}.h = \{\{f,g\}\ ,\ h\} = \{f\ ,\ \{g,h\}\} - \{g\ ,\ \{f,h\}\}$$

$$= \#df.(\#dg.h) - \#dg.(\#df.h) = [\#df\ ,\ \#dg].\ h$$

2.8. Definition

Let M and N be two Poisson manifolds. A differentiable map $\varphi : M \to N$ is called a Poisson morphism if it is such that, for all f and $g \in C^{\infty}(N\ ,\ \mathbb{R})$:

$$\varphi^{\star}\{f,g\} = \{\varphi^{\star}f\ ,\ \varphi^{\star}g\} \quad .$$

The following proposition indicates a usuful property of Poisson morphisms.

2.9. Proposition

Let M and N be two Poisson manifolds, and $\varphi : M \to N$ a Poisson morphism. Then for any $f \in C^\infty(N,\mathbb{R})$, the pair of Hamiltonian vector fields $\#d(f\circ\varphi)$ on M, and $\#df$ on N, is compatible with the map φ; this means that, for all $x \in M$:

$$T_x\varphi(\#d(f\circ\varphi)(x)) = \#df(\varphi(x)) \quad .$$

Proof. Let g be another element of $C^\infty(N,\mathbb{R})$. We have:

$$\begin{aligned}
\langle dg(\varphi(x)), T_x\varphi(\#d(f\circ\varphi)(x))\rangle &= \langle d(g\circ\varphi)(x) , \#d(f\circ\varphi)(x)\rangle \\
&= \{f\circ\varphi , g\circ\varphi\}(x) \\
&= \{\varphi^\star f , \varphi^\star g\}(x) \\
&= \varphi^\star\{f,g\}(x) = \{f,g\}(\varphi(x)) \\
&= \langle dg(\varphi(x) , \#df(\varphi(x))\rangle \quad .
\end{aligned}$$

2.10. Proposition

Let M be a Poisson manifold, and $\varphi : M \to N$ be a surjective submersion of M onto a differentiable manifold N. The two following properties are equivalent.

1. For all f and $g \in C^\infty(M,\mathbb{R})$, the function $\{f\circ\varphi , g\circ\varphi\}$ is constant on any fiber of the fibration $\varphi : M \to N$.

2. There exists a Poisson structure on N such that φ is a Poisson morphism.

When these two equivalent properties are satisfied, the Poisson structure on N for which φ is a Poisson morphism is unique.

Proof. As φ is a surjective submersion, the map:

$$f \mapsto \varphi^\star f = f\circ\varphi$$

is a vector space isomorphism of $C^\infty(N,\mathbb{R})$ onto the vector subspace of $C^\infty(M,\mathbb{R})$ of functions which are constant on each fiber of the fibration $\varphi : M \to N$. This shows that Property 2 implies Property 1. Conversely, if we assume Property 1 satisfied, we can define the Poisson bracket $\{f,g\}$ of two functions f and $g \in C^\infty(N,\mathbb{R})$ as the unique function on N such that:

$$\{f,g\}\circ\varphi = \{f\circ\varphi\ ,\ g\circ\varphi\} \quad .$$

We can check that properties of Definition 2.1 are satisfied. At last the uniqueness of the Poisson structure on N for which φ is a Poisson morphism, is a consequence of Definition 2.8.

2.11. Example

This example is due to P. Libermann (1982). Let (M,Ω) be a symplectic manifold. We first recall some definitions and notations (see for instance Abraham and Marsden, 1978). If x is a point of M and F_x a vector subspace of the tangent space T_xM, the symplectic orthogonal of F_x is the vector subspace of T_xM :

$$\text{orth } F_x = \{v \in T_xM \mid \Omega(x)(v,w) = 0 \text{ for all } w \in F_x\} \quad .$$

The vector space F_x is said coisotropic (resp. isotropic, resp. Lagrangian) if $\text{orth } F_x \subset F_x$ (resp. if $F_x \subset \text{orth } F_x$, resp. if $F_x = \text{orth } F_x$).

Similarly, let F be a vector subbundle of TM. The symplectic orthogonal orth F of F is the vector subbundle of TM, whose fiber, at each point x of TM, is the symplectic orthogonal $\text{orth } F_x$ of the fiber F_x of F at point x. The vector subbundle F is said coisotropic (resp. isotropic, resp. Lagrangian) if $\text{orth } F \subset F$ (resp. if $F \subset \text{orth } F$, resp. if $F = \text{orth } F$).

We now consider the Poisson structure on M underlying its symplectic structure (example 2.3, 1°). Let $\varphi : M \to N$ be a surjective submersion of M onto a differentiable manifold N such that, for each $x \in N$, $\varphi^{-1}(x)$ is connected. The kernel $\ker(T\varphi)$ of the fibre bundle map $T\varphi : TM \to TN$ is a completely integrable vector subbundle of TM, and the manifold N may be looked at as the manifold of leaves of the foliation of M defined by $\ker(T\varphi)$. Then the two equivalent properties of Proposition 2.10 are satisfied if and only if the vector subbundle $\text{orth}(\ker(T\varphi))$ of TM is completely integrable. This property follows from the Frobenius theorem, and from Proposition 2.7.

In particular, when $\ker(T\varphi)$ is a coisotropic subbundle of TM, it can be shown that $\mathrm{orth}(\ker(T\varphi))$ is completely integrable; therefore in that case, the two equivalent properties of Proposition 2.10 are satisfied.

2.12. Remark

Under the hypotheses of the last example, let Λ be the Poisson tensor field of N, defining the Poisson structure on N for which $\varphi : M \to N$ is a Poisson morphism. Let x be a point of N, and 2p the rank of Λ at point x. Then it can be seen that the rank of the 2-form induced on the submanifold $\varphi^{-1}(x)$ by the symplectic 2-form Ω, is constant, equal to $2(p+m-n)$, where 2m is the dimension of M, and n the dimension of N.

3. CHARACTERISTIC FIELD AND INTEGRAL MANIFOLDS OF A POISSON STRUCTURE

In this paragraph M is a Poisson manifold, and Λ its Poisson tensor field. For simplicity, in the following definitions and propositions, all functions, differential forms and vector fields considered are defined on the whole manifold M. The reader will check easily that the results can be extended to the case when these functions, differential forms or vector fields, are defined on open subsets of the manifold M.

3.1. Definitions

1°) A function $f \in C^\infty(M,\mathbb{R})$ is an invariant of the Poisson structure if it is an element of the center of the Lie algebra, that means, if for all $g \in C^\infty(M,\mathbb{R})$, one has :

$$\{f,g\} = 0 \quad .$$

2°) The characteristic field of the Poisson structure is the subset $\mathcal{F}$ of the tangent bundle TM, image of the fibre bundle morphism $(\#) : T^\star M \to TM$.

3°) An integral manifold of the Poisson structure is a connected, immersed submanifold N of M such that, for all $x \in N$:

$$T_x N = \mathcal{F}_x \quad ,$$

where $\mathcal{F}_x$ is the fiber at point x of the characteristic field $\mathcal{F}$.

3.2. Remarks

1°) A function $f \in C^\infty(M,\mathbb{R})$ is an invariant of the Poisson structure if and only if its differential df is a section of the annihilator $\mathcal{F}^0$ of $\mathcal{F}$, that means, if and only if, for each $x \in M$, df(x) belongs to the vector subspace of $T_x^\star M$ of linear forms on T_xM which vanish on the vector subspace $\mathcal{F}_x$.

2°) For each $x \in M$, the fiber $\mathcal{F}_x$ of the characteristic field at point x, is a vector subspace of T_xM, whose dimension is equal to the rank of the skew-symmetric two times contravariant tensor $\Lambda(x)$. But in general, the dimension of $\mathcal{F}_x$ depends on the point $x \in M$; for this reason, $\mathcal{F}$ is not always a vector subbundle of TM.

3.3. Proposition

Let α and β be two Pfaff forms on M, of class C^∞. There exists a Pfaff form γ on M, of class C^∞, such that:

$$[\#\alpha \ , \#\beta] = \#\gamma \quad .$$

Proof. The use of a partition of unity enables us to treat the problem locally, in the domain U of a chart of the manifold M. Let $x^1, \ldots x^m$ be the local coordinates associated with this chart. The Pfaff forms α and β may be written locally as:

$$\alpha = \sum_{i=1}^{m} \alpha_i \, dx^i \quad ; \quad \beta = \sum_{j=1}^{m} \beta_j \, dx^j \quad ,$$

where α_i and β_j are C^∞ functions defined on U. We have:

$$\#\alpha = \sum_{i=1}^{m} \alpha_i \, \#dx^i \quad ; \quad \#\beta = \sum_{j=1}^{m} \beta_j \, \#dx^j \quad ;$$

therefore:

$$[\#\alpha \ , \#\beta] = \sum_{i=1}^{m} \sum_{j=1}^{m} [\alpha_i \, \#dx^i \ , \beta_j \, \#dx^j]$$

or, using well known formulas about the bracket :

$$[\#\alpha\ ,\ \#\beta] = \sum_{i=1}^{m} \sum_{j=1}^{m} \Big(\alpha_i\, \beta_j\, [\#dx^i\ ,\ \#dx^j]$$

$$+ \alpha_i\, \langle d\beta_j\ ,\ \#dx^i \rangle\ \#dx^j$$

$$- \beta_j\, \langle d\alpha_i\ ,\ \#dx^j \rangle\ \#dx^i \Big) \quad .$$

By using Proposition 2.7, we have:

$$[\#dx^i\ ,\ \#dx^j] = \#d\{x^i\ ,\ x^j\} \quad ,$$

and we have:

$$[\#\alpha\ ,\#\beta]\ = \#\gamma \quad ,$$

where:

$$\gamma = \sum_{i=1}^{m} \sum_{j=1}^{m} (\alpha_i \beta_j d\{x^i,x^j\} + \alpha_i (\#dx^i.\beta_j) dx^j - \beta_j (\#dx^j.\alpha_i) dx^i).$$

The last proposition shows that the space of sections of $\mathfrak{F}$ which are images, by the fibre bundle morphism (#), of C^∞ Pfaff forms on M, is invariant by the bracket operation. This property looks like the Frobenius condition, for the complete integrability of a vector subbundle of the tangent bundle. But the classical Frobenius theorem is not applicable here, because $\mathfrak{F}$ is not a vector subbundle of TM. However, we have the following result, due to A. Kirillov (1976):

3.4. Theorem

Let x be any point of the Poisson manifold M. There exists a unique maximal integral manifold N_x of the Poisson structure containing the point x. Any other integral manifold of the Poisson structure which contains x, is a connected, open submanifold of N_x. Moreover, N_x has a unique symplectic structure, whose symplectic 2-form is defined by the following property, valid for all f and g $\in C^\infty(M,\mathbb{R})$ and all $y \in N_x$:

$$(\star) \qquad \Omega_{N_x}(y)\ (\#df(y)\ ,\ \#dg(y)) = \{f,g\}(y) \quad .$$

Finally, the manifold M is partitioned into maximal integral manifolds of its Poisson structure, which are symplectic connected immersed submanifolds of M (in general, not all of the same dimension).

Proof. We will first prove the existence of a local integral manifold N of the Poisson structure containing x. Let 2p be the dimension of $\mathfrak{F}_x$. If p = 0, the result is true, because $N_x = \{x\}$. We assume now p > 0; therefore, there exists a function $f \in C^\infty(M, \mathbb{R})$ such that $\#df(x) \neq 0$. By integration along integral curves of #df, we can define, on an open neighbourhood U of x, a differentiable function g such that:

$$\#df.g = 1 \quad ,$$

or, according to the very definition of #df:

$$\{f,g\} = 1 \quad .$$

By Proposition 2.7, we have on U:

$$[\#df \ , \#dg] = 0 \quad .$$

Moreover, the vector fields #df and #dg are linearly independent at each point of U: if a and b are two scalars and y a point of U such that:

$$a \ \#df(y) + b \ \#dg(y) = 0 \quad ,$$

we have:

$$\#d(af + bg)(y) = 0 \quad ,$$

and:

$$0 = \#d(af + bg).(-bf + ag) \ (y) = \{af + bg \ , -bf + ag\}(y)$$
$$= a^2 + b^2 \quad ,$$

which shows that $a = b = 0$.
By restricting eventually U, we may assume that there exists a surjective submersion:

$$\varphi : U \to W$$

of U onto a manifold W, whose dimension is dim M - 2, such that each leaf of the foliation of U defined by #df and #dg is the inverse image by φ of a point of W. Let h_1 and h_2 be two differentiable functions defined on U, which are constant on each fiber of the fibration $\varphi : U \to W$. We have:

$$\#df.h_i = 0 \quad ; \quad \#dg.h_i = 0 \quad , \quad (i = 1 \text{ or } 2);$$

$$\#df.\{h_1,h_2\} = \{f , \{h_1,h_2\}\} = \#dh_1.(\#df.h_2) - \#dh_2.(\#df.h_1)$$
$$= 0 \quad ,$$

and similarly:

$$\#dg.\{h_1,h_2\} = 0 \; .$$

This shows that $\{h_1,h_2\}$ is constant on each fiber of the fibration $\varphi : U \to W$. By Proposition 2.10, we see that there exists on W a unique Poisson structure for which $\varphi : U \to W$ is a Poisson morphism. The rank of the Poisson tensor field Λ_W at point $\varphi(x)$, is equal to $2(p-1)$. If $p-1 = 0$, $\varphi^{-1}(\varphi(x))$ is an integral manifold of the Poisson structure containing x; the rank of the Poisson tensor field Λ is constant along integral curves of the vector fields #df and #dg, because the integral flows of these vector fields are one-parameter local groups of Poisson morphisms; therefore the rank of Λ is constant along $\varphi^{-1}(\varphi(x))$, and equal to 2, that means, equal to the dimension of this manifold. This shows that $\varphi^{-1}(\varphi(x))$ satisfies the condition defining integral manifolds of the Poisson structure.
Now if $p-1 > 0$, replacing M by W, we can repeat the same argument as above. After a finite number of steps, we can assert the existence of a 2p-dimensional integral manifold N of the Poisson structure, containing the point x.
If N' is another 2p-dimensional integral manifold of the Poisson structure containing x, we see that $N \cap N'$ is open in N and in N'. This shows the local uniqueness of N. Then the existence and uniqueness of a maximal integral manifold of the Poisson structure containing x, is proved by the same procedure as in the case of a foliation of M (Chevalley 1946).
At last, we check that the 2-form Ω_{N_x} defined on N_x by the formula (★) above, is of class C^∞; it is non degenerate by the very definition of $\mathfrak{F}$, and closed because we have, if f, g and h are three differentiable functions, and y a point of N_x:

$$d\Omega_{N_x}(\#df\ ,\ \#dg\ ,\ \#dh)(y) = \underset{(f,g,h)}{\mathfrak{S}} (\#df.\Omega_{N_x}(\#dg\ ,\ \#dh))(y)$$

$$- \underset{(f,g,h)}{\mathfrak{S}} \Omega_{N_x}([\#df\ ,\ \#dg]\ ,\ \#dh)(y)$$

$$= \underset{(f,g,h)}{\mathfrak{S}} \{f\ ,\ \{g,h\}\}(y)$$

$$- \underset{(f,g,h)}{\mathfrak{S}} \{\{f,g\}\ ,\ h\}(y)$$

$$= 0\quad ,$$

by 2.7 and the Jacobi identity. The symbol $\underset{(f,g,h)}{\mathfrak{S}}$ in the formulae above stands for a sum over the three circular permutations of (f,g,h).

3.5. Examples.

1°) In example 2.3, 1°, when (M,Ω) is a symplectic manifold, we have $\mathcal{F} = TM$; therefore the maximal integral manifolds of the Poisson structure are the connected components of M. More generally, when for all $x \in M$, the dimension of $\mathcal{F}_x$ is an even integer 2p which does not depend on x, $\mathcal{F}$ is a completely integrable vector subbundle of TM, whose rank is 2p. The maximal integral manifolds of the Poisson structure are the leaves of the foliation of M defined by $\mathcal{F}$: they are symplectic manifolds, all of the same dimension 2p. In that case, one can prove a version of Darboux theorem (Lichnerowicz 1977, Symes 1980): every point of M has an open neighbourhood, domain of a chart with local coordinates $x^1, \dots x^{2p}, x^{2p+1}, \dots x^m$, such that the expression of the Poisson bracket of two functions f and g is

$$\{f,g\} = \sum_{i=1}^{p} \left(\frac{\partial f}{\partial x^{p+i}} \frac{\partial g}{\partial x^{i}} - \frac{\partial f}{\partial x^{i}} \frac{\partial g}{\partial x^{p+i}}\right) \quad .$$

2°) We consider now example 2.3, 2°, when M is the dual space of a real, finite dimensional Lie algebra $\mathcal{G}$. Let G be a connected Lie group having $\mathcal{G}$ as Lie algebra. Any element X of $\mathcal{G}$ may be looked at as a linear function on $\mathcal{G}^\star$. The corresponding Hamiltonian vector field #dX on $\mathcal{G}^\star$ must satisfy, for all $Y \in \mathcal{G}$ and all $x \in \mathcal{G}^\star$ (the second equality below being a consequence of the very definition of the coadjoint representation $X \mapsto ad_X$ of the Lie algebra $\mathcal{G}$):

$$\langle \# dX(x)\ ,\ Y\rangle = \langle x\ ,\ [X,Y]\rangle = \langle -ad_X^{\star}\, x\ ,\ Y\rangle \quad .$$

Let $\gamma \mapsto Ad_\gamma$ and $\gamma \mapsto Ad_\gamma^{\star}$ (with $\gamma \in G$) be the adjoint and coadjoint representations of the Lie group G. We recall that, for all $\gamma \in G$, $x \in \mathcal{G}^{\star}$, $Y \in \mathcal{G}$:

$$\langle Ad_\gamma^{\star}\, x\ ,\ Y\rangle = \langle x\ ,\ Ad_{\gamma^{-1}}\, Y\rangle \quad ,$$

and that, for all $X \in \mathcal{G}$, $x \in \mathcal{G}^{\star}$, $Y \in \mathcal{G}$:

$$\frac{d}{dt}\, Ad_{\exp(tX)}^{\star}\, x\Big|_{t=0} = ad_X^{\star}\, x \quad ;$$

$$\frac{d}{dt}\, Ad_{\exp(tX)}\, Y\Big|_{t=0} = [X,Y] \quad .$$

We can see now that the maximal integral manifold N_x of the Poisson structure containing a point x of $\mathcal{G}^{\star}$ is the orbit through x of the coadjoint representation of G :

$$N_x = \{Ad_\gamma^{\star}\, x \mid \gamma \in G\} \quad .$$

3°) With the same notations as above, we consider now the case of example 2.3, 3°, when the Poisson structure on $\mathcal{G}^{\star}$ is the modified Kirillov-Kostant-Souriau Poisson structure, associated with a symplectic cocycle Θ. It can be shown that there exists a unique differentiable map $\theta : G \to \mathcal{G}^{\star}$, which has the two properties :

i) for all γ_1 and $\gamma_2 \in G$, one has :

$$\theta(\gamma_1\gamma_2) = Ad_{\gamma_1}^{\star}\, \theta(\gamma_2) + \theta(\gamma_1) \quad ;$$

ii) if $T_e\theta : \mathcal{G} \to \mathcal{G}^{\star}$ is the linear map tangent to θ at the neutral element e of G, one has for all X and Y $\in \mathcal{G}$:

$$\langle T_e\theta(X)\ ,\ Y\rangle = \Theta(X,Y) \quad .$$

We will say that θ is a symplectic 1-cocycle of the Lie group G, with values in $\mathcal{G}^{\star}$, associated with the symplectic cocycle Θ of the Lie algebra $\mathcal{G}$.

Associated with θ, there exists an affine action a_θ of the Lie group G on the dual space $\mathcal{G}^{\star}$ of its Lie algebra, whose linear part is the coadjoint action, defined by :

$$a_\theta(\gamma,x) = \mathrm{Ad}^\star_\gamma\, x + \theta(\gamma) \qquad (\gamma \in G,\ x \in \mathcal{G}^\star) \quad .$$

By the same procedure as in case 2° above, we can now see that the maximal integral manifold N_x of the Poisson structure containing a point x of $\mathcal{G}^\star$, is the orbit through x of the affine action a_θ :

$$N_x = \{\mathrm{Ad}^\star_\gamma\, x + \theta(\gamma) \mid \gamma \in G\} \quad .$$

3.6. Remark (Symes 1980)

Under the hypotheses of examples 2.3, 3°, and 3.5, 3°, we consider the cotangent bundle $T^\star G$. The differential $d\alpha$ of the Liouville 1-form α, is a symplectic 2-form on $T^\star G$. The symplectic cocycle Θ may be looked at as a left invariant differential 2-form on the Lie group G, and the formula (★) of example 2.3, 3°, shows that this 2-form is closed. Let :

$$\Omega_\Theta = d\alpha + q^\star\Theta \quad ,$$

where $q : T^\star G \to G$ is the canonical projection. One can check that Ω_Θ is a symplectic 2-form on $T^\star G$.

Let $\varphi : T^\star G \to \mathcal{G}^\star$ be the map which associates, to each element z of $T^\star G$, the left invariant 1-form whose value at point $q(z) \in G$, is z. We see that φ is a surjective submersion, and that, for all $\xi \in \mathcal{G}^\star$, $\varphi^{-1}(\xi)$ is the graph of the left invariant 1-form ξ. We are now under conditions of example 2.11: the modified Kirillov-Kostant-Souriau Poisson structure on $\mathcal{G}^\star$ associated with the symplectic cocycle Θ, is the unique Poisson structure on $\mathcal{G}^\star$ for which $\varphi : T^\star G \to \mathcal{G}^\star$ is a Poisson morphism (when the Poisson structure on $T^\star G$ is the structure associated with the symplectic structure defined by Ω_Θ).

The above remark applies to examples 2.3, 2° and 3.5, 2° , by making $\Theta = 0$.

3.7. Remark

A differentiable function f defined on a Poisson manifold M is an invariant of the Poisson structure (definition 3.1), if and only if f is constant on each maximal integral manifold of the Poisson structure.

4. AUTOMORPHISMS AND INFINITESIMAL AUTOMORPHISMS OF A POISSON STRUCTURE

In this paragraph M is a Poisson manifold, and Λ its Poisson tensor field.

4.1. Definition

1°) A Poisson automorphism of M is a diffeomorphism $\varphi : M \to M$, which is also a Poisson morphism.

2°) An infinitesimal Poisson automorphism of M is a vector field X on M, whose integral flow φ is such that, for any $t \in \mathbb{R}$, φ_t is a Poisson morphism (from the open subset of M on which φ_t is defined, onto its image).

We can check that when φ and φ' are Poisson automorphisms of M, φ^{-1} and $\varphi' \circ \varphi$ are Poisson automorphisms of M : the set of Poisson automorphisms of M is a subgroup of the group of diffeomorphisms of M.

4.2. Examples

1°) Let (M,Ω) be a symplectic manifold; we look at M as a Poisson manifold, for the underlying Poisson structure. A diffeomorphism $\varphi : M \to M$ is a Poisson automorphism if and only if φ is a symplectomorphism of M, that means, if and only if :

$$\varphi^{\star}\Omega = \Omega \quad .$$

Under the same hypotheses, a vector field X on M is a Poisson infinitesimal automorphism, if and only if X is a locally Hamiltonian vector field on M (see for instance Abraham and Marsden, 1978), that means, if and only if the differential 1-form $i(X)\Omega$ is closed.

2°) Under the hypotheses of examples 2.3,3°, and 3.5, 3°, let γ be an element of the Lie group G. The affine transform of $\mathfrak{G}^{\star}$:

$$x \mapsto \varphi_\gamma(x) = a_\theta(\gamma,x) = \mathrm{Ad}^\star_\gamma\, x + \theta(\gamma)$$

is a Poisson automorphism of $\mathcal{G}^\star$. We have indeed, for f and $g \in C^\infty(\mathcal{G}^\star\ , \mathbb{R})$:

$$\begin{aligned}
\{\varphi^\star_\gamma f\ , \varphi^\star_\gamma g\}_\Theta(x) &= \langle x\ , [d(\varphi^\star_\gamma f)(x)\ , d(\varphi^\star_\gamma g)(x)]\rangle \\
&\quad - \Theta(d(\varphi^\star_\gamma f)(x)\ , d(\varphi^\star_\gamma g)(x)) \\
&= \langle x\ , [\mathrm{Ad}_{\gamma^{-1}}\, df(\varphi_\gamma(x))\ , \mathrm{Ad}_{\gamma^{-1}}\, dg(\varphi_\gamma(x))]\rangle \\
&\quad - \Theta(\mathrm{Ad}_{\gamma^{-1}}\, df(\varphi_\gamma(x))\ , \mathrm{Ad}_{\gamma^{-1}}\, dg(\varphi_\gamma(x))) \\
&= \langle \mathrm{Ad}^\star_\gamma\, x + \theta(\gamma)\ , [df(\varphi_\gamma(x))\ , dg(\varphi_\gamma(x))]\rangle \\
&\quad - \Theta(df(\varphi_\gamma(x))\ , dg(\varphi_\gamma(x))) \\
&= \varphi^\star_\gamma\{f,g\}\ (x) \quad .
\end{aligned}$$

In the above calculation we have used the property, valid for any $y \in \mathcal{G}^\star$:

$$\begin{aligned}
\langle y\ , d(\varphi^\star_\gamma f)(x)\rangle &= \langle \mathrm{Ad}^\star_\gamma\, y\ , df(\varphi_\gamma(x))\rangle \\
&= \langle y\ , \mathrm{Ad}_{\gamma^{-1}}\, df(\varphi_\gamma(x))\rangle \quad .
\end{aligned}$$

We have also used the property which relates the cocycles θ of the Lie group G, and Θ of the Lie algebra $\mathcal{G}$, valid for any X and $Y \in \mathcal{G}$, and any $\gamma \in G$:

$$\langle\theta(\gamma)\ , [X,Y]\rangle = \Theta(X,Y) - \Theta(\mathrm{Ad}_{\gamma^{-1}}\, X\ , \mathrm{Ad}_{\gamma^{-1}}\, Y) \quad .$$

4.3. Proposition

Let X be a vector field on the Poisson manifold M. The three following properties are equivalent.

1°) For all f and $g \in C^\infty(M\ , \mathbb{R})$, we have :

$$X.\{f,g\} = \{X.f\ , g\} + \{f\ , X.g\} \quad .$$

2°) The Lie derivative of the Poisson tensor field Λ, with respect to the vector field X, vanishes :

$$\mathcal{L}(X)\Lambda = 0 \quad .$$

3°) The vector field X is a Poisson infinitesimal automorphism.

Proof. Equivalence of properties 1° and 2° is easy. In order to prove the equivalence of these two properties with property 3°, we remark that if φ is the integral flow of X, we have for all $t_o \in \mathbb{R}$:

$$\begin{aligned}\frac{d}{dt}\varphi^\star_{-t}\{\varphi^\star_t f , \varphi^\star_t g\}\Big|_{t=t_o} = &- \varphi^\star_{-t_o}(X.\{\varphi^\star_{t_o} f , \varphi^\star_{t_o} g\}) \\ &+ \varphi^\star_{-t_o}(\{X.(\varphi^\star_{t_o} f) , \varphi^\star_{t_o} g\}) \\ &+ \varphi^\star_{-t_o}(\{\varphi^\star_{t_o} f , X.(\varphi^\star_{t_o} g)\}) \quad .\end{aligned}$$

It is then easy to see that properties 2° and 3° are equivalent.

The last proposition shows in particular that the set of infinitesimal automorphisms of the Poisson manifold M, is a Lie subalgebra of the Lie algebra of differentiable vector fields on M.

4.4. Example

Let $f \in C^\infty(M , \mathbb{R})$ be a function on the Poisson manifold M. The associated Hamiltonian vector field #df (definition 2.6) is a Poisson infinitesimal automorphism of M. We have indeed, for all g and $h \in C^\infty(M , \mathbb{R})$:

$$\begin{aligned}\#df.\{g,h\} &= \{f , \{g,h\}\} \\ &= \{\{f,g\} , h\} + \{g , \{f,h\}\} \\ &= \{\#df.g , h\} + \{g , \#df.h\} \quad .\end{aligned}$$

The following definition generalizes the definitions of locally and globally Hamiltonian vector fields (Abraham and Marsden, 1978), which are well known for a symplectic manifold. We will see that, on a Poisson manifold, locally Hamiltonian vector fields are infinitesimal automorphisms of the Poisson structure; but infinitesimal Poisson automorphisms may exist, which are not locally Hamiltonian vector fields.

4.5. Definitions

1°) A differential p-form η on the Poisson manifold M is said $\mathfrak{F}$-closed if, for any family $(f_1, \dots f_{p+1})$ of p+1 differentiable functions on M, one has :

$$d\eta(\#df_1, \dots \#df_{p+1}) = 0 \quad .$$

2°) A vector field X on the Poisson manifold M is said locally Hamiltonian if there exists an $\mathfrak{F}$-closed Pfaff form α on M, such that :

$$X = \#\alpha \quad .$$

Any closed 1-form (and, therefore, the differential df of any differentiable function f on M) is $\mathfrak{F}$-closed. Hence a Hamiltonian vector field #df on M, is locally Hamiltonian.

4.6. Proposition

Let α be a Pfaff form on the Poisson manifold M. The vector field $\#\alpha$ is an infinitesimal Poisson automorphism if and only if α is $\mathfrak{F}$-closed, that means, if and only if $\#\alpha$ is locally Hamiltonian.

Proof. Let f and g be two elements of $C^\infty(M,\mathbb{R})$. We have :

$$\begin{aligned}\#\alpha.\{f,g\} &= -\langle\alpha, \#d\{f,g\}\rangle \\ &= -\langle\alpha, [\#df\ ,\ \#dg]\rangle \quad ; \\ \{\#\alpha.f\ ,\ g\} &= \{\Lambda(\alpha,df)\ ,\ g\} \\ &= -\{\langle\alpha\ ,\ \#df\rangle\ ,\ g\} \\ &= \#dg.(\langle\alpha\ ,\ \#df\rangle) \quad ; \\ \{f\ ,\ \#\alpha.g\} &= -\#df.(\langle\alpha\ ,\ \#dg\rangle) \quad .\end{aligned}$$

We obtain:

$$\#\alpha.\{f,g\} - \{\#\alpha.f\ ,\ g\} - \{f\ ,\ \#\alpha.g\} = d\alpha(\#df\ ,\ \#dg) \quad ,$$

and the result follows from this equality .

4.7. Proposition

Let X and Y be two vector fields on the Poisson manifold M.

1°) If X is an infinitesimal Poisson automorphism and Y a locally Hamiltonian vector field, the bracket [X,Y] is locally Hamiltonian; more precisely, if β is an $\mathfrak{F}$-closed Pfaff form such that :

$$Y = \#\beta \quad ,$$

one has :

$$[X,Y] = \#(\mathcal{L}(X)\beta) \ .$$

2°) If X and Y are both locally Hamiltonian, the bracket [X,Y] is a Hamiltonian vector field; more precisely, if α and β are two $\mathfrak{F}$-closed Pfaff forms such that :

$$X = \#\alpha \quad , \qquad Y = \#\beta \quad ,$$

one has :

$$[X,Y] = \#d(i(X)\beta) = -\#d(i(Y)\alpha) \quad .$$

Proof. Let β be an $\mathfrak{F}$-closed Pfaff form such that :

$$Y = \#\beta \ .$$

For any function $f \in C^\infty(M,\mathbb{R})$, we have :

$$\begin{aligned}[X,Y].f &= X.(Y.f) - Y.(X.f) \\ &= X.(\Lambda(\beta,df)) - \Lambda(\beta\ ,\ d(X.f)) \\ &= \Lambda(\mathcal{L}(X)\beta\ ,\ df)\end{aligned}$$

because the Lie derivative $\mathcal{L}(X)\Lambda$ of the Poisson tensor field Λ vanishes (proposition 4.3). Therefore we have, for any function $f \in C^\infty(M,\mathbb{R})$:

$$[X,Y].f = \#(\mathcal{L}(X)\beta).f \quad ,$$

and this shows that :

$$[X,Y] = \#\mathcal{L}(X)\beta \quad .$$

But as X and Y are infinitesimal Poisson automorphisms, their bracket $[X,Y]$ is also an infinitesimal Poisson automorphism; by proposition 4.6, we see that $\mathcal{L}(X)\beta$ is $\mathfrak{F}$-closed, or that $[X,Y]$ is locally Hamiltonian. This completes the proof of 1°.

Under the hypotheses of 2°, by using the formula :

$$\mathcal{L}(X)\beta = d\, i(X)\beta + i(X)\, d\beta \quad ,$$

we obtain :

$$[X,Y] = \#d(i(X)\beta) + \#(i(X)\, d\beta) \quad .$$

But for any function $f \in C^\infty(M,\mathbb{R})$, we have :

$$\begin{aligned} \#(i(X)\, d\beta).f &= \Lambda(i(X)\, d\beta\ ,\ df) \\ &= d\beta(\#df\ ,\ X) \\ &= 0 \quad . \end{aligned}$$

The last equality is due to the facts that β is $\mathfrak{F}$-closed, and that, X being locally Hamiltonian, for any point $x \in M$, there exists a differentiable function h on M such that :

$$X(x) = \#dh(x) \quad .$$

Therefore we have :

$$[X,Y] = \#d(i(X)\beta) \quad ,$$

which completes the proof of 2°.

The reader is referred to the paper of Lichnerowicz (1977) for a much more thorough study of the various Lie algebras of vector fields associated with a Poisson manifold, of their derivations, ideals and deformations. Results indicated above are, for their main part, adapted from the corresponding results established by Lichnerowicz in the particular case when the rank of the Poisson tensor field Λ is constant.

5. FUNCTIONS IN INVOLUTION ON A POISSON MANIFOLD

5.1. Definition

Let M be a Poisson manifold. Two functions f and $g \in C^{\infty}(M,\mathbb{R})$ are said in involution when :

$$\{f,g\} = 0 \quad .$$

The corresponding definition is well known for functions defined on a symplectic manifold. The importance of this concept is related to the classical Liouville theorem (Arnold, 1974, Arnold and Avez, 1967) about completely integrable Hamiltonian systems.

Following the work of Lax (1968) about isospectral deformations, several recent works were devoted to completely integrable Hamiltonian systems : see the papers referred to in the introduction, paragraph 1.5. In these works appear theorems which give conditions under which functions are in involution (see in particular the paper of T. Ratiu, 1980). Some of these theorems may be put under a simpler and more general form, when the concept of Poisson manifold is used. This is the case for the Adler-Kostant-Symes theorem, which may be formulated as follows.

5.2. Theorem (M. Selmi, 1982, and the author)

Let M and N be two Poisson manifolds, and $\varphi : M \to N$ a Poisson morphism, which is also a surjective submersion. Let $s : U \to M$ be a section of φ, that means, a differentiable map from an open subset U of N, into M, such that $\varphi \circ s$ is the identity map of U. We assume that the submanifold s(U) of M satisfies the following property:

Property P : for any pair (h,k) of differentiable functions on M, whose restrictions to s(U) are constant, the Poisson bracket $\{h,k\}$ vanishes on s(U).

Then if f and $g \in C^{\infty}(M,\mathbb{R})$ are two invariants of the Poisson structure of M (definition 3.1), the two functions $f \circ s$ and $g \circ s$, defined on the open subset U of N, are in involution:

$$\{f \circ s \, , g \circ s\} = 0 \quad .$$

Proof. We set :

$$U_1 = s(U) \quad .$$

Let x be a point of U, and $y = s(x)$ the corresponding point of U_1. We have the direct sum decomposition :

$$T_yM = T_yU_1 \oplus \ker T_y\varphi \quad .$$

By duality, we deduce the direct sum decomposition of the cotangent space $T_y^\star M$:

$$T_y^\star M = (\ker T_y\varphi)^\circ \oplus (T_yU_1)^\circ \quad ,$$

where $(\ker T_y\varphi)^\circ$ and $(T_yU_1)^\circ$ are the annihilators, respectively, of $\ker T_y\varphi$ and of T_yU_1 (that means, the vector subspaces of the cotangent space $T_y^\star M$, made of linear forms on T_yM which vanish, respectively, on $\ker T_y\varphi$ and on T_yU_1). The two subspaces $(\ker T_y\varphi)^\circ$ and $(T_yU_1)^\circ$ may be identified, respectively, with the dual spaces $T_y^\star U_1$ of T_yU_1, and $(\ker T_y\varphi)^\star$ of $\ker T_y\varphi$. We note :

$$\pi_1 : T_y^\star M \to (\ker T_y\varphi)^\circ \equiv T_y^\star U_1 \quad ,$$

$$\pi_2 : T_y^\star M \to (T_yU_1)^\circ \equiv (\ker T_y\varphi)^\star$$

the two projections defined by this direct sum decomposition. Let Λ_M be the Poisson tensor field on M. We have :

$$\{f \circ s \, , \, g \circ s\}(x) = \Lambda_M(\pi_1(df(y)) \, , \, \pi_1(dg(y)) \quad ,$$

because $\varphi : M \to N$ is a Poisson morphism. We may write :

$$\begin{aligned}\{f\circ s \, , \, g\circ s\}(x) &= \Lambda_M(df(y) - \pi_2(df(y)) \, , \, dg(y) - \pi_2(dg(y))) \\ &= \langle dg(y) - \pi_2(dg(y)) \, , \, \#df(y)\rangle \\ &\quad + \langle \pi_2(df(y)) \, , \, \#dg(y)\rangle \\ &\quad + \Lambda_M(\pi_2(df(y)) \, , \, \pi_2(dg(y))) \quad .\end{aligned}$$

But the first two terms of this last expression vanish, because as f and g are Poisson invariants, $\#df(y)$ and $\#dg(y)$ are equal to zero. On the other hand, $\pi_2(df(y))$ and $\pi_2(dg(y))$ belong to $(T_yU_1)^\circ$; there exist two functions h and k, constant on U_1, such that :

$$dh(y) = \pi_2(df(y)) \quad ; \quad dk(y) = \pi_2(dg(y)) \quad ,$$

and we have, by Property P :

$$\Lambda_M(\pi_2(df(y)) \, , \, \pi_2(dg(y))) = \{h,k\}(y) = 0 \quad .$$

Finally we have :

$$\{f \circ s \, , \, g \circ s\}(x) = 0 \quad .$$

5.3. Remark

When M is a symplectic manifold, property P of theorem 5.2 means that s(U) is a coisotropic submanifold of M. When M is a Poisson manifold, it seems that submanifolds of M which verify Property P play a part very similar to that of coisotropic submanifolds of a symplectic manifold.

5.4. Application

We give here the usual form of the Adler-Kostant-Symes theorem and we will show how it can be deduced from theorem 5.2.

Let $\mathfrak{g}$ be a real, finite dimensional Lie algebra, $\mathfrak{h}$ and $\mathfrak{k}$ two Lie subalgebras of $\mathfrak{g}$ such that we have the direct sum vector space decomposition :

$$\mathfrak{g} = \mathfrak{h} \oplus \mathfrak{k} \quad .$$

We have, for the dual space $\mathfrak{g}^\star$, the corresponding direct sum decomposition :

$$\mathfrak{g}^\star = \mathfrak{k}^0 \oplus \mathfrak{h}^0 \quad ,$$

where :

$$\mathfrak{k}^0 \equiv \mathfrak{h}^\star \quad ; \quad \mathfrak{h}^0 \equiv \mathfrak{k}^\star \quad .$$

On the spaces $\mathfrak{g}^\star$ and $\mathfrak{h}^\star$, we consider the Kirillov-Kostant-Souriau Poisson structures. Let $\lambda \in \mathfrak{g}^\star$ be such that :

$$(\star) \qquad \langle \lambda \, , \, [\mathfrak{h},\mathfrak{h}] \rangle = 0 \quad ;$$

$$(\star\star) \qquad \langle \lambda , [\mathfrak{k},\mathfrak{k}] \rangle = 0 \ .$$

Let f and g be two functions on $\mathcal{G}^\star$, which are invariants of its Poisson structure. We note f_λ and g_λ the two functions defined on $\mathcal{G}^\star$ by :

$$f_\lambda(x) = f(x + \lambda) \quad , \quad (x \in \mathcal{G}^\star) \ ;$$

$$g_\lambda(x) = g(x + \lambda) \quad , \quad (x \in \mathcal{G}^\star) \ .$$

We note $i : \mathfrak{h}^\star \equiv \mathfrak{k}^0 \to \mathcal{G}^\star$ the canonical injection.

Then the two functions $f_\lambda \circ i$ and $g_\lambda \circ i$, defined on the Poisson manifold $\mathfrak{h}^\star$, are in involution :

$$\{f_\lambda \circ i \ , g_\lambda \circ i\} = 0 \ .$$

In order to deduce this result from theorem 5.2, we take :

$$M = \mathcal{G}^\star \quad ; \quad N = \mathfrak{h}^\star \equiv \mathfrak{k}^0 \ .$$

We define $\varphi : M \to N$, and $s : N \to M$, by :

$$\varphi(y) = \pi_1(y - \lambda) \quad , \quad (y \in \mathcal{G}^\star) \quad ,$$

$$s(x) = x + \lambda \quad , \quad (x \in \mathfrak{k}^0) \quad ,$$

where $\pi_1 : \mathcal{G}^\star = \mathfrak{k}^0 \oplus \mathfrak{h}^0 \to \mathfrak{k}^0$ is the first projection.

Using the property $(\star)$ above, we can check that φ is a Poisson morphism. Similarly, using the property $(\star\star)$, we see that the property P of theorem 5.2 is satisfied by the affine submanifold $s(\mathfrak{k}^0)$ of the Poisson manifold $\mathcal{G}^\star$. We can apply theorem 5.2, and we obtain the result indicated above.

REFERENCES

Abraham, R., and Marsden, J. E.: 1978, Foundations of Mechanics , second edition, Benjamin-Cummings, Reading.

Adler, M.: 1979, 'On a trace functional for formal pseudo-differential operators and the symplectic structure of the Korteweg-de-Vries type equations', Inventiones Math. 50, 219-248.

Adler, M., and Van Moerbeke, P.: 1980a, 'Completely integrable systems, Euclidean Lie algebras, and curves', Advances in

Mathematics 38, 267-317.
Adler, M., and Van Moerbeke, P.: 1980b, 'Linearization of Hamiltonian systems, Jacobi varieties and representation theory', Advances in Mathematics 38, 318-379.
Arnold, V.: 1974, Les méthodes mathématiques de la mécanique classique, Editions Mir, Moscou.
Arnold, V. I., et Avez, A.: 1967, Problèmes ergodiques de la mécanique classique, Gauthier-Villars, Paris.
Chevalley, C.: 1946, Theory of Lie groups, Princeton University Press, Princeton.
Iacob, A., and Sternberg, S.: 1979, 'Coadjoint structure, solitons and integrability', preprint.
Kazhdan, D., Kostant, B., and Sternberg, S.: 1978, 'Hamiltonian group actions and dynamical systems of Calogero type', Comm. in Pure and Appl. Math. 31, 481-507.
Kirillov, A.: 1974, Eléments de la théorie des représentations, Editions Mir, Moscou.
Kirillov, A.: 1976, 'Local Lie algebras', Uspekhi Mat. Nauk 31:4, 57-76 (in Russian); English translation, Russian Math. Surveys 31:4, 55-75
Kostant, B.: 1970, 'Quantization and unitary representations, I: Prequantization', in C. T. Taam (ed.), Lectures in modern analysis and applications III, Lecture notes in Mathematics n° 170, Springer-Verlag, Berlin, pp. 87-208.
Kostant, B.: 1979, 'The solution to a generalized Toda lattice and representation theory', Advances in Mathematics 34, 195-338.
Kuperschmidt, B., and Manin, Yu.: 1977, 'Equations of long waves with a free surface', Funct. Anal. Appl. 11, 31-42.
Lax, P. D.: 1968, 'Integrals of non linear equations of evolution and solitary waves', Comm. in Pure and Appl. Math. 21, 467-490.
Libermann, P.: 1982, 'Problèmes d'équivalence en géométrie symplectique', Séminaire de géométrie du Schnepfenried, 10 au 15 mai 1982, to appear in Astérisque.
Lichnerowicz, A.: 1976, 'Variétés symplectiques, variétés canoniques et systèmes dynamiques', in Topics in Differential Geometry (a volume in honour of E. T. Davis), Academic Press, New York, pp. 57-85.
Lichnerowicz, A.: 1977, 'Les variétés de Poisson et leurs algèbres de Lie associées', J. of Differential Geometry 12, 253-300.
Lichnerowicz, A.: 1979, 'La géométrie des transformations canoniques', Bulletin de la Société mathématique de Belgique 31, 105-135.

Marle, C.-M.: 1982, 'Contact manifolds, canonical manifolds and the Hamilton-Jacobi method in analytical mechanics', IUTAM-ISIMM Symposium on Modern developments in analytical mechanics, Torino (Italy), June 7-11, 1982, to appear.

Miscenko, A. S., and Fomenko, A. T.: 1978, 'Euler equations on finite dimensional Lie groups', Izvestija Akad. Nauk SSSR 42:2 (in Russian); English translation, Math. USSR Izvestija 12:2, 371-389.

Moser, J.: 1975, 'Three integrable Hamiltonian systems connected with isospectral deformations', Advances in Mathematics 16, 197-220.

Olshanetsky, M. A., and Perelomov, A. M.: 1976, 'Completely integrable Hamiltonian systems connected with semi-simple Lie algebras', Inventiones Math. 37, 93-108.

Olshanetsky, M. A., and Perelomov, A. M.: 1979, 'Explicit solutions of classical generalized Toda models', Inventiones Math. 54, 261-269.

Ouzilou, R.: 1981, 'Actions hamiltoniennes sur les variétés de Poisson', in A. Crumeyrolle and J. Grifone (ed.), Symplectic geometry, Pitman Books Ltd, London, to appear.

Ratiu, T.: 1980, 'Involution theorems', in G. Kaiser and J. E. Marsden (ed.), Geometric methods in Mathematical Physics, Lecture notes in Mathematics n° 775, Springer-Verlag, Berlin, pp. 219-257.

Reyman, A. G., and Semenov-Tian-Shansky, M. A.: 1979, 'Reduction of Hamiltonian systems, affine Lie algebras and Lax equations I', Inventiones Math. 54, 81-100.

Reyman, A. G., and Semenov-Tian-Shansky, M. A.: 1981, 'Reduction of Hamiltonian systems, affine Lie algebras and Lax equations II', Inventiones Math. 63, 423-432.

Schouten, J. A.: 1954, 'On the differential operators of first order in tensor calculus', in Convegno Intern. Geometria Differenziale Italia, 1953, Ed. Cremonese, Roma.

Selmi, M.: 1982, Systèmes complètement intégrables, Thèse de troisième cycle, Université Pierre et Marie Curie, Paris.

Souriau, J.-M.: 1969, Structure des systèmes dynamiques, Dunod, Paris.

Symes, W. W.: 1980a, 'Hamiltonian group actions and integrable systems', Physica 1 D, 339-374.

Symes, W. W.: 1980b, 'Systems of Toda type, inverse spectral problems and representation theory', Inventiones Math. 59, 13-51.

Verdier, J.-L.: 1980, 'Algèbres de Lie, systèmes hamiltoniens, courbes algébriques', Exposé au Séminaire Bourbaki n° 556, novembre 1980, Paris.

M.CAHEN and S.GUTT*

THEORY OF DEFORMATIONS AND GROUP REPRESENTATIONS

The setting of analytical mechanics is a differentiable manifold: phase space. Classical observables are functions on this manifold; their evolution in time is decribed by means of the flow of a vector field: the hamiltonian field. The setting for quantum mechanics is a separable Hilbert space: the space os states. Quantum observables are self-adjoint operators which have domain dense in this space; their evolution in time is described by means of a one parameter group of unitary transformations.

The profoundly different nature of the fundamental observables intervening in these two theories poses a certain number of problems and in particular renders the study of the relations between quantum mechanics and classical mechanics difficult in the limit where Planck's constant $\hbar$ tends to zero.

The Weyl correspondence [17,18] is a linear bijection Ω between a certain subspace $H \subset C^{\infty}(\mathbb{R}^{2n})$ and a subspace A of self-adjoint operators of $L^2(\mathbb{R}^{2n})$. More precisely, one considers the phase space $\mathbb{R}^{2n}$ equipped with a system of coordinates $(p_i, q^i, i \leqslant n)$; the symplectic form of $\mathbb{R}^{2n}$ in these coordinates is written $\sum_i dp_i \wedge dq_i$. If f is a function on $\mathbb{R}^{2n}$ with real or complex values we denote its inverse Fourier tranform $\hat{f}$. Thus let ψ be a function on $\mathbb{R}^n$ and let $q^j (j \leqslant n)$ be the coordinates of $\mathbb{R}^n$; we define the operators P_j and Q^j $(j \leqslant n)$ by

C. P. Bruter et al. (eds.), Bifurcation Theory, Mechanics and Physics, 77–98.

$$(P_j\psi)(q) = i\hbar \frac{\partial\psi}{\partial q^j}(q)$$

$$(Q^j\psi)(q) = (q^j\psi)(q)$$

For a function $f \in H$, the Weyl transform $\Omega(f)$ is the operator:

$$\Omega(f) = \int_{\mathbb{R}^{2n}} \hat{f}\,(p,q)\, \exp \frac{(pP + qQ)}{i\hbar}\, dp\, dq \qquad (0.1)$$

Since Moyal [14] it is known that the commutator of the operators $\Omega(f)$ and $\Omega(g)$ is the image under the Weyl transform of a certain deformed bracket (Moyal bracket)

$$[\Omega(f),\Omega(g)] = -i\hbar\, \Omega([f,g]_M) \qquad (0.2)$$

where:

$$[f,g]_M = \sum_{n=0}^{\infty} (-1)^n \frac{\hbar^{2n}}{(2n+1)!} P^{2n+1}(f,g) \qquad (0.3)$$

$P(f,g)$ = Poisson bracket of f and g

$$= \sum_{i=1} \frac{\partial f}{\partial q^i}\frac{\partial g}{\partial p^i} - \frac{\partial f}{\partial p^i}\frac{\partial g}{\partial q^i} = \sum_{a,b=1}^{2n} \Lambda^{ab} \frac{\partial f}{\partial x^a}\frac{\partial g}{\partial x^b}$$

denoting the coordinates $(p_i, q^i; i \leqslant n)$ by x^a $(a \leqslant 2n)$. The symbol of the bidifferential operator P^r is the rth power of the symbol of the operator P.

$$P^r(f,g) = \sum_{\substack{a_1\ldots a_r \\ b_1\ldots b_r}} \Lambda^{a_1b_1}\cdots\cdots\Lambda^{a_rb_r} \frac{\partial^r f}{\partial x^{a_1}\ldots\partial x^{a_r}} \frac{\partial^r g}{\partial x^{b_1}\ldots\partial x^{b_r}}$$

Formula (0.3) shows that the Moyal bracket coincides with the Poisson bracket if f and g are polynomials of degree $\leqslant 2$. One may then inquire whether some modification of the

Weyl transform will give give rise to a homomorphism from the Poisson algebra to the Lie algebra of operators. It follows from a theorem of Van Hove [15] that such a modification is impossible. Thus, we observe that parallel to (0.2) we have:

$$\Omega(f)\,\Omega(g) = \Omega(f *_M g) \qquad (0.4)$$

$$f \underset{M}{*} g = (\exp i\hbar P)(f,g)$$

Flato, Fronstal, Lichnerowicz, and some others [2] have made the suggestion that quatum mechanics be developed within the framework of classical mechanics, that is for example by using as quantum observables certain functions on phase space. From this point of view, the Weyl transformation plays the special role of establishing a comparison with the operational presentation of quantum mechanics. In order to justify this approach it is necessary on the one hand to give an autonomous formulation of quantum mechanics (notably the calculus of spectra) in terms of the classical observables and of the deformation of the associative structure and of the Lie structure of the algebra of these observables, and on the other hand, it is necessary to generalize to phase spaces distinct of R^{2n} both the definition, and eventually the explicit construction, of deformations.

Different spectral calculi have been developed in this deformation theory [2,5,9]. We will not study this aspect. We shall focus essentially on indicating the largest known class at present of manifolds (phase spaces) on which it is possible to demonstrate the existence of a deformation.

We shall also briefly indicate, because this is still in a very informal state, how it is possible to unite to these deformation theories certain aspects of the theory of group representations.

1. Deformations; * products .

Let (M,ω) be a symplectic manifold and let $N=C^{\infty}(M,R)$; we designate by $E(N;\nu)$ the space of formal power series in a complex parameter ν with coefficients in N.

Definition 1: Call a bilinear mapping $N \times N \to E(N,\nu)$ a * product if it has the form $(f,g) \to \sum_{r=0}^{\infty} \nu^r C_r(f,g) \underset{\text{def}}{=} f *_\nu g$ where the cochains C_r satisfy the following hypotheses:

(i) $(f *_\nu g) *_\nu h = f *_\nu (g *_\nu h)$

(ii) $C_0(f,g) = f.g$ This just gives a formal deformation of the associative product

(iii) $C_1(f,g) = \{f,g\}$ the Poisson bracket of f and g associated with the symplectic structure of M. It follows from (i) and (iii) that the formal series $\frac{1}{2\nu}(f*_\nu g - g*_\nu f)$ is a formal deformation of the Lie structure associated to the Poisson bracket

(iv) $C_r(f,g) = (-1)^r C_r(g,f)$. This condition which is satisfied by the Moyal deformation guarantees that if ν is purely imaginary, $g *_\nu f = \overline{f *_\nu g}$ ($\overline{\ }$ denotes complex conjugate).

(v) C_r is local, that is, if f (or g) vanishes on an open set $U \subset M$, $C_r(f,g)$ vanishes on U. Then it follows from a generalization of a theorem of Peetre [3] that the cochain C_r is locally differentiable, that is, it is given in each local chart by a bidifferential operator. If the manifold M is compact, C_r is a bidifferential operator; if the manifold M is not compact the "order" of C_r may not be bounded. To avoid this occurence, which only introduces technical complications, we here make the stipulation that the cochain C_r is a bidifferential operator.

(vi) $C_r(f,g) = 0 (r > 0)$ whenever f or g is a constant function on M. This implies that the constant function 1 is an identity for the * product. The * product extends easily to $E(N,\nu)$ which then becomes an associative algebra. The product on N is constructed by induction. In each order it is a matter of solving the associativity relation. The first two orders are given by (ii) and (iii). The third order has been determined by Lichnerowicz [2], the fourth order by J.Vey [16], the fifth order by one of us [8] (this expression was also known to J.Vey but doesn't appear to have been published). From this, one easily deduces the existence of a sixth term but its explicit expression is not known. The enormous complexity of the cochains, their lack of uniqueness makes it appear illusory to seek a systematic constructive procedure in the general case. Thus we must settle for results with respect to the existence of deformations.

Generally speaking, it is known that the existence of deformations of algebras depends in the associative case on the Hochschild cohomology and in the Lie case on the Chevalley cohomology [6].

2. Hochschild Cohomology.

A local (differentiable) p-cochain is a local multilinear mapping C_p: NxNx...xN → N (differentiable means given by a p-differential operator).

The Hochschild coboundary of C_p is the p+1 cochain $\tilde{\delta}C_p$ defined by:

$$\begin{aligned}\tilde{\delta}C_p(u_0,u_1,\ldots,u_p) &= u_0 C_p(u_1,\ldots,u_p) - C_p(u_0 u_1,\ldots,u_p) + \ldots \\ &+ (-1)^{p-1} C_p(u_0,\ldots,u_{p-1}u_p) \\ &+ (-1)^p\, C_p(u_0,\ldots,u_{p-1})u_p \\ &= \sum_{i=0}^{p-1} (-1)^i [\, u_i C_p(u_0,\ldots,\hat{u}_i,\ldots,u_p) - \\ &\quad - C_p(u_0,\ldots,u_i u_{i+1},\ldots u_p) \\ &\quad + u_{i+1} C_p(u_0,\ldots,\hat{u}_{i+1},\ldots,u_p)] \qquad (2.1)\end{aligned}$$

Here ^ means delete. It follows immediately from this second expression of the coboundary and from the Leibniz rule that if C_p is 1-differentiable (that is, given by a multidifferential operator of order 1 in each argument) $\tilde{\delta}C_p = 0$; one says that C_p is a cocycle.

Since the operator $\tilde{\delta}$ has zero square, if we denote the space of p-cocycles $Z_p(N)$ and the space of p-coboundaries $B_p(N)$ (the latter are the p-cochains which are coboundaries of p-1 cochains) then $Z_p(N) \supset B_p(N)$.

<u>Definition</u>. The <u>p^{th} Hochschild cohomology space (local, differentiable)</u> of N is defined by:

$$\tilde{H}^p_{loc}(N) = Z^p_{loc}(N) \,/\, B^p_{loc}(N)$$

$$\tilde{H}^p_{diff}(N) = Z^p_{diff}(N) \; / \; B^p_{diff}(N)$$

Theorem (Vey) [16] The pth differentiable Hochschild cohomology space of N is isomorphic to the space of fields of completely antisymmetric p-tensors (contravariant) of class C^∞.

This result easily generalizes to the case of local cohomology [3].

Explicitly, one may interpret the theorem of Vey to say that if C is a Hochschild p-cocycle of N and if $u_1,\ldots,u_p \in N$, there exists a completely antisymmetric p-tensor A such that

$$C(u_1,\ldots,u_p) = A(du_1,\ldots,du_p) + \tilde{\delta}E(u_1,\ldots,u_p).$$

In particular, this implies that a p-cocycle is exact if and only if its completely antisymmetric part is zero.

3. Construction of a * product by induction.

Suppose that one had constructed a * product up to order n, the associativity relation being satisfied up to that order. The associativity relation in order n+1 is written:

$$\tilde{\delta}C_{n+1}(u,v,w) = E_{n+1}(u,v,w) \underset{def}{=} \sum_{\substack{r+s=n+1 \\ r,s\geq 1}} C_r(C_s(u,v),w) - C_r(u,C_s(v,w)) \qquad (3.1)$$

It is known [6] that E_{n+1} is a 3-cocycle and to extend an order it is necessary that this 3-cocycle be exact. If n is even, one of the indices r or s is even also and then, taking account of property (iv) the completely antisymmetric part of E_{n+1} is zero. It then follows from the theorem of Vey that E_{n+1} is exact and it is then possible to extend by one order. On the other hand, if n is odd, the completely antisymmetric part is not necessarily zero. If it is, then E_{n+1} is exact and we may extend by one order.

In particular if $\tilde{H}^3_{diff}(N) = 0$ every deformation is prolongable.

This condition can only be verified in dimension 2.

4. <u>Second construction</u> (generalization of Moyal).

On any symplectic manifold (M,ω) there exists a linear connection without torsion Γ, such that the covariant derivative of the form ω is zero relative to this connection: $\nabla\omega = 0$. This connection is by no means unique.

Let (U,ϕ) be a local chart of (M,ω) and let x^a $(a \leq 2n)$ be the local coordinates; we denote the components of the form ω in this chart ω_{ab}, and $-\Lambda^{ab}$ the elements of the inverse matrix.

$$\sum_b \Lambda^{ab}\omega_{bc} = -\delta^a_c$$

These are the components of a tensor field on M.

Let us define the cochain P^r by its local expression in U,

$$P^r(u,v) = \sum_{\substack{i_1\ldots i_r \\ j_1\ldots j_r}} \Lambda^{i_1 j_1}\ldots\Lambda^{i_r j_r}\,\nabla_{i_1\ldots i_r}u\;\nabla_{j_1\ldots j_r}v$$

One sees that $P(u,v)$ is just the Poisson bracket of u and v.

Then set the formal power series:

$$uv + \sum_{r \geq 1} \nu^r P^r(u,v)$$

It is easy to see that this formal power series defines a * product on M iff the linear connection Γ is flat.

5. Third construction (Neroslavsky-Vlassov)

The idea of this construction is to make use of the fact the antisymmetrization of the * product is a deformation of the Poisson bracket. A deformation of the Lie algebra $(N,\{,\})$ depends on the Chevalley cohomology of this algebra with values in N relative to the adjoint representation. This cohomology has been partially determined [12,7]. We shall only make use here of the following lemma.

Lemma: Let $E: N \times N \times \ldots \times N \to N$ be a multilinear mapping which is antisymmetric and which is a differentiable cocycle both of the Hochschild cohomology and of the Chevalley cohomology. Then E is a 1-differentiable cocycle and there exists a closed 3-form A such that

$$E(u,v,w) = A(X_u,X_v,X_w) \quad \text{for all } u,v,w \text{ in } N$$

where $i(X_u)\omega = -du$. Further, E is an exact Chevalley cocycle iff the form A is exact. It then results from the analysis made in #3 and from an analogous analysis in the case of a deformation of $(N,\{,\})$ that for n odd the completely antisymmetric part of E_{n+1} is both a Chevalley and Hochschild cocycle and thus is associated with a closed 3-form A. If this 3-form is exact (A=dB) one may modify C_n by the

1-differentiable cochain $\tilde{B}(u,v) = B(X_u,X_v)$; $C'_n = C_n - \tilde{B}$ which is a Hochschild cocycle; the antisymmetric part of the modified associator E'_{n+1} is zero and so E'_{n+1} is a Hochschild coboundary and one may carry through the induction. Thus we have the

Theorem (Neroslavsky-Vlasov) If the symplectic manifold (M,ω) has third de Rham cohomology space zero, there exists a * product on (M,ω).

6. Fourth construction (Lichnerowicz [11])

The idea here is to make use of the Moyal product which exists on any flat symplectic manifold by inducing a * product on certain quotients of flat manifolds. More precisely, let (M,) be a symplectic manifold on which a * product exists, and let G be a Lie group of transformations of M stabilizing ω and leaving the * product invariant (that is, for all u,v in N and for all g in G, $g^* u * g^* v = g^*(u*v)$). Suppose that $\pi: M \to M/G$ is a submersion and that M/G admits a "compatible" symplectic structure Then there exists an induced * product on M/G. In particular if M is an open set of $\mathbb{K}^{n(n-1)}$ ($\mathbb{K} = \mathbb{R}, \mathbb{C}, \mathbb{H}$) equipped with the restriction of the "canonical" symplectic form of $\mathbb{K}^{n(n-1)}$ one may find a solvable subgroup G of $Gl(n(n-1),\mathbb{K})$ preserving the symplectic form such that M/G is symplectomorphic to the cotangent bundle of $SO(n)$ ($\mathbb{K}\equiv\mathbb{R}$), $SU(n)$ ($\mathbb{K}=\mathbb{C}$), $Sp(n)$ ($\mathbb{K}\equiv\mathbb{H}$). The * product of Moyal on $\mathbb{K}^{n(n-1)}$ being invariant under the group of symplectic affinities, this induces a * product on $T^*SO(n)$, $T^*SU(n)$, $T^*Sp(n)$. These examples are interesting bacause dim $H^3_{\text{de Rham}}(SO(n))$=dim $H^3_{\text{de Rham}}(SU(n))$= dim $H^3_{\text{de Rham}}(Sp(n))$=1 and the existence of a * product does not follow from the theorem of Neroslavsky-Vlassov. This same method provides a * product on the torus $T^{2n}=R^{2n}/Z^{2n}$; now for n > 1 dim $H^3_{\text{de Rham}}(T^{2n})= C^3_{2n}$.

7. Fifth construction [8]

The idea here consists in replacing the de Rham cohomology of a homogeneous symplectic space with the invariant de Rham cohomology. Thus suppose that the symplectic manifold (M,ω) admits a Lie group G of symplectic transformations, and suppose that there is a linear connection Γ such that G is contained in the group of affinities of this connection. Then we have the

Lemma: Let E be a Hochschild 3-cocycle, exact and invariant under G, then there exists a 2-cochain B, also invariant under G, such that $E=\tilde{\delta}B$.

It follows that

Proposition. If (M,ω) admits a Lie group G of symplectic transformations, stabilizing a linear connection Γ and if $H^3_{\text{de Rham, inv}}(N) = 0$ there exists a G invariant * product on M.

This has been used to demonstrate the existence of * products which are invariant on symmetric hermitian spaces [8] and on orbits of the coadjoint representation of the Poincaré group [13].

8. Sixth construction

The existence of a * product on the cotangent bundle of a compact simple classical group leads one to think that there may exist a * product on the cotangent bundle of an arbitrary compact group. Since we know the Moyal product on $\mathbb{R}^{2n}$, and we know that it is invariant under the group of translations it suffices to show the existence of an invariant product on a compact simply connected group in order to deduce the existence of a * product (invariant) on an arbitrary compact connected group. We demonstrate the existence below with some details.

In fact, this result is a particular case of a more general theorem, that says that there exists an invariant * product on the cotangent bundle of an arbitrary connected Lie group. This result, whose proof is independent of the compact case will be published elsewhere [4].

Theorem. Let G be a compact simply connected group and let (T^*G,Ω) be the cotangent bundle equipped with the canonical symplectic structure. Then there exists on (T^*G,Ω) a * product which is invariant under the left action of G on T^*G.

The theorem will follow from a sequence of lemmas. Let us introduce some necessary notation. Let $\Pi: T^*G \to G$ be the canonical projection; let $\{X_i ; i \leq n = \dim G\}$ be a basis for the Lie algebra of G, denoted $\mathfrak{g}$, and let $\tilde{X}_i$ be the left invariant vector field on G associated with X_i.
The $\{\tilde{X}_i ; i \leq n\}$ form at each point x of G a basis of G_x.

Let us designate by $\{\theta^i;\ i \leq n\}$ the left invariant 1-forms which constitute the dual basis $(\theta^i(\tilde{X}_j) = \delta^i{}_j)$. Call $p_i: T^*G \to R$ the function defined by

$$p_i(\xi) = \xi(\tilde{X}_{i\pi(\xi)}) \qquad \xi \in T^*G \tag{6.1}$$

One verifies that the 2n 1-forms on T^*G $\{dp_i, \pi^*\theta^i;\ i \leq n\}$ form a basis of $(T^*G)_\xi$. The dual basis of $(T^*G)_\xi$ is written $\{Z^i, Y_i;\ i \leq n\}$ and the 2n C^∞ vector fields Z^i, Y_i have the properties:

$$dp_i(Z^j) = \delta_i{}^j \qquad dp_i(Y^j) = 0$$

$$\pi^*\theta^i(Z^i) = 0 \qquad \pi^*\theta^i(Y_j) = \delta^i{}_j \tag{6.2}$$

In particular it follows that the vectors Z^i are vertical and that $\pi^* Y_i = \tilde{X}_i$. The Liouville 1-form on T^*G admits a simple global expression

$$\lambda = \sum_{i=1}^{n} p_i \pi^* \theta^i \tag{6.3}$$

and the canonical symplectic 2-form is written:

$$\Omega = \sum_{i=1}^{n} dp_i \wedge \pi^*\theta^i - \frac{1}{2} \sum_{i,j,k=1}^{n} p_i C^i_{jk}\ \pi^*\theta^j \wedge \pi^*\theta^k \tag{6.4}$$

where the C^i_{jk} are the structure constants of the algebra $\mathfrak{g}$ in the basis X_i.

If f is a C^∞ function on T^*G, the corresponding hamiltonian field is :

$$X_f = \sum_{i=1}^{n} (Z^i f) Y_i + \sum_k \Big[\sum_{i,j} p_i C^i_{jk} (Z^j f) - (Y_k f) \Big] Z^k \tag{6.5}$$

In particular:

$$X_{p_i} = Y_i + \sum_{j,k} p_j C^j_{ik} Z^k \qquad (6.6)$$

Right translation on G lifts to an action of G on T^*G defined by:

$$g \circ \xi \underset{\text{def}}{=} R^*_g \xi \qquad \xi \in T^*G \qquad (6.7)$$

The fundamental vector field X_i associated to this action is such that

$$X^*_i(\xi) = \frac{d}{dt}(\exp -t\, X_i . \xi)_{t=0} = \frac{d}{dt}(\xi_\circ R_{\exp -t\, X_i} *)_{t=0}$$

and one verifies that

$$X^*_i(\xi) = Y_i(\xi) + \sum_{j,k} C^j_{ik}\, p_j(\xi)\, Z^k(\xi) = X_{p_i}(\xi) \qquad (6.8)$$

Lemma 1. (Koszul, Chevalley) [10] The third de Rham cohomology space of G has dimension 1. A representative of the non-trivial cohomology class is given by the left invariant 2-form

$$\omega = \frac{1}{3!} \sum_{i,j,r,s} \beta_{ij}\, C^i_{rs}\, \theta^r \wedge \theta^s \wedge \theta^j$$

(The β_{ij} are the components in the basis X_i of the KIlling form of $\mathfrak{g}$). It follows from this that

$$\dim H^3_{\text{de Rham}}(T^*G) = \dim H^3_{\text{de Rham, inv}}(T^*G) = 1$$

where $H^p_{\text{de Rham,inv}}$ is the pth de Rham cohomology space invariant on the left under the action of G on T^*G:
$(g \circ \xi \underset{\text{def}}{=} L^*_{g^{-1}} \xi;\ \xi \in T^*G$

Lemma 2. <u>The 3-form on T^*G</u>:

$$\tilde{\omega} = \frac{1}{3!} \sum_{i,j,r,s} \beta_{ij} C^i_{rs} \pi^* \theta^r \wedge \pi^* \theta^s \wedge \pi^* \theta^j$$

<u>is a representative of the non-trivial cohomology class of</u> $H^3_{\text{de Rham, inv}}(T^*G)$.

<u>Proof</u>: The 3-form $\tilde{\omega}$ is closed and left invariant. Thus it suffices to verify that it is not the differential of a left-invariant 2-form. If γ is a left-invariant 2-form on T^*G (under the action of G), γ is written:

$$\gamma = \frac{1}{2} \sum_{i,j} \alpha_{ij}(p)\pi^* \theta^i \wedge \pi^* \theta^j + \sum_{i,j} \mu^i j(p)\, dp_i \wedge \pi^* \theta^j + \frac{1}{2} \sum_{i,j} \nu^{ij}(p)\, dp_i \wedge dp_j$$

where the $\alpha_{ij}, \mu^i j, \nu^{ij}$ depend only on the functions p_k. The terms in $\pi^* \theta^i \wedge \pi^* \theta^j \wedge \pi^* \theta^k$ of $d\gamma$ all come from the differentiation of the term $\frac{1}{2} \sum_{i,j} \alpha_{ij}(p)\, \pi^* \theta^i \wedge \pi^* \theta^j$.

Let us choose a particular value p_o of the functions p. The 2-form on G :

$$\alpha = \frac{1}{2} \sum_{i,j} \alpha_{ij}(p_o)\, \theta^i \wedge \theta^j$$

must satisfy $d\alpha = \omega$, this is clearly absurd.

<u>Lemma 3</u>. <u>A closed, left invariant 3-form</u> ρ <u>on T^*G is exact</u> if:

$$\rho(X^*_i, X^*_j, X^*_k) = 0 \qquad \text{for all } i,j,k=1,\ldots,n$$

Proof: In fact, if γ is a 2-form on T^*G, left invariant under G, and if

$$(\tilde{\omega}+d\gamma)(X_i^*,X_j^*,X_k^*)=0 \quad \text{for all } i,j,k$$

one has, on the section $p_i=0$,

$$(\tilde{\omega}+d\gamma)(Y_i,Y_j,Y_k)=0 \quad \text{for all } i,j,k$$

this means that the term in $\pi^*\theta^i \wedge \pi^*\theta^j \wedge \pi^*\theta^k$ of the form $(\tilde{\omega}+d\gamma)_{p=0}$ is zero. This is impossible by the reasoning of Lemma 2.

Induction hypotheses (i) The cochains C_r which define the * product are invariant under G, that is:

$$(L^*_{g^{-1}})^* C_r(u,v) = C_r((L^*_{g^{-1}})^*u,\ (L^*_{g^{-1}})^*v)$$

for all $g \in G$ and for all $u,v \in C^\infty(T^*G,\mathbb{R})$. This hypothesis is meaningful bacause $C_0(u,v)=uv$ satisfies it trivially. Further, since the Liouville form of T^*G is G invariant, the same is true of the symplectic form and thus of the Poisson bracket $\{u,v\} = C_1(u,v)$.

We observe that the vector fields Y_i, Z^i are left invariant under G. Let us denote these vector fields U_a $(a=1,\ldots,2n)$. A G-invariant multidifferential operator on T G will be written:

$$C(u_1,\ldots,u_p)= \sum_{\substack{a_1,\ldots,a_{\ell_1}\\ k_1,\ldots,k_{\ell_p}}} C^{a_1\ldots a_{\ell_1}\ldots k_1\ldots k_{\ell_p}} (U_{a_1}\ldots U_{a_{\ell_1}}u_1)\ldots(U_{k_1}\ldots U_{k_{\ell_p}}u_p)$$

where the $C^{a_1\ldots a_{\ell_1},\ldots,k_1\ldots k_{\ell_p}}$ are C^∞ functions depending only on the p_r, and are symmetric in the indices $a_1 \ldots a_{\ell_1}, \ldots$ and in the indices $k_1 \ldots k_{\ell_p}$.

We remark also that the vector fields U_a have the following commutators:

$$[Z^j, Z^k] = [Z^j, Y_k] = 0$$

$$[Y_j, Y_k] = C^{\ell}_{jk} Y_\ell \qquad (6.9)$$

Thus we may imagine that in writing multidifferential operators, the fields Z^j are "the first ones" to operate on functions.

(ii) The cochains C_r are such that

$$C_r(p_i, p_j) = 0 \qquad \text{for all } i,j;\ r > 1$$

We remark that

$$C_0(p_i, p_j) = p_i\ p_j \qquad C_1(p_i, p_j) = C_{ij}^{\ \ k} p_k \ .$$

Lemma 4. Let us call p_i $(i \leq n)$ the natural coordinates of $\mathbb{R}^n$ and let ρ be the representation of $\mathfrak{g}$ in $C^\infty(\mathbb{R}^n)$ defined by:

$$(\rho(X_i)f(p_k) = \sum_{j,k} C^k_{ij}\ p_k\ \frac{\partial f}{\partial p^j}\ (p)$$

If the second Chevalley cohomology space of $\mathfrak{g}$ relative to this representation ρ is zero, the induction hypotheses above are consistent.

Proof. To demonstrate this lemma, we shall follow the induction of the theorem of Neroslavsky Vlassov step by step.

Let us observe right away that $C_r(p_i, p_j) \neq 0$ iff there exists in the expression of $C_r(u,v)$ a term of the form

$$\sum_{i,j} C^{(r)}_{ij}(p)\, Z^i u\, Z^j v$$

Such a term is 1-differentiable and is then a Hochschild cocycle. If then:

$$C'_r(u,v) \underset{def}{=} C_r(u,v) - \sum_{i,j} C^{(r)}_{ij}(p)\, Z^i u\, Z^j v$$

the cochain C'_r is G-invariant if C_r is and:

$$\partial C'_r = \partial C_r$$

Thus suppose that one had constructed a * product by induction up to even order n, satisfying the hypotheses. Since the completely antisymmetric part of E_{n+1} (see notations of #3) is zero, there exists a C_{n+1} such that

$$E_{n+1} = \partial C_{n+1}$$

Since E_{n+1} is G invariant, one may (lemma of #5) suppose that C_{n+1} is G invariant and in virtue of the remark above, one may also suppose that $C_{n+1}(p_i,p_j) = 0$ and the hypotheses are satisfied to order n+1.

The completely antisymmetric part of E_{n+2} corresponds (proof of Neroslavsky Vlassov) to a left invariant 3-form γ, which is closed and which vanishes on the vectors X^*_i.
In fact:

$$\gamma(X^*_i, X^*_j, X^*_k) = \frac{1}{3} \underset{p_i,p_j,p_k}{S} E_{n+2}(p_i,p_j,p_k) = 0$$

where $\underset{u,v,w}{S}$ designates the sum over the circular permutations.

It follows from lemma 3 that γ is exact:

$$\gamma = d\alpha$$

and one may suppose that α is a G invariant 2 form.

Let $\alpha_{ij}(p) \underset{def}{=} \underline{\alpha}(X_i^*, X_j^*)$ and let the mapping

$\mathfrak{g} \times \mathfrak{g} \to C^\infty(\mathbb{R}^n)$ be defined by:

$$\underline{\alpha}(X_i, X_j) = \alpha_{ij}(p)$$

One verifies that:

$(d\alpha)(X_i, X_j, X_k) = 0$ iff $\underline{\alpha}$ is a Chevalley 2-cocycle for the ρ representation.

Suppose that the second Chevalley cohomology space is zero. Then it follows that there is a $\beta: \mathfrak{g} \to C^\infty(\mathbb{R}^n)$ such that

$$\underline{\alpha}(X_i, X_j) = C^s_{ir}\, p_s \frac{\partial \beta_j}{\partial p_r} - C^s_{jr}\, p_s \frac{\partial \beta_i}{\partial p_r} + C_{ij}^{\ \ k}\, \beta_k$$

where $\beta_\ell = \beta(X_\ell)$. The left invariant 1-form on T^*G, $\underline{\beta} = \sum_\ell \beta_\ell \pi^* \theta^\ell$, is such that

$$d\, \underline{\beta}\, (X_i, X_j) = \alpha_{ij}(p)$$

Thus we modify C_{n+1} by $\alpha - d\beta$, that is:

$$C'_{n+1}(u,v) = C_{n+1}(u,v) - \frac{3}{2}(\alpha - d\underline{\beta})(X_u, X_v)$$

where $X_u (X_v)$ is the hamiltonian field associated to u (v). We remark that $\tilde{\delta} C'_{n+1} = \tilde{\delta} C_{n+1}$ because $(\alpha - d\underline{\beta})$ is 1-differentiable. Further, C'_{n+1} is left invariant and vanishes on the functions p_i. So one has :

$$\mathop{\mathfrak{S}}_{u,v,w} E'_{n+2}(u,v,w) = \mathop{\mathfrak{S}}_{u,v,w} E_{n+2}(u,v,w) - 3d\alpha(X_u, X_v, X_w) = 0$$

and then E'_{n+2} is exact; as above we choose C_{n+2} left invariant and zero on the p_i.

Consistency is thus demonstrated.

The second Chevalley cohomology space of $\mathfrak{g}$ relative to a representation is zero if the space of the representation is finite-dimensional. To put ourselves in this setting, we shall add a third inductive hypothesis.

Third induction hypothesis (iii) The cochain $C_r(u,v)$ may be written in the form

$$C_r(u,v) = C_r^{a_1 \dots a_p, b_1 \dots b_q} (U_{a_1} \dots U_{a_p} u)(U_{b_1} \dots U_{b_q} v)$$

where the $C_r^{a_1 \dots a_p, b_1 \dots b_q}$ are polynomials in the variables p_k.

Lemma 5. The third induction hypothesis is consistent.

Proof: In the first place we observe that if the $C_r (r \leq n)$ have polynomial coefficients then E_{n+1} also has polynomial coefficients. In virtue of what preceded E_{n+1} is an exact 3-cocycle and it follows from an analysis of the Hochschild cohomology [4] that $E_{n+1} = \tilde{\delta} C_{n+1}$ where C_{n+1} has polynomial coefficients.

Finally, if a 3-form is exact and has polynomial coefficients one verifies by going back to the case of R^n, that it is the differential of a 2-form with polynomial coefficients.

The existence theorem now follows, on the one hand from the consistency of the induction hypothese, and on the other hand from the fact that at each order E_n is exact.

9. Remarks on the application of * products to group representations.

Two examples have been treated up to now: nilpotent groups and compact groups. In each of these cases it has been shown that it is possible to exhibit all the irreducible unitary representations of these groups by method *.

For nilpotent groups [1] one proves that there exists a * product on each orbit in the coadjoint representation; next one shows that the exponential * is a well defined formal power series and this allows one to define an "embedding" of the group in the space of formal power series Finally, choosing an "agreeable" subspace of $E(N,\lambda)$ one constructs a representation of G which is equivalent to that associated to the orbit by the method of Kirillov.

For compact groups, it has been shown above that there exists a * product on T^*G; in fact it may be shown that one can choose this * product in such a way as to exhibit in $L^2(T^*G)$ on the one hand $L^2(G)$ and on the other hand $L^2(G) \otimes$ the polynomial space in the p-variables. These subspaces are stabilized by the * "adjoint" action and in particular, one verifies that on $L^2(G)$ this action coincides with the action given by the regular representation.

* Chargé de recherches au F.N.R.S.
Université Libre de Bruxelles
Campus Plaine C.P.218
B-1050 Bruxelles

Bibliography

1. D.Arnal:"Moyal product and representations of nilpotent groups".(preprint)
2. F.Bayen,M.Flato,C.Fronsdal,A.Lichnerowicz,D.Sternheimer: "Deformation theory and quantization".Ann.Physics 111,1, 1978,p.61-151.
3. M.Cahen,M.De Wilde,S.Gutt:"Local cohomology of the algebra of C^∞ functions on a symplectic manifold". L.M.P. 4 (1980)p.157-167.
4. M.Cahen,S.Gutt:"Regular * representations of compact Lie groups"(to be published in L.M.P.)
5. C.Fronsdal:"Somme ideas about quantization".Reports in Math.Physics 15 (1978) p.111-145.
6. M.Gerstenhaber:"On the deformations of rings and algebras" Ann.of Math. 79(1969)pp.59-103.
7. S.Gutt:"2e et 3e espace de cohomologie différentiable de l'Algèbre de Lie de Poisson d'une variété symplectique". Ann.I.H.P. 33 (1980) 1-31.
8. S.Gutt. Thèse. Bruxelles 1980.
9. Hansen. preprint (to be published in L.M.P.)
10. J.L.Koszul:"Nombres de Betti d'un groupe de Lie". C.R.Acad.Sc.Paris.
11. A.Lichnerovicz:"Twisted products for cotangent bundles of classical groups". L.M.P. 2(1977)p.133-143.
12. A.Lichnerovicz:"Cohomologie 1-différentiable des algèbres de Lie attachées à une variété symplectique ou de contact".J.Math.p.et appl. 53 (1974) 459-484.
13. Molin: Thèse 3e Cycle (Dijon) 1981
14. J.Moyal:"Quantum mechanics as a statistical theory". Proc.Cambridge Phil.Soc. 45 (1949) p;99-129.
15. Van Hove:"Sur certaines représentations unitaires d'un groupe infini de transformations".Mémoire de l'Acad. Roy.de Belg.26(1951)
16. J.Vey:"Déformation du crochet de Poisson sur une variété symplectique". Com.Math.Helvetici 50(1975)p.421-454.
17. Voros: Thèse Paris 1980
18. H.Weyl:Theory of groups and quantum mechanics.

Jean Leray

APPLICATION TO THE SCHRÖDINGER ATOMIC EQUATION OF AN EXTENSION OF FUCHS THEOREM

(Translation of : Actes du 6ème Congrès du Groupement des Mathématiciens d'Expression Latine, Gauthier-Villars(1982) p. 179-182).

1. HISTORY

Schrödinger's equation, concerning the atom with several electrons, has been the subject of much research. The existence of its discontinuous spectrum and the properties of its continuous spectrum were established by Kato, Zislin and Sigalov, Simon ; see the monographs by Kato [5], Jörgens and Weidmann [4], Simon [7]. The speed of decreasing at infinity of their eigenfunctions has been determined by Simon [8] and several other authors. The calculation of a few eigenvalues of two - electron atoms was done with a great precision using Ritz method by Hylleraas, Schiff Pekeris and Lifson [6] ; Bazley [1] has rigorously evaluated the error, using a method by A. Weinstein.

2. THE PROBLEM.

Assume the atomic mass of the nucleus to be infinite. Call $N = (n + 1)/3$ the number of electrons. Their position is defined by a vector $x \in \mathbb{R}^{n+1}$; the atomic nucleus is the origin of $\mathbb{R}^{n+1}$. Schrödinger's equation is elliptic and its coefficients are holomorph except when the distance r_j between an electron and the nucleus or the distance r_{jk} between two electrons vanishes ; thus the solutions of that equation have for singular support the cone where

$$(1) \qquad \prod_j r_j \prod_{j<k} r_{jk} = 0 .$$

We shall soon publish formulae describing the behaviour of those solutions at the vertex of this cone, that is at the origin of $\mathbb{R}^{n+1}$. Here we merely describe the outline of those formulae and the very classical nature of the proof.

C. P. Bruter et al. (eds.), Bifurcation Theory, Mechanics and Physics, 99–108.

3. MOTIVATION.

i) The functions used for the applications of Ritz method to the calculation of eigenvalues can only give precise results if they approximatively verify Schrödinger's equation ; but the functions used were polynomial-exponentials unable to approximate the solution till the order 2 at the origin ; this defect has been compensated by the use of polynomial-exponentials depending on a considerable number of parameters, as large as 560.

ii) In the case of the fundamental state of the two-electron atom, Fock [3] has proposed an expansion of the solution at the origin ; but its justification is merely formal and the calculation of its terms is very complicated.

iii) The case of the one-electron atom is entirely solved by the application of Fuchs theorem.

The extension of this theorem to the case of several electrons is possible. It gives expansions of the solutions of the equation, convergent on a neighbourhood of the origin. It gives at the origin a limited expansion of arbitrary order of any solution of the equation ; this expansion is a function defined on $\mathbb{R}^{n+1}$, exponentially decreasing at infinity. Such expansions could give approximations of the eigenfunctions of Schrödinger's equation.

4. NOTATIONS.

Since the equation of the singular support is (1), use polar coordinates in $\mathbb{R}^{n+1}$: let S^n be the sphere of $\mathbb{R}^{n+1}$ with center at the origin and radius 1 ; denote $x \in \mathbb{R}^{n+1}$ by

$$x = ry, \text{ where } y \in S^{n+1} \text{ and } r = |x| \in \mathbb{R}_+ ; r^2 = \sum_j r_j^2 . \tag{2}$$

Denote by $\mathcal{H}$ (and $\mathcal{H}'$) the Hilbert space of square integrable functions $S^n \to \mathbb{R}$ (and square integrable gradients). Denote by $\mathcal{H}_\ell$ the space of spherical harmonics of degree ℓ.

$$\mathcal{H}_\ell \subset \mathcal{H}' \subset \mathcal{H} .$$

The $\mathcal{H}_\ell$ are orthogonal to each other in $\mathcal{H}$ and $\mathcal{H}'$. The $\mathcal{H}_\ell$ sustains $\mathcal{H}$ and $\mathcal{H}'$.

Denote by B, B' the open balls in R^{n+1} with center at the origin and respective radii b, b' < b .

5. RESULTS.

When the potentials and the energy appearing in SCHRÖDINGER'S equation vanish, that is, when that equation reduces to LAPLACE'S equation $\Delta u = 0$, then our results reduce to the following : every harmonic function u defined in B, has an expansion

$$u(x) = \sum_{\ell \in \mathbb{N}} r^{\ell} h_{\ell}(y) \text{ , where } h_{\ell} \in \mathcal{H}_{\ell} \text{ ;}$$

it is its TAYLOR'S expansion ; $r^{\ell} h_{\ell}$ is a harmonic and homogeneous polynomial of degree ℓ .

We extend that property to SCHRÖDINGER'S equation as follows. We construct a linear operator $\mathcal{U}_{\ell}$ mapping each $h_{\ell} \in \mathcal{H}_{\ell}$ into a solution $u = \mathcal{U}_{\ell}(h_{\ell})$ of SCHRÖDINGER'S equation ; u is a square-integrable function with square-integrable gradient defined in a ball B independent of ℓ; the function $x \to u(x) - r^{\ell} h_{\ell}(x)$ vanishes at least $(\ell + 1)$ times at the origin.

Any solution u of SCHRÖDINGER'S equation, square integrable with square integrable gradient, defined in a neighbourhood of the origin, has in a sufficiently small ball B' the expansion

$$(3) \qquad u = \sum_{\ell \in \mathbb{N}} \mathcal{U}_{\ell} (h_{\ell}) \text{, where } h_{\ell} \in \mathcal{H}_{\ell} \text{ .}$$

Each operator $\mathcal{U}_{\ell}$ is defined by a series of operators $\mathcal{U}_{\ell,m}$:

$$(4) \qquad \mathcal{U}_{\ell}(h_{\ell}) = \sum_{m \in \mathbb{N}} \mathcal{U}_{\ell,m} (h_{\ell}) \text{ ,}$$

the function $\mathcal{U}_{\ell,m} (h_{\ell})$ vanishing at least $(\ell+m)$ times at the origin ;

$$\mathcal{U}_{\ell,0}(h_{\ell}) = r^{\ell} h_{\ell} \text{ .}$$

Thus, modulo the functions vanishing k times at the origin, any solution of SCHRÖDINGER'S equation, square integrable with square integrable gradient in a neighbourhood of the origin, belongs to the finite dimensional vector space whose elements are the functions

$$(5) \qquad \sum_{\ell + m \leq k} \mathcal{U}_{\ell,m}(h_\ell) .$$

In order to show the analytical structure of those functions, let us use the notations introduced in section 2 and define

$$(6) \qquad F(x) = \sum_j r_j - \frac{1}{2Z} \sum_{j<k} r_{j,k} ,$$

Z being the atomic number ; thus $\frac{1}{2}\Delta F$ is the potential; since $N \leqq 2Z$, we have

$$F(x) \geqq (1 - \frac{N-1}{2Z}) \sum_j r_j \geqq (1 - \frac{N-1}{2Z}) r \geqq 0 .$$

Any solution u of SCHRÖDINGER'S equation, square integrable with square integrable gradient in a neighbourhood of the origin, has the following expression :

$$(7) \qquad u(x) = e^{-F(x)} U(r, \log r, y) ,$$

where $U_{\ell m}(r,\rho,.)$ means the value, belonging to $\mathcal{H}'$, of a function of r and ρ , holomorphic for $|r| <$ const. $\rho \in \mathbb{C}$; in other words it is an entire function of ρ .

For any constant $c \in \mathbb{C}$, the function

$$x \mapsto e^{-F(x)} U(r, c + \log r, y)$$

is a solution of the same SCHRÖDINGER'S equation.

The value at x of the general term of the series (4) is :

$$\mathcal{U}_{\ell,m}(h_\ell)(x) = e^{-F(x)} r^{\ell+m} U_{\ell,m}(\log r, y) ;$$

$\mathcal{U}_{\ell,m}(\rho,.)$ is the value, belonging to $\mathcal{H}'$, of a polynomial in ρ of degree m .

For any c , ℓ', m' such that $1' + m' \leqq k$, the function

$$x \mapsto e^{-F(x)} r^{\ell' + m'} U_{\ell',m'}(c + \log r, y)$$

belongs to the vector space whose elements are the functions (5) .

Thus, its elements are functions defined on $\mathbb{R}^{n+1}$ and are exponentially decreasing at infinity. If k is so small that those functions can be obtained in a closed analytic form, then some of them could perharps be interesting approximations of the eigenfunctions of SCHRÖDINGER'S operator.

6. CONSTRUCTION OF THE OPERATORS $\mathcal{U}_\ell$.

By (4), this construction is the construction of the operators $\mathcal{U}_{\ell,m}$;that last construction reduces to a simple use of the operators $\mathcal{V}_\ell$ which are described by section 8 : they invert the Laplacian Δ of $\mathbb{R}^{n+1}$.

Their expressions are

$$(8) \qquad \mathcal{V}_\ell = \sum_{\gamma=-1}^{\infty} \mathcal{K}_{\ell,\gamma} \left(\frac{\partial}{\partial\rho}\right)^{\gamma} ,$$

the $\mathcal{K}_{\ell,\gamma}$ being the coefficients of Laurent expansion of the resolvent of the spherical Laplacian Δ_s

$$\mu \mapsto [\Delta_s + \mu\,(n-1+\mu)]^{-1} ,$$

at its simple poles $\mu = \ell \in \mathbb{N}$; if $\mu \notin \mathbb{N}$ and $n-1+\mu \notin \mathbb{N}$, then the value of that resolvent is a bounded operator $\mathcal{H} \to \mathcal{H}'$.

Let us note by $\mathcal{P}_\ell$ the orthogonal projection of $\mathcal{H}$ into $\mathcal{H}_\ell$; the $\mathcal{K}_{\ell,\gamma}$ are easily expanded in series

$$\mathcal{K}_{\ell,\gamma} = \sum_{m \in N} C_{\ell,\gamma,m}\, \mathcal{P}_m, \text{ where } C_{\ell,\gamma,m} \in \mathbb{R} .$$

The kernel p_ℓ of the operator $\mathcal{P}_\ell$ is the derivative of a TCHEBYCHEV (resp. LEGENDRE) polynomial if n is odd (resp even). Thus the operator $\mathcal{K}_{\ell,\gamma}$ has the kernel

$$k_{\ell,\gamma} = \sum_{m \in N} C_{\ell,\gamma,m}\, p_m ;$$

that series rapidly converges if γ is large ; it converges only in a space of distributions if γ is small.

But a closed form of $k_{\ell,\gamma}$ can be obtained as much easily as γ is smaller : the $k_{\ell,\gamma}$ are indeed the Laurent coefficients of the spherical Green function ; that function is given by derivations of the Green function of S^1 if n is odd and of the Green function of S^2 if n is even ; and those two special Green functions are respectively given by trigonometric functions and by Legendre functions.

The above derivations are analogous to those that J. CHEEGER and M. TAYLOR [2] use in their formula (2.4).

Note.- We have not yet completely applied our results to the fundamental state of helium, i.e, to obtaining an expansion equivalent to FOCK'S result [3].

7. EXTENSION TO SOME PARTIAL DIFFERENTIAL EQUATIONS OF FUCH'S THEOREM (i.e. : the theory of regular singularities of ordinary differential equation).

Let us recall that (6) has defined a function $F : \mathbb{R}^{n+1} \to \mathbb{R}_+$. The multiplication by e^F of the unknown u of SCRÖDINGER'S equation transforms this equation into an equation of the following type :

$$(9) \qquad (\Delta + A)\, u = 0$$

where

$$A = A_0 + \sum_{\nu=1}^{n+1} A_\nu \frac{\partial}{\partial x_\nu}$$

is a differential operator of first order, whose coefficients are bounded and homogeneous measurable functions, obviously of degree zero : $\mathbb{R}^{n+1} \to \mathbb{C}$.

It is the general equation of this type (9) that we first study. Its properties are those stated in section 5 , where F has to be cancelled. Let us specify that the radius of the ball B where all the $\mathcal{U}_\ell(h_\ell)$ are defined is

$$b = 1/4 \ \sup_x \left[\sum_\nu |A_\nu(x)|^2 \right]^{1/2}$$

Those results are an extension of FUCHS' theorem on regular singularities of ordinary differential equations. That would be more obvious if we had assumed that $r A_0, A_1 \ldots A_{n+1}$ are holomorphic functions of r at the origin, whose coefficients are bounded and homogeneous, obviously of degree zero, functions of x .

8. THE CONSTRUCTION OF $\mathcal{V}_\ell$.

The inverse operator $\mathcal{V}_\ell$ of Δ has been used in section 6 where its expression (8) is given. This operator is defined for all $\ell \in \mathbb{N}$; it maps any polynomial of ρ ,

$$\rho \mapsto V(\rho,.) \in \mathcal{H} ,$$

into a polynomial with a degree increased by one

$$\rho \mapsto U(\rho,.) \in \mathcal{H}' ,$$

such that the functions u and v defined by the convention (2) and the formulae

$$v(x) = r^{\ell-2}\, V(\log r\, ,\, y)\ ,\ u(x) = r^{\ell}\, U(\log r\, ,\, y)$$

verify in $\mathbb{R}^{n+1}$ the equation

$$\Delta u = v\ .$$

The construction of $\mathcal{V}_\ell$ makes use of the decomposition of any element of $\mathcal{H}$ and $\mathcal{H}'$ into a sum of spherical harmonics, i.e of elements of the orthogonal subspaces $\mathcal{H}_\ell$ of those two Hilbert spaces.

9. CONSTRUCTIONS OF THE EXPANSION (3) OF A SOLUTION OF (9) IN NEIGHBOURHOOD OF THE ORIGIN.

Notations : Let $f : B \to \mathbb{C}$ and $q \in \mathbb{R}\setminus \mathbb{Z}$. Let us denote by σ the measure of $\mathbb{S}^n$ and, using the notation (2), put

$$\|f(r,.)\|^2 = \int_{\mathbb{S}^n} |f(r,y)|^2 \sigma\ ;$$

$$\|f\|^2_{B,q} = b^{2q} \int_0^b r^{-2q-1} \|f(r,.)\|^2\, dr\ ;$$

$$|||f|||^2_{B,q} = b^{2q} \int_0^b \left[\left(\frac{n-1}{2}\right)^2 r^{-2q-1} \|f(r,.)\|^2 + r^{-2q+1} \left\|\frac{\partial f}{\partial x}(r,.)\right\|^2 \right] dr\ ;$$

let $\mathcal{H}_{B,q}$ and $\mathcal{K}_{B,q}$ be the Hilbert spaces of respective norms

$$\|\cdot\|_{B,q} \text{ and } |||\cdot|||_{B,q}\ .$$

They decrease when q increases.

For $\ell \in \mathbf{Z}$, define

(10) $\mathcal{H}_{B,\ell} = \bigcap_q \mathcal{H}_{B,q}$, $\mathcal{H}'_{B,\ell} = \bigcap_q \mathcal{H}'_{B,q}$ where $q < \ell$.

THE OPERATOR $\mathcal{W}_q$. For every $q \in \mathbb{R}\backslash\mathbf{Z}$ such that $q > -n$, by decomposition of the elements of $\mathcal{H}$ and $\mathcal{H}'$ in sums of spherical harmonics one can build a linear operator, inverse of Δ

(11) $\mathcal{W}_q : \mathcal{H}_{B,q-1} \ni v \mapsto u \in \bigcap_p \mathcal{H}'_{B,p}$ where $p < q+1$,

such that

(12) $\Delta u = v$.

The projection $\mathfrak{X}_\ell$ into $\mathcal{H}_\ell$ of the solutions of (9) belonging to $\mathcal{H}'_{B,q}$.

Assume $q > -n$. Denote by ℓ the integer such that

$$q < \ell < q + 1$$

Let $u \in \mathcal{H}'_{B,q}$ be a solution of (9); by (9) $\Delta u = -Au \in \mathcal{H}_{B,q-1}$; by (11) and (12)

(13) $w = u - \mathcal{W}_q \Delta u$

is harmonic and satisfies $w \in \mathcal{H}'_{B,q}$; hence , using the notation (10)

$$w \in \mathcal{H}'_{B,\ell} ,$$

and therefore the existence of a unique element $h_\ell \in \mathcal{H}_\ell$ such that

(14) $w - r^\ell h_\ell \in \mathcal{H}'_{B,\ell+1}$; $h_\ell = 0$ if $\ell < 0$.

The operator

$$\mathfrak{X}_\ell : u \to h_\ell \in \mathcal{H}_\ell$$

has, by (11), (13) and (14), the following property :

$$u - r^\ell \mathfrak{X}_\ell u \in \bigcap_p \mathcal{H}'_{B,p} \text{ , where } p < q+1 .$$

But, by (4)

$$\mathcal{U}_\ell(\mathfrak{X}_\ell u) - r^\ell \mathfrak{X}_\ell u \in \mathcal{H}'_{B,\ell+1} .$$

Therefore

(15) $u - \mathcal{U}_\ell(\mathfrak{X}_\ell u) \in \bigcap_p \mathcal{H}'_{B,p}$ where $p < q+1$;

hence, a fortiori

$$u \in \mathcal{H}'_{B,\ell} .$$

Thus, if u is solution of (9), and $u \in \mathcal{H}'_{B,q}$ then $u \in \mathcal{H}'_{B,\ell}$.

Consequently since $\mathcal{U}_\ell(\mathfrak{X}_\ell u)$ is a solution of (9), the formula (15) shows that

(16) $$u - \mathcal{U}_\ell(\mathfrak{X}_\ell u) \in \mathcal{H}'_{B,\ell+1} .$$

If $\ell < 0$, $\mathfrak{X}_\ell u = 0$; therefore each solution u of (9) satisfying $u \in \mathcal{H}'_{B,q}$ where $q > -n$ satisfies $u \in \mathcal{H}'_{B,0}$.

<u>CONCLUSION</u> : If $\ell \in \mathbb{N}$, define the linear operator $\mathcal{Y}_\ell$ by $\mathcal{Y}_\ell u = u - \mathcal{U}(\mathfrak{X}_\ell u)$.

Thus we have put each solution u of (9) belonging to $\mathcal{H}'_{B,\ell}$ under the form :

$$u = \mathcal{U}_\ell(\mathfrak{X}_\ell u) + \mathcal{Y}_\ell u \ , \quad \text{where} \quad \mathfrak{X}_\ell u \in \mathcal{H}_\ell \text{ and } \mathcal{Y}_\ell u \in \mathcal{H}'_{B,\ell+1} ;$$

$\mathcal{Y}_\ell u$ satisfies (9). Hence, by an obvious induction, the following result : each solution u of (9) satisfying $u \in \mathcal{H}'_{B,q}$, where $q > -n$, admits the following expression, where each term verifies (9).

$$u = \mathcal{U}_0(h_0) + \mathcal{U}_1(h_1) + \ldots + \mathcal{U}_\ell(h_\ell) + \mathcal{Y}_\ell \mathcal{Y}_{\ell-1} \ldots \mathcal{Y}_0 u ,$$

where

$$\ell \in \mathbb{N}, h_0 = \mathfrak{X}_0 u \in \mathcal{H}_0 , h_j = \mathfrak{X}_j \mathcal{Y}_{j-1} \ldots \mathcal{Y}_0 \in \mathcal{H}_j \quad \text{if} \quad j > 0 ,$$

$$\mathcal{Y}_\ell \mathcal{Y}_{\ell-1} \ldots \mathcal{Y}_0 u \in \mathcal{H}'_{B,\ell+1} .$$

That result and sufficiently precise estimates of the operators $\mathcal{U}_\ell$, $\mathcal{W}_q$, $\mathfrak{X}_\ell$ and $\mathcal{Y}_\ell$ lead to (3).

BIBLIOGRAPHY

[1] BAZLEY, N. Proc. Nat. Acad. Sci., Vol. 45, N°6, p.850-853 (1959).
BAZLEY, N. Phys. Rev., Vol. 124, n°2, p.483-492 (1961).

[2] CHEEGER, J. and TAYLOR, M. On the diffraction of waves by conical singularities (to appear).

[3] FOCK, V.A., Izv. Akad. Nauk, SSSR, Ser. Fiz. 18, p. 1961 (1954).

[4] JÖRGENS, K. and WEIDMANN, J. Spectral Properties of Hamiltonian Operators, Lecture Notes in Mathematics, 313, Springer (1973).

[5] KATO, T., Perturbation Theory for Linear Operators, Grundlehren der mathematischen Wissenschaften, 132, Second Edition, Springer (1980)

[6] SCHIFF, B., PEKERIS C.L. and LIFSON, H., Fine structure of the 2^3P and 3^3P States of Helium, Phys. Rev., 137, A, p. 1672 (1965).

SCHIFF, B., LIFSON, H., PEKERIS, C.L. and RABINOWITS P., $2^{1,3}P$, $3^{1,3}P$ and $4^{1,3}P$ States of He and the 2^1P State of Li^+, Phys. Rev., 140, A, p. 1104 (1965).

[7] SIMON, B., Quantum Mechanics for Hamiltonians Defined as Quadratic Forms, Princeton University Press (1971).

[8] SIMON, B., Pointwise Bounds on Eigenfunctions and Wave Packets in N-Body Quantum Systems ; Proc. Amer. Math. Soc. 42, p. 395 (1974); 45, p.454 (1974); Trans. Amer. Math. Soc. 208, p. 317 (1975).

CARMONA, R. and SIMON, B., Pointwise Bounds on Eigenfunctions and Wave Packets in N-Body Quantum Systems, Lower Bounds and Path Integrals (to appear).

*

The present paper is supplemented by

LERAY, J., La fonction de Green de la sphère S^n et l'application effective à l'équation de Schrödinger atomique d'une extension du théorème de Fuchs, METHODS OF FUNCTIONAL ANALYSIS AND THEORY OF ELLIPTIC OPERATORS, NAPOLI, 1982.

Collège de France
F - 75231 Paris Cedex 05
FRANCE

Shih Wei Hui

Egalité de consommation pour tous les êtres humains.

ON THE CAUCHY PROBLEM FOR THE EQUATION OF A GENERAL FLUID

1. The purpose of this work is to study the Cauchy problem for the equation which describes the movement of a general fluid [1]. It is given by the following system (I) = (1,1',1").

$$\frac{\partial \rho}{\partial t} + \rho \sum_{k=1}^{3} \frac{\partial v_k}{\partial x_k} + \sum_{k=1}^{3} v_k \frac{\partial \rho}{\partial x_k} = 0 \quad , \tag{1}$$

for $j = 1,2,3$

$$\rho\left[\frac{\partial v_j}{\partial t} + \sum_{k=1}^{3} v_k \frac{\partial v_j}{\partial x_k}\right] = -\frac{\partial p}{\partial x_j} + \sum_{k=1}^{3} \frac{\partial \sigma_{jk}}{\partial x_k} + F_j(x,t) \ , \tag{1)'$$

and

$$\rho T\left[\frac{\partial s}{\partial t} + \sum_{k=1}^{3} v_1 \frac{\partial s}{\partial x_k}\right] = \sum_{k=1}^{3} \frac{\partial}{\partial x_k}\left(\chi \frac{\partial T}{\partial x_k}\right) +$$

$$+ \frac{\eta}{2}\left\{\sum_{i,k=1}^{3}\left[\frac{\partial v_i}{\partial x_k} + \frac{\partial v_k}{\partial x_i} - \frac{2}{3}\delta_{ik}\left(\sum_{\ell=1}^{3}\frac{\partial v_\ell}{\partial x_\ell}\right)\right]^2\right\} + \tag{1)''$$

$$+ \zeta\left[\sum_{i,k=1}^{3} \frac{\partial v_i}{\partial x_i} \cdot \frac{\partial v_k}{\partial x_k}\right] ,$$

with

$$\sigma_{jk} = \eta\left[\frac{\partial v_j}{\partial x_k} + \frac{\partial v_k}{\partial x_j}\right] + \left(\zeta - \frac{2}{3}\eta\right)\delta_{jk}\left(\sum_{\ell=1}^{3}\frac{\partial v_\ell}{\partial x_\ell}\right) .$$

This system (I) is parametrized by the five strictly positive real analytic functions χ, η, ζ, p, s of two variables

C. P. Bruter et al. (eds.), Bifurcation Theory, Mechanics and Physics, 109–138.

$(\rho, T) \in \mathbb{R}^2$ and the three real analytic functions $F_i(x,t)$ of the space-time variables $(x,t) = (x_1,x_2,x_3,t) \in \mathbb{R}^4$, where the unknown functions are $v_i(x,t)$, $i = 1,2,3$ and $\rho(x,t) > 0$, $T(x,t)$.

We shall study the following problems for the system (I).

I.1 For which analytic hypersurface "Σ" in $\mathbb{R}^4$ and initial condition "C" on Σ such that $(\Sigma, \emptyset)$ form, for the system (I) a well posed Cauchy problem ? In other words show the existence and the uniqueness of an analytic solution with the given initial condition, and the stability of the solution with respects to a small perturbation of (Σ, C) (cf. Theorem 1).

I.2. (I.1) being fulfilled express the solution in terms of (Σ, C) (cf. Theorem 2).

Here, we shall not follow the idea of S. Lie (find a transformation which changes (I) into an expression of Cauchy-Kowaleska type) nor the general theory of integrability for a differential system of Cartan-Kähler. In fact we shall follow a method which has been already established for several years, but only partially summarized in a recent note in Comptes-Rendus[5], i.e. we completely determine the "système stratification canonique" and the "équations secondaires" of the system (I) from which our results follow. Let us recall that in the classical French school [3][6], it is usual to give the solutions of a partial differential equation in terms of convergent series. On one hand, this method enables us to verify directly that the expression given in Theorem 2 is, in fact, a solution of (I). On the other hand, to get the explicit form of the solution may have some practical advantage on a mere existence theorem in some space of functions. The detailed proof [4] of this work will appear elsewhere in the near future.

2. The symmetry of (1)" assures that it is sufficient to study the following two types of analytic hypersurface in $\mathbb{R}^4$ parametrized by $\xi = (\xi_1, \xi_2, \xi_4) \in \mathbb{R}^3$.

$$x_1 = \xi_1 \ , \quad x_2 = \xi_2 \ , \quad x_3 = f(\xi_1,\xi_2,\xi_4) \ , \quad t = \xi_4 \qquad (\Sigma_f)$$

and $\xi = (\xi_1,\xi_2,\xi_3) \in \mathbb{R}^3$

$$x_1 = \xi_1 \ , \quad x_2 = \xi_2 \ , \quad x_3 = \xi_3 \ , \quad t = g(\xi_1,\xi_2,\xi_3) \qquad (\Sigma_g)$$

where f and g are real analytic functions of ξ . We deduce from (1) and the Cartan ideal I, of the first Ehresmann space $J^1(\mathbb{R}^4, \mathbb{R}^4 \times \mathbb{R}^1_+)$, where $\mathbb{R}_+$ is the set of strictly positive real numbers, that an admissible initial condition on Σ_f is given by the analytic functions

$$v^i_o(\xi) \quad (i=1,2,3) \; , \; \rho_o(\xi) > 0 \; , \; T_o(\xi)$$
$$p^1_3(\xi) \; , \; p^2_3(\xi) \; , \; p^4_3(\xi) \; , \; p^5_3(\xi) \qquad (C_f)$$

and on Σ_g by

$$v^i_o(\xi) \quad (i = 1,2,3) \; , \; \rho_o(\xi) > 0 \; , \; T_o(\xi) \; , \qquad (C_g)$$
$$p^i_4(\xi) \; , \; p^5_4(\xi) \; .$$

The Cauchy problem for Σ_f (resp. Σ_g) consists of finding analytic functions $v^i(x,t)$, $i = 1,2,3$, $\rho(x,t) > 0$, $T(x,t)$ satisfying (I), and defined in a tubular neighborhood of Σ_f and (resp. Σ_g) uniquely determined by the conditions given by C_f

$$v^i|_{\Sigma f} = v^i_o \; , \; \rho|_{\Sigma f} = \rho_o \quad T|_{\Sigma f} = T_o \; , \; i = 1,2,3$$

$$\frac{\partial v^1}{\partial x_3}\Big|_{\Sigma f} = p^1_3 \; , \; \frac{\partial v^2}{\partial x_3}\Big|_{\Sigma f} = p^2_3 \; , \; \frac{\partial \rho}{\partial x_3}\Big|_{\Sigma f} = p^4_3 \; , \; \frac{\partial T}{\partial x_3}\Big|_{\Sigma f} = p^5_3 \; ,$$

(resp. given by C_g)

$$v^i|_{\Sigma_g} = v^i_o \; , \; \frac{\partial v^i}{\partial t}\Big|_{\Sigma_g} = p^i_4 \; , \; i = 1,2,3$$

$$\rho|_{\Sigma_g} = \rho_o \; , \; T|_{\Sigma_g} = T_o \; , \; \frac{\partial T}{\partial t}\Big|_{\Sigma_g} = p^5_4 \; .$$

Then we have

Theorem 1 : The initial conditions (Σ_f, C_f) is a well posed Cauchy problem for the system (I) if and only if they satisfy the following inequation on partial derivatives

$$v_o^3(\xi) - v_o^1(\xi)\cdot\frac{\partial f}{\partial\xi_1} - v_o^2(\xi)\frac{\partial f}{\partial\xi_2} - \frac{\partial f}{\partial\xi_4} \neq 0 \quad .$$

Theorem 1 bis : The initial condition (Σ_g, C_g) is a well posed Cauchy problem for the system (I), if and only if they satisfy the following inequation on partial derivatives

$$\left(\frac{\partial g}{\partial\xi_1}\right)^2 + \left(\frac{\partial g}{\partial\xi_2}\right)^2 + \left(\frac{\partial g}{\partial\xi_3}\right)^2 \neq 0$$

$$v_o^1(\xi)\frac{\partial g}{\partial\xi_1} + v_o^2(\xi)\frac{\partial g}{\partial\xi_2} + v_o^3(\xi)\frac{\partial g}{\partial\xi_3} \neq 1 \quad .$$

As a consequence [2], there do no exist any initial condition on the hyperplane $t = 0$ for which the corresponding Cauchy problem is well posed. And the condition of a "well posed problem" is independent of the eight parameter functions χ , η , ζ , p , s , and F_i , $i = 1,2,3$.

3. In this paragraph we shall construct two sequences of analytic functions of three variables

$$A^i_{q^\lambda,\ell} \quad , \quad \tilde{A}^i_{q^\lambda,\ell} \quad , \quad i = 1,2,3,4,5$$

$q^\lambda = q_1^{\lambda_1},\ldots,q_\mu^{\lambda_\mu}$, $1 \leq q_1 < q_2 < \ldots < q_\mu \leq 4$, $\lambda_i \geq 1$, $\ell = \sum_{i=1}^{\mu} \lambda_i$, associated to (Σ_f, C_f) and (Σ_g, C_g) respectively - well posed initial condition of (I) as indicated in Theorem 1 and bis. We shall denote for each given function K of the set of variables $x = (x_1,x_2,x_3,x_4)$, $X = \{X_{q^\lambda,\ell}\}$, $\ell \leq m$, the j- function of Ehresmann of K , $1 \leq j \leq 4$

$$E_j K(x,X) = \frac{\partial K}{\partial x_j} + \Sigma \frac{\partial K}{\partial X_{q^\lambda,\ell}} \cdot X_{jq^\lambda,\ell+1} \quad ,$$

function of the set of variable $x = (x_1,x_2,x_3,x_4)$, $X = \{X_{q^\lambda,\ell}\}$, $\ell \leq m + 1$. Then the sequence of function $A^i_{q^\lambda,\ell}$ associated to (Σ_f, C_f) is given as follows. Denote by $\delta_i = \dfrac{\partial f}{\partial\xi_i}$, $i = 1,2,4$ (cf. Theorem 1) and by

$$\tilde{p}_1 = \frac{\partial p}{\partial \rho}, \quad \tilde{p}_2 = \frac{\partial p}{\partial T}, \quad \eta_1 = \frac{\partial \eta}{\partial \rho}, \quad \eta_2 = \frac{\partial \eta}{\partial T}, \quad \zeta_1 = \frac{\partial \zeta}{\partial \rho}$$

$$s_1 = \frac{\partial s}{\partial \rho}, \quad s_2 = \frac{\partial s}{\partial T}, \quad \chi_1 = \frac{\partial \chi}{\partial \rho}, \quad \chi_2 = \frac{\partial \chi}{\partial T}, \quad \zeta_2 = \frac{\partial \zeta}{\partial T}$$

$$a_1 = \frac{4}{3}\frac{\partial \eta}{\partial \rho} + \frac{\partial \zeta}{\partial \rho}, \qquad a_2 = \frac{4}{3}\frac{\partial \eta}{\partial T} + \frac{\partial \zeta}{\partial T}$$

$$b_1 = \frac{\partial \zeta}{\partial \rho} - \frac{2}{3}\frac{\partial \eta}{\partial \rho}, \qquad b_2 = \frac{\partial \zeta}{\partial T} - \frac{2}{3}\frac{\partial \eta}{\partial} .$$

Consider the following analytic functions of three variables in $\xi = (\xi_1, \xi_2, \xi_4)$:

$$\begin{aligned}\phi^1 = {} & p^1_1[a_1 p^4_1 + a_2 p^5_1] + [p^1_2 + p^2_1][\eta_1 p^4_2 + \eta_2 p^5_2] + [p^1_3 + p^3_1][\eta_1 p^4_3 + \eta_2 p^5_3] + \\ & + [p^2_2 + p^3_3][b_1 p^4_1 + b_2 p^5_1] - [\tilde{p}_1 p^4_1 + \tilde{p}_2 p^5_1] - \\ & - \rho_o[v^1_o p^1_1 + v^2_o p^1_2 + v^3_o p^1_3 + p^1_4] + F_1 \quad ;\end{aligned}$$

$$\begin{aligned}\phi^2 = {} & p^2_2[a_1 p^4_2 + a_2 p^5_2] + [p^1_2 + p^2_1][\eta_1 p^4_1 + \eta_2 p^5_1] + [p^2_3 + p^3_2][\eta_1 p^4_3 + \eta_2 p^5_3] + \\ & + [p^1_1 + p^3_3][b_1 p^4_2 + b_2 p^5_2] - [\tilde{p}_1 p^4_2 + \tilde{p}_2 p^5_2] - \\ & - \rho_o[v^1_o p^2_1 + v^2_o p^2_2 + v^3_o p^2_3 + p^2_4] + F_2 \quad ;\end{aligned}$$

$$\begin{aligned}\phi^3 = {} & p^3_3[a_1 p^4_3 + a_2 p^5_3] + [p^1_3 + p^3_1][\eta_1 p^4_1 + \eta_2 p^5_1] + [p^2_3 + p^3_2][\eta_1 p^4_2 + \eta_2 p^5_2] + \\ & + [p^1_1 + p^2_2][b_1 p^4_3 + b_2 p^5_3] - [\tilde{p}_1 p^4_3 + \tilde{p}_2 p^5_3] - \\ & - \rho_o[v^1_o p^3_1 + v^2_o p^3_2 + v^3_o p^3_3 + p^3_4] + F_3 \quad ;\end{aligned}$$

$$\phi^4 = \chi_1[p_1^4p_1^5+p_2^4p_2^5+p_3^4p_3^5]+\chi_2[(p_1^5)^2+(p_2^5)^2+(p_3^5)^2] +$$

$$+ \frac{\eta}{2} \sum_{i,k=1}^{3} [p_k^i+p_i^k - \frac{2}{3}\delta_{ik} \sum_{\ell=1}^{3} p_\ell^\ell]^2 + \zeta \sum_{i,k=1}^{3} p_i^i p_k^k -$$

$$- \rho_0 T_0 [s_1(v_0^1 p_1^4+v_0^2 p_2^4+v_0^3 p_3^4+p_4^4) + s_2(v_0^1 p_1^5+v_0^2 p_2^5+v_0^3 p_3^5+p_4^5)];$$

$$\phi^5 = p_3^4[p_1^1+p_2^2+p_3^3]+p_3^1p_1^4+p_3^2p_2^4+p_3^3p_3^4 \quad ,$$

where the p_j^i which are not in (C_f) are given by

$$p_1^i = v^i(1) - p_3^i \cdot \delta_1 \qquad p_2^i = v^i(2) - p_3^i \cdot \delta_2$$

$$p_4^i = v^i(4) - p_3^i \cdot \delta_4 \quad , \qquad i = 1,2,3,4,5$$

where

$$p_3^3 = p_3^1\delta_1 + p_3^2\delta_2 - [v_0^3 - v_0^1\delta_1 - v_0^2\delta_2 - \delta_4]p_3^4/\rho_0 -$$

$$[v^1(1)+v^2(2)] - [v_0^1\rho_0(1)+v_0^2\rho_0(2)+\rho_0(4)]/\rho_0$$

with

$$v^i(\ell) = \frac{\partial v_0^i}{\partial \xi_\ell} \quad , \qquad \rho_0(\ell) = \frac{\partial \rho_0}{\partial \xi_\ell} \quad , \ \ell = 1,2,4.$$

Now let us denote by

$$c_2^1 = -\phi^1+\{[\tfrac{4}{3}\eta+\zeta][\delta_1 p_3^1(1)-p_1^1(1)]+\eta[\delta_2 p_3^1(2)-p_2^1(2)]+$$

$$+ [\tfrac{1}{3}\eta+\zeta][\delta_1 \, p_3^2(2)-p_2^2(1)-p_3^3(1)]\} \quad ;$$

$$C_2^2 = -\phi^2+\{[\tfrac{4}{3}\eta+\zeta][\delta_2 p_3^2(2)-p_2^2(2)]+\eta[\delta_1 p_3^2(1)-p_1^2(1)] +$$

$$+ [\tfrac{1}{3}\eta+\zeta][\delta_1 p_3^1(2)-p_2^1(1)-p_3^3(2)]\} \quad ;$$

$$C_2^3 = -\phi^3+\{\eta[\delta_1 p_3^3(1)+\delta_2 p_3^3(2)-p_1^3(1)-p_2^3(2)]-$$

$$- [\tfrac{1}{3}\eta+\zeta][p_3^1(1)+p_3^2(2)]\} \quad ;$$

$$C_2^4 = -\phi^4+\chi[\delta_1 p_3^5(1)+\delta_2 p_3^5(2)-p_1^5(1)-p_2^5(2)] \quad ;$$

$$C_2^5 = -\phi^5-\rho_o[p_3^1(1)-p_3^2(2)]-[v_o^1 p_3^4(1)+v_o^2 p_3^4(2)+p_3^4(4)] \ ;$$

where

$$p_j^i(k) = \frac{\partial p_j^i}{\partial \xi_k} \quad , \quad k = 1,2,4 \quad ; \quad i = 1,2,3,4,5, \ j = 1,2,3,4.$$

Now let us define

$$A_{o,o}^i(\xi) = v_o^i(\xi) \quad , \qquad i = 1,2,3 \ ,$$

$$A_{o,o}^4(\xi) = \rho_o(\xi) \quad , \qquad A_{o,o}^5(\xi) = T_o(\xi) \quad ,$$

for $\ell = 1$, by

$$A_{j,1}^i = A_{j,1}^i(\xi) = p_j^i(\xi) \quad , \quad i = 1,2,3,4,5 \ , \quad j = 1,2,3,4;$$

and for $\ell = 2$, $q^\lambda = 3^2$ by

$$A_{3^2,2}^1 = \frac{1}{\eta[\frac{4}{3}\eta+\zeta][1+\delta_1^2+\delta_2^2]^2}\cdot\{[\eta\delta_1^2+(\tfrac{4}{3}\eta+\zeta)\delta_2^2+(\tfrac{4}{3}\eta+\zeta)]C_2^1 -$$

$$- [\tfrac{1}{3}\eta+\zeta]\delta_1\delta_2 C_2^2]+[\tfrac{1}{3}\eta+\zeta]\delta_1 C_2^3\} \quad ;$$

$$A^{2}_{3^2,2} = \frac{1}{\eta[\frac{4}{3}\eta+\zeta][1+\delta_1^2+\delta_2^2]^2}\cdot\{-[\frac{1}{3}\eta+\zeta]\delta_1\delta_2 C_2^1 +$$

$$+ [(\frac{4}{3}\eta+\zeta)\delta_1^2+\eta\delta_2^2+(\frac{4}{3}\eta+\zeta)]C_2^2+[\frac{1}{3}\eta+\zeta]\delta_2 C_2^3\} \quad ;$$

$$A^{3}_{3^2,2} = \frac{1}{\eta[\frac{4}{3}\eta+\zeta][1+\delta_1^2+\delta_2^2]^2}\cdot\{[\frac{1}{3}\eta+\zeta]\delta_1 C_2^1+[\frac{1}{3}\eta+\zeta]\delta_2 C_2^2 +$$

$$+ [(\frac{4}{3}\eta+\zeta)\delta_1^2+(\frac{4}{3}\eta+\zeta)\delta_2^2+\eta]C_2^3\} \quad ;$$

$$A^{4}_{3^2,2} = \frac{1}{[\frac{4}{3}\eta+\zeta][1+\delta_1^2+\delta_2^2][v_o^3-v_o^1\delta_1-v_o^2\delta_2-\delta_4]}\cdot\{\rho_o[\delta_1 C_2^1+\delta_2 C_2^2-C_2^3]+$$

$$+ [\frac{4}{3}\eta+\zeta][1+\delta_1^2+\delta_2^2]C_2^5\} \quad ;$$

$$A^{5}_{3^2,2} = \frac{1}{\chi[1+\delta_1^2+\delta_2^2]}C_2^4 \quad .$$

We obtain the other q^λ with $\ell = 2$ by the following formula

$$A^i_{1q_j,2} = p^i_j(1) - A^i_{q_j3,2}\cdot\delta_1$$

$$A^i_{2q_j,2} = p^i_j(2)-A^i_{q_j3,2}\cdot\delta_2$$

$$A^i_{4q_j,2} = p^i_j(4)-A^i_{q_j3,2}\cdot\delta_4 \quad .$$

Now in the case when $A^i_{3^\ell,\ell}$, $\ell \geq 3$, we have

$$A^{1}_{3^{\ell},\ell} = \frac{1}{\eta[\frac{4}{3}\eta+\zeta][1+\delta_1^2+\delta_2^2]^2}\cdot\{[\eta\delta_1^2+(\frac{4}{3}\eta+\zeta)\delta_2^2+(\frac{4}{3}\eta+\zeta)]C^1_\ell -$$

$$-[\frac{1}{3}\eta+\zeta]\delta_1\delta_2 C^2_\ell + [\frac{1}{3}\eta+\zeta]\delta_1 C^3_\ell\} \qquad ;$$

$$A^{2}_{3^{\ell},\ell} = \frac{1}{\eta[\frac{4}{3}\eta+\zeta][1+\delta_1^2+\delta_2^2]^2}\cdot\{-[\frac{1}{3}\eta+\zeta]\delta_1\delta_2 C^1_\ell +[(\frac{4}{3}\eta+\zeta)\delta_1^2 +$$

$$+\eta\delta_2^2+(\frac{4}{3}\eta+\zeta)]C^2_\ell + [\frac{1}{3}\eta+\zeta]\delta_2 C^3_\ell\} \qquad ;$$

$$A^{3}_{3^{\ell},\ell} = \frac{1}{\eta[\frac{4}{3}\eta+\zeta][1+\delta_1^2+\delta_2^2]^2}\cdot\{[\frac{1}{3}\eta+\zeta]\delta_1 C^1_\ell+[\frac{1}{3}\eta+\zeta]\delta_2 C^2_\ell +$$

$$+[(\frac{4}{3}\eta+\zeta)\delta_1^2+(\frac{4}{3}\eta+\zeta)\delta_2^2+\eta]C^3_\ell\} \qquad ;$$

$$A^{4}_{3^{\ell},\ell} = \frac{1}{[\frac{4}{3}\eta+\zeta][1+\delta_1^2+\delta_2^2][v^3_o-v^1_o\delta_1-v^2_o\delta_2-\delta_4]}\cdot\{\rho_o[\delta_1 C^1_\ell+$$

$$+ \delta_2 C^2_\ell - C^3_\ell] + [\frac{4}{3}\eta+\zeta][1+\delta_1^2+\delta_2^2]C^5_\ell\};$$

$$A^{5}_{3^{\ell},\ell} = \frac{1}{\chi[1+\delta_1^2+\delta_2^2]}\,C^4_\ell \qquad ;$$

where the C_ℓ are given by

$$C^1_\ell = -\{E_3^{\ell-2}(\phi^1)+E_3^{\ell-3}(((\frac{4}{3}\eta_1+\zeta_1)p_3^4+(\frac{4}{3}\eta_2+\zeta_2)p_3^5)\cdot A^1_{1^2,2} +$$

$$+ (A^1_{2^2,2}+A^1_{3^2,2})(\eta_1 p_3^4+\eta_2 p_3^5)+$$

$$+ (A^{2}_{12,2}+A^{3}_{13,2})((\tfrac{1}{3}\eta_1+\zeta_1)p^{4}_{3}+(\tfrac{1}{3}\eta_2+\zeta_2)p^{5}_{3})) +$$

$$+ E^{\ell-4}_{3}(A^{1}_{1^{2}3,3}\cdot((\tfrac{4}{3}\eta_1+\zeta_1)p^{4}_{3}+(\tfrac{4}{3}\eta_2+\zeta_2)p^{5}_{3}) +$$

$$+ (A^{1}_{2^{2}3,3}+A^{1}_{3^{3},3})(\eta_1 p^{4}_{3}+\eta_2 p^{5}_{3}) + (A^{2}_{123,3}+A^{3}_{13^{2},3})$$

$$\cdot((\tfrac{1}{3}\eta_1+\zeta_1)p^{4}_{3}+(\tfrac{1}{3}\eta_2+\zeta_2)p^{5}_{3}))+\ldots+$$

$$+ E_3(A^{1}_{1^{2}3^{\ell-4},(\ell-2)}((\tfrac{4}{3}\eta_1+\zeta_1)p^{4}_{3}+(\tfrac{4}{3}\eta_2+\zeta_2)p^{5}_{3}) +$$

$$+ (A^{1}_{2^{2}3^{\ell-4},(\ell-2)}+A^{1}_{3^{\ell-2},(\ell-2)})(\eta_1 p^{4}_{3}+\eta_2 p^{5}_{3}) +$$

$$+ (A^{2}_{123^{\ell-4},(\ell-2)}+A^{3}_{13^{\ell-3},(\ell-2)})((\tfrac{1}{3}\eta_1+\zeta_1)p^{4}_{3}+(\tfrac{1}{3}\eta_2+\zeta_2)p^{5}_{3}))+$$

$$+ [A^{1}_{1^{2}3^{\ell-3},(\ell-1)}((\tfrac{4}{3}\eta_1+\zeta_1)p^{4}_{3}+(\tfrac{4}{3}\eta_2+\zeta_2)p^{5}_{3}) +$$

$$+ (A^{1}_{2^{2}3^{\ell-3},(\ell-1)}+A^{1}_{3^{\ell-1},(\ell-1)})(\eta_1 p^{4}_{3}+\eta_2 p^{5}_{3}) +$$

$$+ (A^{2}_{123^{\ell-3},(\ell-1)}+A^{3}_{13^{\ell-2},(\ell-1)})((\tfrac{1}{3}\eta_1+\zeta_1)p^{4}_{3} +$$

$$+ (\tfrac{1}{3}\eta_2+\zeta_2)p^{5}_{3})]\} + [\tfrac{4}{3}\eta+\zeta][\delta_1 A^{1}_{3^{\ell-1},(\ell-1)}(1)-A^{1}_{13^{\ell-2},(\ell-1)}(1)]$$

$$+\eta[\delta_2 A^1_{3^{\ell-1},(\ell-1)}(2)-A^1_{23^{\ell-2},(\ell-1)}(2)] \quad +$$

$$+\ [\tfrac{1}{3}\eta+\zeta][\delta_1 A^2_{3^{\ell-1},(\ell-1)}(2)-A^2_{23^{\ell-2},(\ell-1)}(1)-A^3_{3^{\ell-1},(\ell-1)}(1)];$$

$$C^2_\ell = -\{E_3^{\ell-2}(\phi^2)+E_3^{\ell-3}(A^2_{2^2,2}((\tfrac{4}{3}\eta_1+\zeta_1)p_3^4+(\tfrac{4}{3}\eta_2+\zeta_2)p_3^5 +$$

$$+\ (A^2_{1^2,2}+A^2_{3^2,2})(\eta_1 p_3^4+\eta_2 p_3^5) +$$

$$+\ (A^1_{12,2}+A^3_{23,2})((\tfrac{1}{3}\eta_1+\zeta_1)p_3^4 + (\tfrac{1}{3}\eta_2+\zeta_2)p_3^5)) +$$

$$+\ E_3^{\ell-4}(A^2_{2^23,3}((\tfrac{4}{3}\eta_1+\zeta_1)p_3^4+(\tfrac{4}{3}\eta_2+\zeta_2)p_3^5) +$$

$$+\ (A^2_{1^23,3}+A^2_{3^3,3})(\eta_1 p_3^4+\eta_2 p_3^5) + (A^1_{123,3}+A^3_{23^2,3})$$

$$((\tfrac{1}{3}\eta_1+\zeta_1)p_3^4+(\tfrac{1}{3}\eta_2+\zeta_2)p_3^5))+\ldots+$$

$$+\ E_3(A^2_{2^23^{\ell-4},(\ell-2)}((\tfrac{4}{3}\eta_1+\zeta_1)p_3^4+(\tfrac{4}{3}\eta_2+\zeta_2)p_3^5) +$$

$$+\ (A^2_{1^23^{\ell-4},(\ell-2)}+A^2_{3^{\ell-2},(\ell-2)})(\eta_1 p_3^4+\eta_2 p_3^5) +$$

$$+\ (A^1_{123^{\ell-4},(\ell-2)}+A^3_{23^{\ell-3},(\ell-2)})((\tfrac{1}{3}\eta_1+\zeta_1)p_3^4+(\tfrac{1}{3}\eta_2+\zeta_2)p_3^5))+$$

$$+ [A^{2}_{2^{2}3^{\ell-3},(\ell-1)}((\tfrac{4}{3}\eta_1+\zeta_1)p_3^4+(\tfrac{4}{3}\eta_2+\zeta_2)p_3^5 +$$

$$+ (A^{2}_{1^{2}3^{\ell-3},(\ell-1)}+A^{2}_{3^{\ell-1},(\ell-1)})(\eta_1 p_3^4+\eta_2 p_3^5) +$$

$$+ (A^{1}_{123^{\ell-3},(\ell-1)}+A^{3}_{23^{\ell-2},(\ell-1)})((\tfrac{1}{3}\eta_1+\zeta_1)p_3^4 +$$

$$+ (\tfrac{1}{3}\eta_2+\zeta_2)p_3^5)]\} + [\tfrac{4}{3}\eta+\zeta][\delta_2 A^{2}_{3^{\ell-1},(\ell-1)}(2)-A^{2}_{23^{\ell-2},(\ell-1)}(2)]+$$

$$+ \eta[\delta_1 A^{2}_{3^{\ell-1},(\ell-1)}(1)-A^{3}_{13^{\ell-2},(\ell-1)}(1)] +$$

$$+ [\tfrac{1}{3}\eta+\zeta][\delta_1 A^{1}_{3^{\ell-1},(\ell-1)}(2)-A^{1}_{23^{\ell-2},(\ell-1)}(1)-A^{3}_{3^{\ell-1},(\ell-1)}(2)];$$

$$C_\ell^3 = - \{E_3^{\ell-2}(\phi^3)+E_3^{\ell-3}(A^{3}_{3^2,2}((\tfrac{4}{3}\eta_1+\zeta_1)p_3^4+(\tfrac{4}{3}\eta_2+\zeta_2)p_3^5) +$$

$$+ (A^{3}_{1^2,2}+A^{3}_{2^2,2})(\eta_1 p_3^4+\eta_2 p_3^5)+(A^{1}_{13,2}+A^{2}_{23,2})((\tfrac{1}{3}\eta_1+\zeta_1)p_3^4 +$$

$$+ (\tfrac{1}{3}\eta_2+\zeta_2)p_3^5)) + E_3^{\ell-4}(A^{3}_{3^3,3}((\tfrac{4}{3}\eta_1+\zeta_1)p_3^4+(\tfrac{4}{3}\eta_2+\zeta_2)p_3^5)+$$

$$+ (A^{3}_{1^2 3,3}+A^{3}_{2^2 3,3})(\eta_1 p_3^4+\eta_2 p_3^5) +$$

$$+ (A^{1}_{13^2,3}+A^{2}_{23^2,3})((\tfrac{1}{3}\eta_1+\zeta_1)p_3^4+(\tfrac{1}{3}\eta_2+\zeta_2)p_3^5))+\ldots+$$

$$+ E_3(A^3_{3^{\ell-2},(\ell-2)}((\tfrac{4}{3}\eta_1+\zeta_1)p_3^4+(\tfrac{4}{3}\eta_2+\zeta_2)p_3^5) +$$

$$+ (A^3_{1^2 3^{\ell-4},(\ell-2)} + A^3_{2^2 3^{\ell-4},(\ell-2)})(\eta_1 p_3^4+\eta_2 p_3^5)+$$

$$+ (A^1_{13^{\ell-3},(\ell-2)} + A^2_{23^{\ell-3},(\ell-2)})((\tfrac{1}{3}\eta_1+\zeta_1)p_3^4+(\tfrac{1}{3}\eta_2+\zeta_2)p_3^5))+$$

$$+ [A^3_{3^{\ell-1},(\ell-1)}((\tfrac{4}{3}\eta_1+\zeta_1)p_3^4+(\tfrac{4}{3}\eta_2+\zeta_2)p_3^5) +$$

$$+ (A^3_{1^2 3^{\ell-3},(\ell-1)} + A^3_{2^2 3^{\ell-3},(\ell-1)})(\eta_1 p_3^4+\eta_2 p_3^5) +$$

$$+ (A^1_{13^{\ell-2},(\ell-1)} + A^2_{23^{\ell-2},(\ell-1)})((\tfrac{1}{3}\eta_1+\zeta_1)p_3^4+(\tfrac{1}{3}\eta_2+\zeta_2)p_3^5)]\} +$$

$$+\eta[\delta_1 A^3_{3^{\ell-1},(\ell-1)}(1)+\delta_2 A^3_{3^{\ell-1},(\ell-1)}(2)-A^3_{13^{\ell-2},(\ell-1)}(1)-$$

$$- A^3_{23^{\ell-2},(\ell-1)}(2)] - [\tfrac{1}{3}\eta+\zeta][A^1_{3^{\ell-1},(\ell-1)}(1)+A^2_{3^{\ell-1},(\ell-1)}(2)];$$

$$C^4_\ell = -\{E_3^{\ell-2}(\phi^4)+E_3^{\ell-3}((A^5_{1^2,2}+A^5_{2^2,2}+A^5_{3^2,2})(\chi_1 p_3^4+\chi_2 p_3^5)) +$$

$$+ E_3^{\ell-4}((A^5_{1^2 3,3}+A^5_{2^2 3,3}+A^5_{3^3,3})(\chi_1 p_3^4+\chi_2 p_3^5))+ \dots +$$

$$+ E_3((A^5_{1^2 3^{\ell-4},(\ell-2)}+A^5_{2^2 3^{\ell-4},(\ell-2)}+A^5_{3^{\ell-2},(\ell-2)})(\chi_1 p_3^4+\chi_2 p_3^5))+$$

$$+[(A^5_{1^2 3^{\ell-3},(\ell-1)}+A^5_{2^2 3^{\ell-3},(\ell-1)}+A^5_{3^{\ell-1},(\ell-1)})(\chi_1 p_3^4+\chi_2 p_3^5)]\}+$$

$$+\chi[\delta_1 A^5_{3^{\ell-1},(\ell-1)}(1)+\delta_2 A^5_{3^{\ell-1},(\ell-1)}(2)-A^5_{13^{\ell-2},(\ell-1)}(1) -$$

$$- A^5_{23^{\ell-2},(\ell-1)}(2)] \quad ;$$

$$C^5_\ell = -\{ E_3^{\ell-2}(\phi^5)+E_3^{\ell-3}(p_3^4(A^1_{13,2}+A^2_{23,2}+A^3_{3^2,2})+p_3^1 A^4_{13,2}+$$

$$+ p_3^2 A^4_{23,2}+p_3^3 A^4_{3^2,2}) + E_3^{\ell-4}(p_3^4(A^1_{13^2,3}+A^2_{23^2,3}+A^3_{3^3,3})+$$

$$+ p_3^1 A^4_{13^2,3}+p_3^2 A^4_{23^2,3}+p_3^3 A^4_{3^3,3})+\ldots+$$

$$+ E_3(p_3^4(A^1_{13^{\ell-3},(\ell-2)}+A^2_{23^{\ell-3},(\ell-2)}+A^3_{3^{\ell-2},(\ell-2)}) +$$

$$+ p_3^1 A^4_{13^{\ell-3},(\ell-2)}+p_3^2 A^4_{23^{\ell-3},(\ell-2)}+p_3^3 A^4_{3^{\ell-2},(\ell-2)}) +$$

$$+ [p_3^4(A^1_{13^{\ell-2},(\ell-1)}+A^2_{23^{\ell-2},(\ell-1)}+A^3_{3^{\ell-1},(\ell-1)}) +$$

$$+ p_3^1 A^4_{13^{\ell-2},(\ell-1)}+p_3^2 A^4_{23^{\ell-2},(\ell-1)}+p_3^3 A^4_{3^{\ell-1},(\ell-1)}]\} -$$

$$- \rho_o[A^1_{3^{\ell-1},(\ell-1)}(1)+A^2_{3^{\ell-1},(\ell-1)}(2)] -$$

$$- [v_o^1 A^4_{3^{\ell-1},(\ell-1)}(1)+v_o^2 A^4_{3^{\ell-1},(\ell-1)}(2)+A^4_{3^{\ell-1},(\ell-1)}(4)];$$

Finally the analytic functions $A^i_{q^\lambda,\ell}(\xi)$, $\xi = (\xi_1,\xi_2,\xi_3)$ are given by :

$$A^i_{1q_2^{\lambda_2}\cdots q_\ell^{\lambda_\ell},\ell} = A^i_{q_2^{\lambda_2}\cdots q_\ell^{\lambda_\ell},(\ell-1)}(1) - \delta_1 \cdot A^i_{q_2^{\lambda_2}\cdots q_\ell^{\lambda_\ell}3,\ell} \ .$$

$$A^i_{2q_2^{\lambda_2}\cdots q_\ell^{\lambda_\ell},\ell} = A^i_{q_2^{\lambda_2}\cdots q_\ell^{\lambda_\ell},(\ell-1)}(2) - \delta_2 \cdot A^i_{q_2^{\lambda_2}\cdots q^{\lambda_\ell}3,\ell} \ .$$

$$A^i_{q_2^{\lambda_2}\cdots q^{\lambda_\ell}4,\ell} = A^i_{q_2^{\lambda_2}\cdots q_\ell^{\lambda_\ell},(\ell-1)}(4) - \delta_4 \cdot A^i_{q_2^{\lambda_2}\cdots q^{\lambda_\ell}3,\ell} \ ,$$

where

$$A^i_{q_2^{\lambda_2}\cdots q_\ell^{\lambda_\ell},(\ell-1)}(k) = \frac{\partial A^i_{q_2^{\lambda_2}\cdots q_\ell^{\lambda_\ell},(\ell-1)}}{\partial \xi_k} \quad (k = 1,2,4) \ .$$

Now we consider the case (Σ_g, C_g) as in Theorem 1›bis. Let $\xi = (\xi_1,\xi_2,\xi_3) \in \mathbb{R}^3$, $\delta_i = \dfrac{\partial g}{\partial \xi_i}$, $i = 1,2,3$; the sequence of analytic functions $\tilde{A}^i_{q^\lambda,\ell}$ in ξ is given as follows.

Let us consider the following analytic function of ξ :

$$G^1 = p_1^1[a_1 p_1^4 + a_2 p_1^5] + [p_2^1 + p_1^2][\eta_1 p_2^4 + \eta_2 p_2^5] + [p_3^1 + p_1^3][\eta_1 p_3^4 + \eta_2 p_3^5] +$$

$$+ [p_2^2 + p_3^3][b_1 p_1^4 + b_2 p_1^5] - [\tilde{p}_1 p_1^4 + \tilde{p}_2 p_1^5] - \rho_o[v_o^1 p_1^1 + v_o^2 p_2^1 +$$

$$+ v_o^3 p_3^1 + p_4^1] + F_1 \;;$$

$$G^2 = p_2^2[a_1 p_2^4 + a_2 p_2^5] + [p_2^1 + p_1^2][\eta_1 p_1^4 + \eta_2 p_1^5] + [p_3^2 + p_2^3][\eta_1 p_3^4 + \eta_2 p_3^5] +$$

$$+ [p_1^1 + p_3^3][b_1 p_2^4 + b_2 p_2^5] - [\tilde{p}_1 p_2^4 + \tilde{p}_2 p_2^5] - \rho_o[v_o^1 p_1^2 + v_o^2 p_2^2 + v_o^3 p_3^2 +$$

$$+ p_4^2] + F_2 \;;$$

$$G^3 = p_3^3[a_1 p_3^4 + a_2 p_3^5] + [p_3^1 + p_1^3][\eta_1 p_1^4 + \eta_2 p_1^5] + [p_3^2 + p_2^3][\eta_1 p_2^4 + \eta_2 p_2^5] +$$

$$+ [p_1^1 + p_2^2][b_1 p_3^4 + b_2 p_3^5] - [\tilde{p}_1 p_3^4 + \tilde{p}_2 p_3^5] - \rho_o[v_o^1 p_1^3 + v_o^2 p_2^3 + v_o^3 p_3^3 +$$

$$+ p_4^3] + F_3 \;;$$

$$G^4 = \chi_1[p_1^4 p_1^5 + p_2^4 p_2^5 + p_3^4 p_3^5] + \chi_2[(p_1^5)^2 + (p_2^5)^2 + (p_3^5)^2] +$$

$$+ \frac{\eta}{2} \sum_{i,k=1}^{3} [p_k^i + p_i^k - \frac{2}{3}\delta_{ik} \sum_{\ell=1}^{3} p_\ell^\ell]^2 + \zeta \sum_{i,k=1}^{3} p_i^i p_k^k -$$

$$-\rho_o T_o[s_1(v_o^1 p_1^4 + v_o^2 p_2^4 + v_o^3 p_3^4 + p_4^4) + s_2(v_o^1 p_1^5 + v_o^2 p_2^5 + v_o^3 p_3^5 + p_4^5)] \;;$$

$$G^5 = p_4^4[p_1^1 + p_2^2 + p_3^3] + p_4^1 p_1^4 + p_4^2 p_2^4 + p_4^3 p_3^4 \;,$$

where

$$p_1^i = v^i(1) - p_4^i \cdot \delta_1 \;, \quad p_2^i = v^i(2) - p_4^i \cdot \delta_2 \;, \quad p_3^i = v^i(3) - p_4^i \delta_3$$

$i = 1,2,3,4,5$, and where

$$p_4^4 = \frac{1}{[1-v_o^1\delta_1-v_o^2\delta_2-v_o^3\delta_3]}\{\rho_o[\delta_1 p_4^1+\delta_2 p_4^2+\delta_3 p_4^3-v^1(1)-v^2(2)-v^3(3)]-$$

$$- [v_o^1\rho_o(1)+v_o^2\rho_o(2)+v_o^3\rho_o(3)]\} ,$$

with

$$v^i(\ell) = \frac{\partial v_o^i}{\partial \xi_\ell} , \quad \rho_o(\ell) = \frac{\partial \rho_o}{\partial \xi_\ell} , \quad \ell = 1,2,3 .$$

We denote by :

$$B_2^1 = -G^1 -\{[\tfrac{4}{3}\eta+\zeta]p_1^1(1)+\eta[p_2^1(2)+p_3^1(3)]+[\tfrac{1}{3}\eta+\zeta][p_2^2(1)+p_3^3(1)]\}+$$

$$+ \delta_1\{[\tfrac{4}{3}\eta+\zeta]p_4^1(1)+[\tfrac{1}{3}\eta+\zeta][p_4^2(2)+p_4^3(3)]\} + \eta[\delta_2 p_4^1(2)+$$

$$+\delta_3 p_4^1(3)] \quad ;$$

$$B_2^2 = -G^2-\{[\tfrac{4}{3}\eta+\zeta]p_2^2(2)+\eta[p_1^2(1)+p_3^2(3)]+[\tfrac{1}{3}\eta+\zeta][p_2^1(1)+p_3^3(2)]\}+$$

$$+ \delta_1\{[\tfrac{1}{3}\eta+\zeta]p_4^1(2)+\eta p_4^2(1)\}+\delta_2\{[\tfrac{4}{3}\eta+\zeta]p_4^2(2)+[\tfrac{1}{3}\eta+\zeta]p_4^3(3)\}+$$

$$+ \delta_3\eta p_4^2(3) \ ;$$

$$B_2^3 = -G^3-\{[\tfrac{4}{3}\eta+\zeta]p_3^3(3) + \eta[p_1^3(1)+p_2^3(2)]+[\tfrac{1}{3}\eta+\zeta][p_3^1(1)+p_3^2(2)]\}+$$

$$+ \delta_1\{[\tfrac{1}{3}\eta+\zeta]p_4^1(3)+\eta p_4^3(1)\}+ \delta_2\{[\tfrac{1}{3}\eta+\zeta]p_4^2(3)+\eta p_4^3(2)\} +$$

$$+ \delta_3[\tfrac{4}{3}\eta+\zeta]p_4^3(3) \quad ;$$

$$B_2^4 = -G^4 - \chi[p_1^5(1)+p_2^5(2)+p_3^5(3)+\delta_1 p_4^5(1)+\delta_2 p_4^5(2)+\delta_3 p_4^5(3)] \;;$$

$$B_2^5 = -G^5 - \{\rho_o[p_4^1(1)+p_4^2(2)+p_4^3(3)] + v_o^1 p_4^4(1)+v_o^2 p_4^4(2)+v_o^3 p_4^4(3)\};$$

where

$$p_j^i(k) = \frac{\partial p_j^i}{\partial \xi_k} \quad , \quad i = 1,2,3,4,5 \;;\; j = 1,2,3,4 \;; \quad k = 1,2,3 \quad .$$

Now let us define

$$\tilde{A}^i_{o,o}(\xi) = v_o^i(\xi) \;,\quad i = 1,2,3,\; \tilde{A}^4_{o,o}(\xi) = \rho_o(\xi) \;,$$

$$\tilde{A}^5_{o,o}(\xi) = T_o(\xi)$$

for $\ell = 1$, by

$$\tilde{A}^i_{j,1} = \tilde{A}^i_{j,1}(\xi) = p_j^i(\xi) \quad , \quad i = 1,2,3,4,5 \;, \quad j = 1,2,3,4 \;,$$

and for $\ell = 2$, $q^\lambda = 4^2$ by :

$$\tilde{A}^1_{4^2,2} = \frac{1}{\eta[\frac{4}{3}\eta+\zeta][\delta_1^2+\delta_2^2+\delta_3^2]^2}\{[\eta\delta_1^2+(\tfrac{4}{3}\eta+\zeta)\delta_2^2+(\tfrac{4}{3}\eta+\zeta)\delta_3^2]B_2^1 - $$

$$- [\tfrac{1}{3}\eta+\zeta]\delta_1\delta_2 B_2^2 - [\tfrac{1}{3}\eta+\zeta]\delta_1\delta_3 B_2^3\} \quad .$$

$$\tilde{A}^2_{4^2,2} = \frac{1}{\eta[\frac{4}{3}\eta+\zeta][\delta_1^2+\delta_2^2\;+\delta_3^2]^2}\cdot\{-[\tfrac{1}{3}\eta+\zeta]\delta_1\delta_2 B_2^1+[(\tfrac{4}{3}\eta+\zeta)\delta_1^2+\eta\delta_2^2+$$

$$+ (\tfrac{4}{3}\eta+\zeta)\delta_3^2]B_2^2-[\tfrac{1}{3}\eta+\zeta]\delta_2\delta_3B_2^3\} \quad ;$$

$$\tilde{A}^3_{4^2,2} = \frac{1}{\eta[\frac{4}{3}\eta+\zeta][\delta_1^2+\delta_2^2+\delta_3^2]^2}\cdot\{-[\tfrac{1}{3}\eta+\zeta]\delta_1\delta_3B_1^1-[\tfrac{1}{3}\eta+\zeta]\delta_2\delta_3B_2^2 +$$

$$+ [(\tfrac{4}{3}\eta+\zeta)\delta_1^2+(\tfrac{4}{3}\eta+\zeta)\delta_2^2+\eta\delta_3^2]B_2^3\} \;;$$

$$\tilde{A}^4_{4^2,2} = \frac{1}{[\frac{4}{3}\eta+\zeta][1-v_o^1\delta_1-v_o^2\delta_2-v_o^3\delta_3][\delta_1^2+\delta_2^2+\delta_3^2]}\cdot$$

$$\{\rho_o[\delta_1B_2^1+\delta_2B_2^2+\delta_3B_2^3] + [\tfrac{4}{3}\eta+\zeta][\delta_1^2+\delta_2^2+\delta_3^2]B_2^5\} \quad ;$$

$$\tilde{A}^5_{4^2,2} = \frac{1}{\chi[\delta_1^2+\delta_2^2+\delta_3^2]}\cdot B_2^4 \quad .$$

We obtain the other q^λ for $\ell = 2$ by

$$\tilde{A}^i_{1q_j,2} = p_j^i(1)-A^i_{q_j4,2}\cdot\delta_1 \quad ,$$

$$A^i_{2q_j,2} = p_j^i(2)-A^i_{q_j4,2}\cdot\delta_2 \quad ,$$

$$\tilde{A}^i_{3q_3,2} = p_j^i(3) - \tilde{A}^i_{q_j^4,2}\cdot\delta_3 \quad .$$

Finally for $\tilde{A}^i_{4^\ell,\ell}$, $\ell \geq 3$, we have

$$\tilde{A}^1_{4^\ell,\ell} = \frac{1}{\eta[\frac{4}{3}\eta+\zeta][\delta_1^2+\delta_2^2+\delta_3^2]^2}\cdot\{[\eta\delta_1^2+(\tfrac{4}{3}\eta+\zeta)\delta_2^2+(\tfrac{4}{3}\eta+\zeta)\delta_3^2]B_\ell^1 \;-$$

$$- [\tfrac{1}{3}\eta+\zeta]\delta_1\delta_2 B^2_\ell - [\tfrac{1}{3}\eta+\zeta]\delta_1\delta_3 B^3_\ell\} \quad ;$$

$$\tilde{A}^2_{4^\ell,\ell} = \frac{1}{\eta\,[\frac{4}{3}\eta+\zeta][\delta_1^2+\delta_2^2+\delta_3^2]^2}\cdot\{-[\tfrac{1}{3}\eta+\zeta]\delta_1\delta_2 B^1_\ell +$$

$$+ [(\tfrac{4}{3}\eta+\zeta)\delta_1^2+\eta\delta_2^2+(\tfrac{4}{3}\eta+\zeta)\delta_3^2]B^2_\ell - [\tfrac{1}{3}\eta+\zeta]\delta_2\delta_3 B^3_\ell\} \quad ;$$

$$\tilde{A}^3_{4^\ell,\ell} = \frac{1}{\eta[\frac{4}{3}\eta+\zeta][\delta_1^2+\delta_2^2+\delta_3^2]^2}\cdot\{-[\tfrac{1}{3}\eta+\zeta]\delta_1\delta_3 B^1_\ell -$$

$$- [\tfrac{1}{3}\eta+\zeta]\delta_2\delta_3 B^2_\ell + [(\tfrac{4}{3}\eta+\zeta)\delta_1^2+(\tfrac{4}{3}\eta+\zeta)\delta_2^2+\eta\delta_3^2]B^3_\ell\} \quad ;$$

$$\tilde{A}^4_{4^\ell,\ell} = \frac{1}{[\frac{4}{3}\eta+\zeta][1-v_o^1\delta_1-v_o^2\delta_2-v_o^3\delta_3][\delta_1^2+\delta_2^2+\delta_3^2]}\cdot$$

$$\cdot\{\rho_o[\delta_1 B^1_\ell+\delta_2 B^2_\ell+\delta_3 B^3_\ell]+[\tfrac{4}{3}\eta+\zeta][\delta_1^2+\delta_2^2+\delta_3^2]B^5_\ell\} \quad ;$$

$$\tilde{A}^5_{4^\ell,\ell} = \frac{1}{\chi[\delta_1^2+\delta_2^2+\delta_3^2]}B^4_\ell \quad ;$$

where the B^i_ℓ are given by :

$$B^1_\ell = -\{E_4^{\ell-2}(G^1)+E_4^{\ell-3}(\tilde{A}^1_{1^2,2}((\tfrac{4}{3}\eta_1+\zeta_1)p_4^4+(\tfrac{4}{3}\eta_2+\zeta_2)p_4^5) +$$

$$+ (\tilde{A}^1_{2^2,2}+\tilde{A}^1_{3^2,2})(\eta_1 p_4^4+\eta_2 p_4^5)+(\tilde{A}^2_{12,2}+\tilde{A}^3_{13,2})((\tfrac{1}{3}\eta_1+\zeta_1)p_4^4+$$

$$+ (\tfrac{1}{3}\eta_2+\zeta_2)p_4^5))+E_4^{\ell-4}(\tilde{A}^1_{1^24,3}((\tfrac{4}{3}\eta_1+\zeta_1)p_4^4+(\tfrac{4}{3}\eta_2+\zeta_2)p_4^5 +$$

$$+ (\tilde{A}^1_{2^2 4,3} + \tilde{A}^1_{3^2 4,3})(\eta_1 p_4^4 + \eta_2 p_4^5) +$$

$$+ (\tilde{A}^2_{124,3} + \tilde{A}^3_{134,3})((\tfrac{1}{3}\eta_1+\zeta_1)p_4^4 + (\tfrac{1}{3}\eta_2+\zeta_2)p_4^5)) + \dots +$$

$$+ E_4(\tilde{A}^1_{1^2 4^{\ell-4},\ell-2}((\tfrac{4}{3}\eta_1+\zeta_1)p_4^4+(\tfrac{4}{3}\eta_2+\zeta_2)p_4^5)+$$

$$+ (\tilde{A}^1_{2^2 4^{\ell-4},(\ell-2)} + \tilde{A}^1_{3^2 4^{\ell-4},(\ell-2)})(\eta_1 p_4^4+\eta_2 p_4^5) +$$

$$+ (\tilde{A}^2_{124^{\ell-4},(\ell-2)} + \tilde{A}^3_{134^{\ell-4},(\ell-2)})((\tfrac{1}{3}\eta_1+\zeta_1)p_4^4+$$

$$+ (\tfrac{1}{3}\eta_2+\zeta_2)p_4^5)) + [\tilde{A}^1_{1^2 4^{\ell-3},(\ell-1)}((\tfrac{4}{3}\eta_1+\zeta_1)p_4^4 +$$

$$+ (\tfrac{4}{3}\eta_2+\zeta_2)p_4^5)+(\tilde{A}^1_{2^2 4^{\ell-3},(\ell-1)} + \tilde{A}^1_{3^2 4^{\ell-3},(\ell-1)})$$

$$\cdot (\eta_1 p_4^4+\eta_2 p_4^5) + (\tilde{A}^2_{124^{\ell-3},(\ell-1)}+\tilde{A}^3_{134^{\ell-3},(\ell-1)})$$

$$\cdot ((\tfrac{1}{3}\eta_1+\zeta_1)p_4^4+(\tfrac{1}{3}\eta_2+\zeta_2)p_4^5)]\} - \{[\tfrac{4}{3}\eta+\zeta]\,\tilde{A}^1_{14^{\ell-2},(\ell-1)}(1)+$$

$$+ \eta[\tilde{A}^1_{24^{\ell-2},(\ell-1)}(2)+\tilde{A}^1_{34^{\ell-2},(\ell-1)}(3)] +$$

$$+ [\tfrac{1}{3}\eta+\zeta][\tilde{A}^2_{24^{\ell-2},(\ell-1)}(1)+\tilde{A}^3_{34^{\ell-2},(\ell-1)}(1)]\}+$$

$$+ \delta_1\{[\tfrac{4}{3}\eta+\zeta]\tilde{A}^1_{4^{\ell-1},(\ell-1)}(1)+[\tfrac{1}{3}\eta+\zeta][\tilde{A}^2_{4^{\ell-1},(\ell-1)}(2)+$$

$$+ \tilde{A}^3_{4^{\ell-1},(\ell-1)}(3)]\} + \eta[\delta_2\tilde{A}^1_{4^{\ell-1},(\ell-1)}(2)+\delta_3\tilde{A}^1_{4^{\ell-1},(\ell-1)}(3)];$$

$$B_\ell^2 = -\{E_4^{\ell-2}(G^2)+E_4^{\ell-3}(\tilde{A}^2_{2^2,2}((\tfrac{4}{3}\eta_1+\zeta_1)p_4^4+(\tfrac{4}{3}\eta_2+\zeta_2)p_4^5) +$$

$$+ (\tilde{A}^2_{1^2,2}+\tilde{A}^2_{3^2,2})(\eta_1 p_4^4+\eta_2 p_4^5) + (\tilde{A}^1_{12,2}+\tilde{A}^3_{23,2})\cdot$$

$$\cdot((\tfrac{1}{3}\eta_1+\zeta_1)p_4^4+(\tfrac{1}{3}\eta_2+\zeta_2)p_4^5)) +$$

$$+ E_4^{\ell-4}(\tilde{A}^2_{2^2 4,3}((\tfrac{4}{3}\eta_1+\zeta_1)p_4^4+(\tfrac{4}{3}\eta_2+\zeta_2)p_4^5) +$$

$$+ (\tilde{A}^2_{1^2 4,3}+\tilde{A}^2_{3^2 4,3})(\eta_1 p_4^4+\eta_2 p_4^5) +$$

$$+ (\tilde{A}^1_{124,3}+\tilde{A}^3_{234,3})((\tfrac{1}{3}\eta_1+\zeta_1)p_4^4 + (\tfrac{1}{3}\eta_2+\zeta_2)p_4^5))+..+$$

$$+ E_4(\tilde{A}^2_{2^2 4^{\ell-4},(\ell-2)}((\tfrac{4}{3}\eta_1+\zeta_1)p_4^4+(\tfrac{4}{3}\eta_2+\zeta_2)p_4^5)+$$

$$+ (\tilde{A}^2_{1^2 4^{\ell-4},(\ell-2)}+\tilde{A}^2_{3^2 4^{\ell-4},(\ell-2)})(\eta_1 p_4^4+\eta_2 p_4^5) +$$

$$+ (\tilde{A}^1_{124^{\ell-4},(\ell-2)}+\tilde{A}^3_{234^{\ell-4},(\ell-2)})((\tfrac{1}{3}\eta_1+\zeta_1)p_4^4+(\tfrac{1}{3}\eta_2+\zeta_2)p_4^5))+$$

$$+ [\tilde{A}^2_{2^2 4^{\ell-3},(\ell-1)}((\tfrac{4}{3}\eta_1+\zeta_1)p_4^4+(\tfrac{4}{3}\eta_2+\zeta_2)p_4^5) +$$

$$+ (\tilde{A}^2_{1^2 4^{\ell-3},(\ell-1)}+\tilde{A}^2_{3^2 4^{\ell-3},(\ell-1)})(\eta_1 p_4^4+\eta_2 p_4^5) +$$

$$+ (\tilde{A}^1_{124^{\ell-3},(\ell-1)} + \tilde{A}^3_{234^{\ell-4},(\ell-2)})((\frac{1}{3}\eta_1+\zeta_1)p^4_4 +$$

$$+ (\frac{2}{3}\eta_2+\zeta_2)p^5_4)]\} -$$

$$-\{[\frac{4}{3}\eta+\zeta]A^2_{24^{\ell-2},(\ell-1)}(2) + \eta[\tilde{A}^2_{14^{\ell-2},(\ell-1)}(1) + \tilde{A}^2_{34^{\ell-2},(\ell-1)}(3)] +$$

$$+ [\frac{1}{3}\eta+\zeta][\tilde{A}^1_{24^{\ell-2},(\ell-1)}(1) + \tilde{A}^3_{34^{\ell-2},(\ell-1)}(2)]\} +$$

$$+ \delta_1\{[\frac{1}{3}\eta+\zeta]\tilde{A}^1_{4^{\ell-1},(\ell-1)}(2) + \eta\tilde{A}^2_{4^{\ell-1}(\ell-1)}(1)\} +$$

$$+ \delta_2\{[\frac{4}{3}\eta+\zeta]\tilde{A}^2_{4^{\ell-1},(\ell-1)}(2) + [\frac{1}{3}\eta+\zeta]\ \tilde{A}^3_{4^{\ell-1},(\ell-1)}(3)\} +$$

$$+ \eta\delta_3\tilde{A}^2_{4^{\ell-1},(\ell-1)}(3) \qquad ;$$

$$B^3_\ell = -\{E^{\ell-2}_4(G^3) + E^{\ell-3}_4(\tilde{A}^3_{3^2,2}((\frac{4}{3}\eta_1+\zeta_1)p^4_4 + (\frac{4}{3}\eta_2+\zeta_2)p^5_4) +$$

$$+ (\tilde{A}^3_{1^2,2} + \tilde{A}^3_{3^2,2})(\eta_1 p^4_4 + \eta_2 p^5_4) +$$

$$+ (\tilde{A}^1_{13,2} + \tilde{A}^2_{23,2})((\frac{1}{3}\eta_1+\zeta_1)p^4_4 + (\frac{1}{3}\eta_2+\zeta_2)p^5_4)) +$$

$$+ E^{\ell-4}_4(\tilde{A}^3_{3^24,3}((\frac{4}{3}\eta_1+\zeta_1)p^4_4 + (\frac{4}{3}\eta_2+\zeta_2)p^5_4) +$$

$$+ (\tilde{A}^3_{1^2 4,3} + \tilde{A}^3_{2^2 4,3}) \cdot (\eta_1 p_4^4 + \eta_2 p_4^5) +$$

$$+ (\tilde{A}^1_{134,3} + \tilde{A}^2_{234,3}) \cdot ((\tfrac{1}{3}\eta_1 + \zeta_1) p_4^4 + (\tfrac{1}{3}\eta_2 + \zeta_2) p_4^5)) + \ldots +$$

$$+ E_4 (\tilde{A}^3_{3^2 4^{\ell-4},(\ell-2)} \cdot ((\tfrac{4}{3}\eta_1 + \zeta_1) p_4^4 + (\tfrac{4}{3}\eta_2 + \zeta_2) p_4^5) +$$

$$+ (\tilde{A}^3_{1^2 4^{\ell-4},(\ell-2)} + \tilde{A}^3_{2^2 4^{\ell-4},(\ell-2)})(\eta_1 p_4^4 + \eta_2 p_4^5) +$$

$$+ (\tilde{A}^1_{134^{\ell-4},(\ell-2)} + \tilde{A}^2_{234^{\ell-4},(\ell-2)})((\tfrac{1}{3}\eta_1 + \zeta_1) p_4^4 +$$

$$+ (\tfrac{1}{3}\eta_2 + \zeta_2) p_4^5)) + [\tilde{A}^3_{3^2 4^{\ell-3},(\ell-1)}((\tfrac{4}{3}\eta_1 + \zeta_1) p_4^4 +$$

$$+ (\tfrac{4}{3}\eta_2 + \zeta_2) p_4^5) + (\tilde{A}^3_{1^2 4^{\ell-3},(\ell-1)} + \tilde{A}^3_{2^2 4^{\ell-3},(\ell-1)})(\eta_1 p_4^4 + \eta_2 p_4^5) +$$

$$+ (\tilde{A}^1_{134^{\ell-3},(\ell-1)} + \tilde{A}^2_{234^{\ell-3},(\ell-1)})((\tfrac{1}{3}\eta_1 + \zeta_1) p_4^4 + (\tfrac{1}{3}\eta_2 + \zeta_2) p_4^5)]\} -$$

$$- \{[\tfrac{4}{3}\eta + \zeta]\tilde{A}^3_{34^{\ell-2},(\ell-1)}(3) + \eta[\tilde{A}^3_{14^{\ell-2},(\ell-1)}(1) + \tilde{A}^3_{24^{\ell-2},(\ell-1)}(2)] +$$

$$+ [\tfrac{1}{3}\eta + \zeta][\tilde{A}^1_{34^{\ell-2},(\ell-1)}(1) + \tilde{A}^2_{34^{\ell-2},(\ell-1)}(2)]\} +$$

$$+ \delta_1\{[\tfrac{1}{3}\eta + \zeta]\tilde{A}^1_{4^{\ell-1},(\ell-1)}(3) + \eta\tilde{A}^3_{4^{\ell-1},(\ell-1)}(1)\} +$$

$$+ \delta_2\{[\tfrac{1}{3}\eta+\zeta]\tilde{A}^2_{4^{\ell-1},(\ell-1)}(3)+\eta\tilde{A}^3_{4^{\ell-1},(\ell-1)}(2)\} +$$

$$+ [\tfrac{4}{3}\eta+\zeta]\delta_3\tilde{A}^3_{4^{\ell-1},(\ell-1)}(3) \quad ;$$

$$B^4_\ell = -\{E_4^{\ell-2}(G^4)+E_4^{\ell-3}((\tilde{A}^5_{1^2,2}+\tilde{A}^5_{2^2,2}+\tilde{A}^5_{3^2,2})(\chi_1 p_4^4+\chi_2 p_4^5)) +$$

$$+ E_4^{\ell-4}((\tilde{A}^5_{1^2 4,3}+\tilde{A}^5_{2^2 4,3}+\tilde{A}^5_{3^2 4,3})(\chi_1 p_4^4+\chi_2 p_4^5))+ \ldots \ldots +$$

$$+ E_4((\tilde{A}^5_{1^2 4^{\ell-4},(\ell-2)}+\tilde{A}^5_{2^2 4^{\ell-4},(\ell-2)}+\tilde{A}^5_{3^2 4^{\ell-4},(\ell-2)})\cdot$$

$$\cdot(\chi_1 p_4^4+\chi_2 p_4^5)) + [(\tilde{A}^5_{1^2 4^{\ell-3},(\ell-1)}+\tilde{A}^5_{2^2 4^{\ell-3},(\ell-1)}+$$

$$+ A^5_{3^2 4^{\ell-3};(\ell-1)})(\chi_1 p_4^4+\chi_2 p_4^5)]\} -$$

$$-\chi[\tilde{A}^5_{14^{\ell-2},(\ell-1)}(1)+\tilde{A}^5_{24^{\ell-2},(\ell-1)}(2)+\tilde{A}^5_{34^{\ell-2},(\ell-1)}(3)]+$$

$$+ \chi[\delta_1\tilde{A}^5_{4^{\ell-1},(\ell-1)}(1)+\delta_2\tilde{A}^5_{4^{\ell-1},(\ell-1)}(2)+\delta_3\tilde{A}^5_{4^{\ell-1},(\ell-1)}(3)],$$

$$B^5_\ell = -\{E_4^{\ell-2}(G^5)+E_4^{\ell-3}(p_4^4(\tilde{A}^1_{14,2}+\tilde{A}^2_{24,2}+\tilde{A}^3_{34,2}) +$$

$$+ p_4^1\tilde{A}^4_{14,2}+p_4^2\tilde{A}^4_{24,2}+p_4^3\tilde{A}^4_{34,2}) + \quad ,$$

$$+ E_4^{\ell-4}(p_4^4(\tilde{A}^1_{14^2,3} + \tilde{A}^2_{24^2,3} + \tilde{A}^3_{34^2,3}) +$$

$$+ p_4^1\tilde{A}^4_{14^2,2} + p_4^2\tilde{A}^4_{24^2,3} + p_4^3\tilde{A}^4_{34^2,3}) + \ldots \ldots +$$

$$+ E_4(p_4^4(\tilde{A}^1_{14^{\ell-3},(\ell-2)} + \tilde{A}^2_{24^{\ell-3},(\ell-2)} + \tilde{A}^3_{34^{\ell-3},(\ell-2)}) +$$

$$+ p_4^1\tilde{A}^4_{14^{\ell-3},(\ell-2)} + p_4^2\tilde{A}^4_{24^{\ell-3},(\ell-2)} + p_4^3\tilde{A}^4_{34^{\ell-3},(\ell-2)}) +$$

$$+ [p_4^4(\tilde{A}^1_{14^{\ell-2},(\ell-1)} + \tilde{A}^2_{24^{\ell-2},(\ell-1)} + \tilde{A}^3_{34^{\ell-2},(\ell-1)}) +$$

$$+ p_4^1\tilde{A}^4_{14^{\ell-2},(\ell-1)} + p_4^2\tilde{A}^4_{24^{\ell-2},(\ell-1)} + p_4^3\tilde{A}^4_{34^{\ell-2},(\ell-1)}]\}-$$

$$-\{\rho_o[\tilde{A}^1_{4^{\ell-1},(\ell-1)}(1) + \tilde{A}^2_{4^{\ell-1},(\ell-1)}(2) + \tilde{A}^3_{4^{\ell-1},(\ell-1)}(3)]$$

$$+ v_o^1\tilde{A}^4_{4^{\ell-1},(\ell-1)}(1) + v_o^2\tilde{A}^4_{4^{\ell-1},(\ell-1)}(2) + v_o^3\tilde{A}^4_{4^{\ell-1},(\ell-1)}(3)\};$$

Finally, the analytic functions $\tilde{A}^i_{q^\lambda,\ell}(\xi)$, $\xi = (\xi_1,\xi_2,\xi_3)$ are given by

$$\tilde{A}^i_{1q_2^{\lambda_2}\ldots q_\ell^{\lambda_\ell},\ell} = \tilde{A}^i_{q_2^{\lambda_2}\ldots q_\ell^{\lambda_\ell},(\ell-1)}(1) - \delta_1\tilde{A}^i_{q_2^{\lambda_2}\ldots q_\ell^{\lambda_\ell}4,\ell} ;$$

$$\tilde{A}^i_{2q_2^{\lambda_2}\dots q^{\lambda_\ell},\ell} = \tilde{A}^i_{q^{\lambda_2}\dots q^{\lambda_\ell},(\ell-1)}(2)-\delta_2\cdot\tilde{A}^i_{q_2^{\lambda_2}\dots q^{\lambda_\ell}4,\ell} \quad ;$$

$$\tilde{A}^i_{3q_2^{\lambda_2}\dots q_\ell^{\lambda_\ell},} = \tilde{A}^i_{q_2^{\lambda_2}\dots q^{\lambda_\ell},(\ell-1)}(3)-\delta_3\cdot\tilde{A}^i_{q_2^{\lambda_2}\dots q_\ell^{\lambda_\ell}4,\ell} \quad .$$

where

$$\tilde{A}^i_{q_2^{\lambda_2}\dots q_\ell^{\lambda_\ell}}(k) = \frac{\partial\tilde{A}^i_{q_2^{\lambda_2}\dots q_\ell^{\lambda_\ell}}}{\partial\xi_k} \quad , \qquad k = 1,2,3 \quad .$$

4. Given a well posed initial condition (Σ_f,C_f) as in Theorem 1 and a fixed point $x^o = (x_1^o,x_2^o,x_3^o,x_4^o)$ on the hypersurface Σ_f, let $\xi^o = (\xi_1^o,\xi_2^o,\xi_4^o) \in \mathbb{R}^3$ be such that $f(\xi^o) = x_3^o$, consider the values of the sequence of functions $A^i_{q^\lambda,\ell}$ constructed in § 3 at ξ^o

$$a^i_{q^\lambda,\ell} = A^i_{q^\lambda,\ell}(\xi^o) \quad , \qquad i = 1,2,3,4,5 \quad .$$

Then we have

<u>Theorem 2</u> : <u>The unique analytic solution of the system</u> (I) <u>on a tubular neighborhood of the hypersurface</u> Σ_f , <u>and near</u> x^o <u>associated to the well posed initial condition</u> (Σ_f,C_f) <u>is given by the convergent series</u>

$$v^i(x_1,x_2,x_3,x_4) = \Sigma\,(\lambda_1!\dots\lambda_\mu!)^{-1} a^i_{q^\lambda,\ell}\,(x_{q_1}-x^o_{q_1})^{\lambda_1}\dots (x_{q_\mu}-x^o_{q_\mu})^{\lambda_\mu}$$

$i = 1,2,3,$

$$\rho(x_1,x_2,x_3,x_4) = \Sigma\,(\lambda_1!..\lambda_\mu!)^{-1} a^4_{q^\lambda,\ell}\,(x_{q_1}-x^o_{q_1})^{\lambda_1} \ldots (x_{q_\mu}-x^o_{q_\mu})^{\lambda_\mu}$$

$$T(x_1,x_2,x_3,x_4) = \Sigma\,(\lambda_1!..\lambda_\mu!)^{-1} a^5_{q^\lambda,\ell}\,(x_{q_1}-x^o_{q_1})^{\lambda_1} \ldots (x_{q_\mu}-x^o_{q_\mu})^{\lambda_\mu} .$$

Now, for the well posed initial condition (Σ_g, C_g) of the theorem 1 bis we set

$$b^i_{q^\lambda,\ell} = \tilde{A}^i_{q^\lambda,\ell}\,(\xi^o) \quad , \qquad i = 1,2,3,4,5 ,$$

and we have

<u>Theorem 2 bis</u> : <u>The unique analytic solution of the system (I) on a tubular neighborhood of the hypersurface</u> Σ_g <u>and near</u> x^o <u>associated to the well posed initial condition</u> <u>(Σ_g, C_g)</u> <u>is given by the convergent series</u> : $i = 1,2,3$

$$v^i(x_1,x_2,x_3,x_4) = \Sigma\,(\lambda_1!\ldots\lambda_\mu!)^{-1} b^i_{q^\lambda,\ell}\,(x_{q_1}-x^o_{q_1})^{\lambda_1} \ldots (x_{q_\mu}-x^o_{q_\mu})^{\lambda_\mu}$$

$$\rho(x_1,x_2,x_3,x_4) = \Sigma\,(\lambda_1!\ldots\lambda_\mu!)^{-1} b^4_{q^\lambda,\ell}\,(x_{q_1}-x^o_{q_1})^{\lambda_1} \ldots (x_{q_\mu}-x^o_{q_\mu})^{\lambda_\mu}$$

$$T(x_1,x_2,x_3,x_4) = \Sigma\ (\lambda_1!\ldots\lambda_\mu!)^{-1} b^5_{q^\lambda,\ell}\ (x_{q_1}-x^o_{q_1})^{\lambda_1}\ldots$$

$$(x_{q_\mu}-x^o_{q_\mu})^{\lambda_\mu}$$

Finally, we remark that it is clear that the inequations mentionned in Theorem 1 and Theorem 1 bis have real analytic solutions ; hence the space of local real analytic solutions of the system (I) at a point $x^o \in \mathbb{R}^4$ contains the space

$$(V_3)^7 \times (M_3 \cdot M_3) \times (V_{3,+})^2$$

where V_3 is the ring of germs of real analytic functions at the origin of $\mathbb{R}^3$, M_3 its maximal ideal, and $V_{3,+}$ the subspace of V_3 consisting of all the strictly positive germs. The different type of inequations appearing in Theorem 1 and 1 bis seem to indicate that the "classical symbol or characteristic" is not quite sufficient to answer the two problems of the beginning of this work.

References :

[1] Landau, L. et Lifschitz, E. : Physique Théorique Tome 6, éditions Mir. 1954.

[2] Nash, J. : Le problème de Cauchy pour les équations différentielles d'un fluide général. Bull. Soc. Math. France, 90, 1962. p. 487.

[3] Picard, E. : Leçons sur quelques types simples d'équations aux dérivées partielles, avec des applications à la physique mathématique, 1925, Gauthier-Villars, Paris.

[4] Shih, W.H. : Sur le problème de Cauchy pour les équations d'un fluide général. Thèse Université de Paris VII, à paraître.

[5] Shih, W. : Quelques notions élémentaires sur les équations aux dérivées partielles. C.R. Acad. Sci. Paris, t. 292, 1981, série I, p. 901.

[6] Valiron, G. : Cours d'analyse mathématique. II Equations fonctionnelles. 1950, Masson.

中 国

上海工业大学

施 惟 慧

SHIH Wei Hui

Shanghai Technological University
People's Republic of China

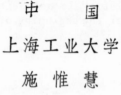

Jean LERAY

THE MEANING OF W.H. SHIH'S RESULT

W.H. Shih(*) comments upon its theorem 1bis as follows : << There does not exist any initial condition on the hyperplane t = 0 for which the corresponding Cauchy problem is well posed.>>

The theory asserts that the problem in question describes the movement of a general fluid, whose state at an initial time is given. If that problem should not be well posed, then the fluid mechanics would be a failure.

W.H. Shih does not establish such a failure.

First of all, W.H. Shih does perhaps change the meaning of that problem by using notions of his own, which were never clearly and explicity published : his (5).

Next, let us assume he means by <<a well posed Cauchy problem>> what J. Hadamard (**) meant when he introduced that notion , pointing out how it depends essentially on the functional space in which the solution is chosen. J. Hadamard did insist on the fact that a space of analytic functions - the only kind of spaces W.H. Shih takes into consideration - is inadequate for most of the problems arising in mathematical physics : it is a too narrow space. That is clear for all the problems where discontinuities or wave propagations can happen. It is also true for many other problems for instance the heat propagation in the euclidian space is governed by the equation :

$$\frac{\partial u(t,x)}{\partial t} = \sum_{k=1}^{3} \frac{\partial^2 u(t,x)}{\partial x_k^{\,2}}$$

and the Cauchy datum : u(o,x) ; a century ago that Cauchy problem was already known to be well posed in many functional spaces, but not in the space of the analytic functions.

C. P. Bruter et al. (eds.), Bifurcation Theory, Mechanics and Physics, 139–140.

Now, in the system governing the fluid mechanics, the equation W.H. Shih (*) denotes by (1') contains the heat operator

$$\rho \frac{\partial}{\partial t} - \eta \sum_{k=1}^{3} \frac{\partial^2}{\partial x_k^2}$$

which plays, explicity or not, a fundamental rôle in the solution (***) of that system with Cauchy data for $t = 0$. Therefore anybody familiar to fluid mechanics exends without hesitation from heat mechanics to fluid mechanics the assertion : <<The spaces of analytic functions are not the spaces in which that Cauchy problem can be solved>>.

The above assertion is perhaps what Shih (*) should state in order to be clear ans precise.

BIBLIOGRAPHY

(*) W.H. Shih, On the Cauchy problem for the equation of a general fluid, this volume, p. 109.

(**) J. Hadamard, any of the treatises about differential equations, for instance the last one, published in China (1950).

(***) A. Matsumura, T. Nishida, The initial value problem for the equations of motion of viscous and heat - conductive gases, J. Math. Kyoto Univ. 20(1080), n° 1, 67-104.

JEAN-FRANCOIS POMMARET

THE GENESIS OF THE GALOIS THEORY FOR SYSTEMS OF PARTIAL DIFFERENTIAL EQUATIONS

INTRODUCTION

It is well known that the development of physics (including mechanics) has always paralled that of mathematics. The only question left is to know whether the tool has motivated its application (tensor analysis for general relativity, the work of Cartan for Gauge theory) or if the contrary has arisen (study of crystals for tensor analysis, continuum mechanics for differential geometry). In the case of the Galois theory for systems of partial differential equations (PDE) or simply in the sequel, differential Galois theory[6] , our purpose is to illustrate the combination of these two processes. Namely, we shall sketch, with the least amount of mathematics, how physical motivations have oriented our work and how it will conversely interfere with future physics.

IN PARTICULAR WE SHOULD LIKE TO POINT OUT THAT A CRUCIAL FACT HAS BEEN TO OVERCOME THREE DEEP MISUNDERSTANDINGS THAT HAVE EXISTED FOR MORE THAN FIFTY YEARS AND THAT WILL STILL BE DONE AS FAR AS NOBODY WILL INVEST HIS TIME IN THE STUDY OF THE MODERN FORMAL THEORY OF PDE IN CONJUNCTION WITH ITS ALGEBRAIC COUNTERPART.

Here are these three striking results, each one being related with the following one as we shall see later on :

1) The classical Galois theory is not at all a theory of field automorphisms but only a particular case of the study of principal homogenous spaces (PHS) for finite algebraic groups, namely it just deals with permutation groups.
2) The Picard-Vessiot theory as developed by E. Picard and E. Vessiot, then by E.R. Kolchin during the last forty years, for the study of linear ordinary differential equations (ODE) is absolutely incorrect. The misunderstanding is provided by the use of maximum differential ideals instead of prime differential ideals in the diffe-

C. P. Bruter et al. (eds.), Bifurcation Theory, Mechanics and Physics, 141–147.

rential algebraic background and the lack of an appropriate formalism.

3) The Cartan structure equations for Lie pseudogroups involving differential forms on jet spaces are not at all[5] the proper generalizations of the Maurer-Cartan equations for Lie groups. The correct structure equations have been provided as early as in 1903 by Vessiot in a brilliant paper [7] and then completely forgotten.

We hope that this lecture will constitute a warning against the way to consider mathematics as a "logical chess game", by cutting it out from its "intuitive roots" and their historical developments. As a byproduct, only the intuitive way has allowed to create links between topics at first sight completely disconnected as the three above points, the method being the use of aesthetic pictorial arguments like diagram chasing (differential geometry) and reversed arrows (algebraic geometry).

I/ FIRST MISUNDERSTANDING :

We sketch the evolution of the Galois theory, presenting this in a few steps, each one clarifying the preceding one.

A) Before 1830 : One must never forget that the initial problem was to "know" about the roots of an irreducible polynomial with coefficients in some field K. We recall that a field is a set of elements such that :

$a,b \in K \Rightarrow a+b \in K,\ ab \in K,\ a \neq 0 \Rightarrow \exists\, b \in K,\ ab = 1$

as we shall always suppose that $\mathbb{Q} \in K$ and that a polynomial P is irreducible if it cannot be written as a product of polynomials also defined over K.

Of course a first question was to explain the words "to know" in a mathematical manner. For most of the people the key machinery was the use of square, cubic,... roots but the reader must not forget that even J. Cardan in this formulas (first discovered by N. Tartaglia, but this is another story !) was not possessing the concept of "zero". Shortly after the case of degree three was known by the above formulas, L. Ferrari similarly solved the case of degree four by bringing it back to the latter situation. Then it was conjectured that, for equations of order five or more, it was not possible to express the roots by formulas involving radicals.

B) 1830-1930 : Most of the great mathematicians attacked this problem (N.H. Abel) in vain. Knowing about the work of N.H. Abel, E. Galois was able to solve this problem by introducing for the first time the word and concept of "group" in mathematics. Though each of his three manuscripts was short (about thirty pages), the French Academy of Sciences, through its referees, refused the first (Cauchy), lost the second (Fourier) and found the third ununderstandable (Poisson). Hence Galois died, exactly 150 years ago, with the only personal belief that he was introducing new concepts in mathematics.

In 1843, that is to say 11 years after Galois's death, Liouville announced the importance of the results but it seems sure that, first nobody knows how Liouville got these papers, second that Liouville did not understand them (Betti) and third that Cauchy was involved in this affair (in a bad way according to certain letters !) as he published at once many papers on permutation groups. In any case this is the starting point of our story and a boost for studying this problem took place in the second half of the nineteenth century.

Let $\eta^1,\ldots,\eta^m$ be the m (distinct) roots of a polynomial $P \equiv (y)^m - \omega^1 (y)^{m-1} + \ldots + (-1)^m \omega^m \in K[y]$ the ring of polynomials in y with coefficients $\omega^1,\ldots,\omega^m$ belonging to K. The trick is to construct an equation, of degree $m\,!$, called the resolvent equation, in such a way that $\eta^1,\ldots,\eta^m$ may be expressed as rational functions of any one root ζ of this resolvent equation and any permutation of $\eta^1,\ldots\eta^m$ exchanges the roots of the resolvent equation . As one may choose $\zeta = \alpha_1 \eta^1 + \ldots + \alpha_m \eta^m$ with $\alpha_1,\ldots,\alpha_m \in K$, we express this by saying that the field $L = K(\eta^1,\ldots\eta^m)$ of rational functions in $\eta^1,\ldots,\eta^m$ is equal to the field $K(\zeta)$ of rational functions in the primitive element ζ . The crucial fact is to notice that the resolvent equation may be reducible over K when $\omega^1,\ldots,\omega^m$ are not algebraically independent. Moreover any irreducible factor of the resolvent equation is invariant by a subgroup of the group of permutations of m objects and these groups are isomorphic. One of these is called the Galois group Γ of P and we may define an algebraic extension L/K to be a Galois extension by the following property :

Definition 1 : L is obtained from K by adding all the roots of a polynomial defined over K.

The work done by Galois was to relate the search for the roots of P with the search for certain subgroups of Γ .

C) 1930-1980 : The main improvement of the theory has been done by E. Artin [1] after the work of the german school (Dedekind, Kronecker).

Any automorphism of L over K, that is to say any $\varphi : L \longrightarrow L$, $\varphi(a+b) = \varphi(a) + \varphi(b), \varphi(ab) = \varphi(a)\,\varphi(b)$ preserving K, preserves P and thus transforms any root η^k into another say η^ℓ. Hence the group aut(L/K) thus defined may be considered as a group of permutations. A delicate fact proved by Artin using tricky methods of linear algebra, is that we may identify aut(L/K) with Γ and that, moreover, a definition equivalent to the preceding may be :

Definition 2 : The field of invariants of aut(L/K) in L is just K.

In a condensed way $K = \{ a \in L \mid \varphi(a) = a, \forall \varphi \in \text{aut}(L/K)\}$. Next, this nice setting allows one to obtain a bijective order reversing correspondence between the intermediate fields and the subgroups of the Galois Group namely with any K' between K and L we associate $\Gamma' = \text{aut}(L/K')$. Moreover, if $K \subset K' \subset L$ then K'/K is again a Galois extension if and only if aut(L/K') is normal in aut(L/K). This property is known as the normal behaviour of Galois extensions.

D) After 1980 : At the end of the last century and the beginning of this, some mathematicians (F. Klein[3], J. Drach[2], E. Vessiot[7]) had the feeling that it was possible to generalize the Galois theory and in particular the Galois correspondence by using systems of algebraic PDE instead of equations, solutions instead of roots and Lie pseudogroups, that is to say groups of transformations solutions of systems of PDE, instead of permutation groups.

Of course, we may endow K with an additional differential structure by introducing commuting derivations ∂_i that is to say $\partial_i : K \longrightarrow K$ with $\partial_i(a+b) = \partial_i(a) + \partial_i(b)$, $\partial_i(ab) = (\partial_i a)b + a(\partial_i b)$ $\forall i, \forall a,b \in K$ and we say that K becomes a differential field.

However the first of the preceding definitions has no chance to be used because there is now an infinite number of solutions.

As for the second definition, the problem is more subtle. Using the case $m = 3$ that can easily be generalized to an arbitrary m, let us write out m times the same polynomial equation $P \equiv (y)^3 - \omega^1 (y)^2 + \omega^2 y - \omega^3 = 0$ with m different indeterminates $y^1, \ldots, y^m$. We get, with $\omega^1, \omega^2, \omega^3 \in K$

$$\begin{cases} P_1 \equiv (y^1)^3 - \omega^1 (y^1)^2 + \omega^2 y^1 - \omega^3 = 0 \\ P_2 \equiv (y^2)^3 - \omega^1 (y^2)^2 + \omega^2 y^2 - \omega^3 = 0 \\ P_3 \equiv (y^3)^3 - \omega^1 (y^3)^2 + \omega^2 y^3 - \omega^3 = 0 \end{cases}$$

which is a linear system for $\omega^1, \omega^2, \omega^3$ the value δ of the determinant $\Delta \equiv (y^1-y^2)(y^1-y^3)(y^2-y^3)$ being the discriminant of P. As we may suppose with no loss of generality that P is irreducible, we must have $\delta \neq 0$ with δ^2 a rational function of $\omega^1, \omega^2, \omega^3$.

The Cramer's resolution becomes just :

$$\left\{\begin{array}{lll} \Phi^1 \equiv y^1 + y^2 + y^3 & = & \omega^1 \\ \Phi^2 \equiv y^1 y^2 + y^1 y^3 + y^2 y^3 & = & \omega^2 \\ \Phi^3 \equiv y^1 y^2 y^3 & = & \omega^3 \end{array}\right.$$

where the Φ are the symmetric functions of the y generating all the other ones. Because $\delta \neq 0$, the roots are different and this algebraic system defines a principal homogenous space (PHS) for the group of permutations in 3 objects which is thus acting freely and transitively by definition.

This zero dimensional algebraic variety is not irreducible in general as the ideal generated by P_1, P_2, P_3 in $K[y^1, y^2, y^3]$ may be not be prime. Indeed, if $\omega^1 = 0$, $\omega^2 = 3$, $\omega^3 = 1$ then $\Delta^2 - (9)^2 = (\Delta - 9)(\Delta + 9) \in K[y]$ belongs to this ideal. As we are dealing with permutation groups, each irreducible component is again a PHS for some group isomorphic to the Galois group of P. Hence we discover that the Galois theory is nothing else than a theory of irreducible PHS for finite algebraic groups. But some of these groups, for example multiplications by cube roots of unity, do not provide automorphisms and we can use no longer the second definition.

This approach of the Galois mechanism has never been used. Now as L is a finite dimensional vector space over K we may construct the tensor product $L \otimes_K L$ and we get an equivalent definition :

Definition 3 : $L \otimes_K L \simeq L \oplus \ldots \oplus L$ ($\dim_K L$ terms)
Only this last definition must be used in a systematic way, even in the case where the algebraic group does not split completely over K. The use of diagram chasing in this framework is quite a new language and the reader understands that not a lot is left from Artin's point of view as aut(L/K) is of no use at all in this situation.

II/ SECOND MISUNDERSTANDING :

As was discovered by Vessiot, the need for fields, and thus irreducible PHS was essential.

A grave error was made by J. Drach in his thesis. Indeed he defined a system of algebraic PDE to be irreducible if it was not possible to add new compatible PDE to the system. This means that the corresponding differential ideal, that is to say an ideal stable by derivations $\partial_1, \ldots, \partial_n$ playing the role of $\partial/\partial x^1, \ldots, \partial/\partial x^n$, must be maximum. In fact, as we need only fields, we need only prime differential ideals and not maximum ones. This deep misunderstanding directed subsequent research in a bad direction.

As an example, let us consider the Picard-Vessiot theory for the linear ODE :

$$P \equiv y_{xxx} - \omega^1 y_{xx} + \omega^2 y_x - \omega^3 y = 0$$

where $\omega^1, \omega^2, \omega^3 \in K$ a differential field. We may repeat the last construction with the Wronskian instead of the discriminant and we discover that we have an irreducible PHS for the general linear group $GL(3)$ as P is linear. Hence the Galois group of such an ODE of order m is $GL(m)$ and not an algebraic subgroup.

This fact is invalidating most of the work on these topics.

III/ THIRD MISUNDERSTANDING :

As we have increased the generality of both the systems and the groups involved in the two preceding paragraphs, we may think with Vessiot that the ultimate step is to deal with algebraic PDE and Lie pseudogroups.

Starting with the knowledge of mechanics and general relativity, we did believe that to each pseudogroup one is able to associate a geometric object, that is to say an object whose transformations are known whenever one effects a change of local coordinates. Then this geometric object (field in the sense of physicists but also in the sens of differential algebra) must satisfy partial differential conditions (Maurer-Cartan equations, constant Riemannian curvature...) depending on some constants related by quadratic identities.[5,6]

It was a surprise for us to discover that this program was first sketched by Vessiot, in complete opposition to the methods of Cartan and more recently Spencer.[4]

As a complementary work, essential for the normal behaviour of the Galois theory, we discovered that the concept of normalizer of some pseudogroup Γ, that is to say the biggest Lie pseudogroup in which Γ is normal, is just the natural generalization of canonical transformations in mechanics.

CONCLUSION

We hope to have convinced the reader that many aspects of algebra and geometry may be incorporated in a new mathematical architecture. As a byproduct it is not possible to separate one aspect from another one.

Exactly as the Galois theory is essential for the study of algebraic equations, we do believe that the differential Galois theory will become fundamental for the study of PDE. Now the fact that the structure equations of Cartan must be replaced by those of Vessiot will lead physicists to change the differential framework of gauge theory. However the situation will not evolve until someone invests some time in understanding the formal theory of PDE. Finally, for the same reason, the quadratic identities for the structure constants we have spoken about in the last paragraph, are nothing else than the generalization of the Jacobi's identities. This fact proves that the recent program for using super algebras (supergravity, supergauge), must be revisited in this framework as it cannot be excluded from differential geometry.

BIBLIOGRAPHY

1) E. ARTIN : Galois Theory, Notre Dame University publications, 1942.
2) J. DRACH : Sur l'intégration logique..., thèse, Ann. Ec. Norm. Sup., 1898.
3) F. KLEIN : Programme d'Erlangen, Collection Discours de la Méthode, Gauthier-Villars Paris, 1974.
4) A. KUMPERA, D.C. SPENCER : Lie equations, Princeton University Press, 1972, 290p.
5) J.F. POMMARET : Systems of partial differential equations, and Lie pseudogroups, Gordon and Breach, 1978.
6) J.F. POMMARET : Differential Galois Theory, Gordon and Breach, 1982, 760 p.
7) E. VESSIOT : Sur la théorie de Galois et ses diverses généralisations, Ann. Ec. Norm. Sup., 1904.

Silviu Guiasu

INFORMATION THEORY AND A STOCHASTIC MODEL FOR EVOLUTION

1. INTRODUCTION

The initial aim of information theory was the construction of a mathematical model for the transmission of messages through noisy or noiseless communication channels. Later on its domain became larger by including in mathematical models a variety of applications from statistical inference, decision theory, pattern-recognition and classification theory, etc. The main reason for this increase of its area of applicability must be looked for in the fact that information theory deals with quantitative measures for very general and abstract notions like the amount of uncertainty contained a priori by an arbitrary experiment having several possible outcomes and the global connection between the subsystems of a given system. In fact Shannon's information entropy is the best measure for the amount of uncertainty known today while Watanabe's measure gives us the possibility to evaluate numerically the amount of global connection between the subsystems of a system. At the same time, by using Kullback's divergence we can globally compute how much two probability distributions of the same dimension differ one from another.

The aim of the present paper is to discuss the possibility we have today, due to information theory, to build up a stochastic model for evolution when we know only some mean values of some random variables representing some state functions of a given system and some correlations between some state functions of the system at different time-moments. More explicitely, let us suppose that we know the mean values

$$E(f(t_1)),\ldots E(f(t_m))$$

at the time-moments $t_1 < t_2 < \ldots < t_m$ and the covariances

$$C(f(t_j),f(t_{j+1}))$$

corresponding to the successive time-moments

C. P. Bruter et al. (eds.), Bifurcation Theory, Mechanics and Physics, 149–167

$$t_j, t_{j+1} \qquad (j=1,\ldots,m-1)$$

attached to the state functions

$$f(t_1),\ldots,f(t_m)$$

corresponding to an arbitrary dynamical system. What is the most suitable model for the system's evolution compatible with these data or, which is the same thing, what is the most suitable probability distribution on the set of all possible trajectories of the system compatible with the data mentioned above? For answering this question we have to precise first of all the criterion or the criteria with respect to which we use the superlative "most suitable". The criteria and the instrument for solving this problem consist of the Principle of Maximum Entropy (PEM) and the Principle of Minimum Interdependence (PIM). Both these principles are necessary and complement one another. By applying PEM we determine, at each moment t_j, $(j=1,\ldots,m)$, the "largest" probability distribution on the set of all possible states of the system compatible with the mean value

$$E(f(t_j)), \tag{1}$$

i.e. the probability distribution of maximum uncertainty, which does not ignore any possibility or which is the most uniform compatible with the constraint given by the mean value (1) of the random variable f_j. By applying PIM we determine the "largest" product probability distribution on the set of all successions of states of the system at the time-moments

$$t_1 < t_2 < \ldots < t_m \tag{2}$$

i.e. the product probability distribution which is the nearest one to the independence of the system's states at the successive time-moments (2) compatible with the data we have on the interdependence between the system's states given quantitatively by the covariances

$$C(f(t_1),f(t_2)),\ C(f(t_2),f(t_3)),\ldots,C(f(t_{m-1}),f(t_m)). \tag{3}$$

Thus, a maximum principle (PEM) and a minimum one (PIM) permit us to built up the "largest" stochastic model for the system's evolution, i.e. the model which does not impose any new constraints on the system's evolution other than the ones given a priori by the mean values (1) for $j=1,\ldots,m$ and by the covariances (3) attached to the successive state functions of the system corresponding to the time-moments (2).

PEM was formulated by E.T.Jaynes (1957) as a deeper variant of Laplace's Principle of Insufficient Reason, while PIM is a special case of Kullback's Principle of Minimum Divergence (Kullback, 1959), considered by Guiasu (1978) like a minimization of Watanabe's measure of interdependence (Watanabe, 1969) slightly generalized. In the construction of the stochastic model for evolution given here we try to show that PEM and PIM are complementary and complete one another.

Therefore, by investigating the information transmission through noisy and noiseless communication channels C.E.Shannon (1948) introduced the information entropy as the measure of either the amount of information supplied a posteriori by a probabilistic experiment or the amount of uncertainty we have a priori on the outcome of the respective probabilistic experiment. The source of inspiration has been the H-function introduced by L. Boltzmann (1896) as a measure of desorder in the study of the molecules of a gas. Shannon's entropy was used by S. Watanabe (1969) for defining the global measure of the interdependence between the subsystems of a given system. On the other hand, S.Kullback (1959) introduced the entropic measure of the divergence between two probability distributions. By using these entropic measures it is possible to formulate two methods for the construction of probability distributions:

1) The construction of the probability distribution compatible with the mean value of some random variables which maximizes the entropy (PEM) ;

2) The construction of the product probability distribution of minimum interdependence (or of minimum divergence from independence) compatible with some mixed moments (like covariances) given a priori (PIM).

It is possible to combine these two methods for constructing the stochastic evolution which does not ignore any possibility compatible with the given data. Thus, if at each time-moment t_j we know the mean value (1) of a state function f_j and for each pair of successive time-moments $t_j < t_{j+1}$ we know the covariance

$$C(f(t_j), f(t_{j+1})) \tag{4}$$

between two state functions, PEM is applied for determining at each moment t_j the probability distribution $p^*(t_j)$ of the possible states of the system while PIM is used for the construction of the product probability distribution

$$\pi(t_1,\ldots,t_m) \tag{5}$$

of the possible trajectories of the evolution corresponding to the successive time-moments (2). Freezing the time, at each time-moment t_j, the probability distribution on the possible states of the system is the most uncertain one compatible with the mean value (1). In this way we obtain at each moment t_j the most uniform probability distribution on the possible states of the system compatible with the constraint given by the mean value (1). According to PEM we introduce at each time-moment t_j such a description of the state space with respect to which the discrimination between the possible states of the system is minimum compatible with the constraint (1). Because the state functions $f(t_j)$, $j=1,\ldots,m$, are not independent, the product probability distribution (5) giving the probabilities for all possible trajectories is not completely determined by the probability distributions

$$p^*(t_j), \quad j=1,\ldots,m \tag{6}$$

According to PIM we introduce the product probability distribution on the system's trajectories which is the nearest one to the independence, compatible with the constraints given by the covariances (3). Because the probability distributions (6) determined by applying PEM are only approximations of what happens at the time moments $t_1,\ldots,t_m$, we have no reason at all to impose them as being the marginal probability distributions of the product probability distribution (5).Therefore by applying PIM-PEM theory we construct the stochastic evolution characterized at each moment of time t_j by the most uniform probability distribution on the system's states compatible with the mean value (1) and, from the dynamic point of view, by the most independent trajectories compatible with the covariances (3). PIM-PEM theory implies the most uniform static description of the system's possible states and the most independent description of the possible trajectories of the system. PEM is necessary because for applying PIM we need the probability distributions (6). On the other hand,because we have no reason to impose (6) as the marginal probability distributions of the product probability distribution (5), PIM does not reduce itself to PEM. In this sense PIM and PEM complete one another for giving the stochastic model of evolution based on a double prudence in attaching probabilities both to the system's states, at each time-moment, and to the system's trajectories.

2. SHANNON'S INFORMATION ENTROPY, WATANABE'S MEASURE OF INTERDEPENDENCE, KULLBACK'S DIVERGENCE

Let us consider a probabilistic experiment whose outcome is one of the possible results $a_1,\ldots,a_n$ with the probabilities

$$p_k \geqq 0, \qquad \sum_{k=1}^{n} p_k = 1 \tag{7}$$

When we perform this experiment only one result from the set mentioned above can occur. The amount of uncertainty contained a priori by this experiment is measured by Shannon's information entropy (Shannon, 1948):

$$H(p) = H_n(p_1,\ldots,p_n) = -\sum_{k=1}^{n} p_k \ln p_k \tag{8}$$

where p is the vector $p=(p_1,\ldots,p_n)$.

If $p = (p_1,\ldots,p_n)$ and $q = (q_1,\ldots,q_n)$ are two probability distributions with positive components, the quantity

$$I(p:q) = \sum_{k=1}^{n} p_k \ln \frac{p_k}{q_k} \tag{9}$$

measures the divergence of the probability distribution p from the "initial" probability distribution q and was introduced by S.Kullback (1959).

Let us consider now a product probability distribution

$$\pi_{k_1\ldots k_m} \geqq 0, \qquad \sum_{k_1}\ldots\sum_{k_m} \pi_{k_1\ldots k_m} = 1 \tag{10}$$

$$\pi = (\pi_{k_1\ldots k_m})$$

where $k_1=1,\ldots,n_1;\ldots;\ k_m=1,\ldots n_m$.

The marginal probability distributions are

$$p_{(j)} = (p_{k_j})_{1 \leqq k_j \leqq n_j} \qquad (j=1,\ldots,m)$$

$$p_{k_j} = \sum_{k_1}\ldots\sum_{k_{j-1}}\sum_{k_{j+1}}\ldots\sum_{k_m} \pi_{k_1\ldots k_{j-1}k_j k_{j+1}\ldots k_m} \tag{11}$$

According to S. Watanabe (1969), the interdependence or the

amount of connection between the marginal probability distributions

$$p_{(1)}, \dots, p_{(m)} \tag{12}$$

with respect to the product probability distribution (10) is measured by

$$W = W(\pi; p_{(1)}, \dots, p_{(m)}) = \sum_{j=1}^{m} H(p_{(j)}) - H(\pi) \tag{13}$$

We have

$$W(\pi; p_{(1)}, \dots, p_{(m)}) \geq 0 \tag{14}$$

with equality if and only if the marginal probability distributions (12) are independent with respect to the joint probability π , i.e. if and only if

$$\pi = p_{(1)} \cdots p_{(m)} \tag{15}$$

where

$$p_{(1)} \cdots p_{(m)} = (p_{k_1} \cdots p_{k_m})_{1 \leq k_1 \leq n_1, \dots, 1 \leq k_m \leq n_m} \tag{16}$$

For m=2, W becomes just the information rate introduced by Shannon (1948). At the same time (13) gives us

$$\begin{aligned} W(\pi; p_{(1)}, \dots, p_{(m)}) &= \\ &= \sum_{k_1} \dots \sum_{k_m} \pi_{k_1 \dots k_m} \ln \frac{\pi_{k_1 \dots k_m}}{p_{k_1} \cdots p_{k_m}} = \\ &= I(\pi : p_{(1)} \cdots p_{(m)}) \end{aligned} \tag{17}$$

measuring the divergence of the product probability π from the independent product probability distribution

$$p_{(1)} \cdots p_{(m)} \tag{18}$$

generated by its marginal probability distributions (11). Let us take m arbitrary probability distributions

$$p^*_{(j)} = (p^*_{k_j})_{1 \leq k_j \leq n_j} \qquad (j=1, \dots, m) \tag{19}$$

having strictly positive components

$$p^*_{k_j} > 0, \quad \sum_{k_j=1}^{n_j} p^*_{k_j} = 1, \quad (j=1,\ldots,m) \tag{20}$$

Let us introduce a slight but essential generalization of the Watanabe's measure W by introducing

$$W^* = W^*(\pi; p^*_{(1)},\ldots,p^*_{(m)}) = \sum_{j=1}^{m} H(p^*_{(j)}) - H(\pi) =$$

$$= \sum_{k_1} \ldots \sum_{k_m} \pi_{k_1 \ldots k_m} \ln \frac{\pi_{k_1 \ldots k_m}}{p^*_{k_1} \ldots p^*_{k_m}} = \tag{21}$$

$$= I(\pi : p^*_{(1)} \ldots p^*_{(m)})$$

which is just the divergence of the arbitrary product probability distribution π from the independent product

$$p^*_{(1)} \ldots p^*_{(m)} \tag{22}$$

of the arbitrary probability distributions (19)-(20).

THEOREM 1: *We have*

$$W^*(\pi; p^*_{(1)},\ldots,p^*_{(m)}) \geqq 0, \tag{23}$$

with equality if and only if

$$\pi = p^*_{(1)} \ldots p^*_{(m)}$$

in which case $p^*_{(1)},\ldots,p^*_{(m)}$ *are just the marginal probability distributions of the probability distribution* π, *i.e.*

$$p_{(j)} = p^*_{(j)} \quad (j=1,\ldots,m) \tag{24}$$

PROOF: Taylor's formula applied to the function

$$\phi(t) = t \ln t$$

gives

$$\phi(t) = \phi(1) + \phi'(1)\,(t-1) + \frac{1}{2}\phi''(\tau)(t-1)^2 = (t-1) + $$

$$+ \frac{1}{2\tau}(t-1)^2 \tag{25}$$

for t > 0 with   between 1 and t. Thus

$$W^* = \sum_{k_1} \dots \sum_{k_m} p^*_{k_1} \dots p^*_{k_m} \frac{\pi_{k_1 \dots k_m}}{p^*_{k_1} \dots p^*_{k_m}} \ln \frac{\pi_{k_1 \dots k_m}}{p^*_{k_1} \dots p^*_{k_m}} =$$

$$= \sum_{k_1} \dots \sum_{k_m} p^*_{k_1} \dots p^*_{k_m} \left(\frac{\pi_{k_1 \dots k_m}}{p^*_{k_1} \dots p^*_{k_m}} - 1\right) +$$

$$+ \sum_{k_1} \dots \sum_{k_m} p^*_{k_1} \dots p^*_{k_m} \frac{1}{2\tau_{k_1 \dots k_m}} \left(\frac{\pi_{k_1 \dots k_m}}{p^*_{k_1} \dots p^*_{k_m}} - 1\right)^2$$

where $\tau_{k_1 \dots k_m}$ is a number between 1 and

$$\frac{\pi_{k_1 \dots k_m}}{p^*_{k_1} \dots p^*_{k_m}} .$$

Thus we have (23) with equality if and only if we have the equality

$$\pi_{k_1 \dots k_m} = p^*_{k_1} \dots p^*_{k_m} \tag{26}$$

for all $k_1, \dots, k_m$, i.e.

$$\pi = p^*_{(1)} \dots p^*_{(m)} \tag{27}$$

q.e.d.

The comparison between the measures W and W* is given in the following theorem:

THEOREM 2: *We have*

$$W^*(\pi; p^*_{(1)}, \dots, p^*_{(m)}) \geq W(\pi; p_{(1)}, \dots, p_{(m)}) \tag{28}$$

with equality if and only if

$$p^*_{(j)} = p_{(j)} , \quad (j=1,\dots,m) \tag{29}$$

i.e. if and only if $p^*_{(j)}$, j=1,...,m *are marginal probability* distributions of the product probability distribution

PROOF: If we have two arbitrary probability distributions with positive components

$$\tilde{p} = (\tilde{p}_1,\ldots,\tilde{p}_n), \qquad \tilde{q} = (\tilde{q}_1,\ldots,\tilde{q}_n)$$

we have the elementary inequality (Guiasu, 1977, p.173)

$$-\sum_{k=1}^{n} \tilde{p}_k \ln \tilde{p}_k \leqq -\sum_{k=1}^{n} \tilde{p}_k \ln \tilde{q}_k \tag{30}$$

From the equalities (17) and (21) by applying (30) we get

$$W^*(\pi;p^*_{(1)},\ldots,p^*_{(m)}) = -\sum_{j=1}^{m}\sum_{k_j=1}^{n_j} p_{k_j} \ln p^*_{k_j} - H(\pi) \geqq$$

$$\geqq -\sum_{j=1}^{m}\sum_{k_j=1}^{n_j} p_{k_j} \ln p_{k_j} - H(\pi) =$$

$$= \sum_{j=1}^{m} H(p_{(j)}) - H(\pi) = W(\pi;p_{(1)},\ldots,p_{(m)})$$

with equality if and only if

$$p^*_{(j)} = p_{(j)}, \quad (j=1,\ldots,m).$$

q.e.d.

3. THE PRINCIPLE OF MAXIMUM ENTROPY

When the probability distribution p is given, we can compute without any difficulty its entropy H(p) measuring the amount of uncertainty contained by p. The converse problem was proposed by E.T.Jaynes (1957) who formulated the Principle of Maximum Entropy (PEM): Find the probability distribution

$$p = (p_1,\ldots,p_n),$$

$$p_k > 0, \qquad \sum_{k=1}^{n} p_k = 1 \tag{31}$$

containing the maximum uncertainty

$$H = H(p) = - \sum_{k=1}^{n} p_k \ln p_k \qquad (32)$$

compatible with (in the most elementary and common case) the mean value of the random variable f,

$$E(f) = \sum_{k=1}^{n} f_k p_k \qquad (33)$$

By introducing the Lagrange's multipliers α and β and by applying the equality (25) we get

$$- H + \alpha.1 + \beta E(f) = \sum_{k=1}^{n} p_k \ln(p_k e^{\alpha+\beta f_k}) =$$

$$= \sum_{k=1}^{n} e^{-\alpha-\beta f_k}(p_k e^{\alpha+\beta f_k}) \ln(p_k e^{\alpha+\beta f_k}) = \qquad (34)$$

$$= \sum_{k=1}^{n} e^{-\alpha-\beta f_k}(p_k e^{\alpha+\beta f_k} - 1) + \sum_{k=1}^{n} e^{-\alpha-\beta f_k} \frac{1}{2\tau_k}(p_k e^{\alpha+\beta f_k} - 1) \geqq$$

$$\geqq \sum_{k=1}^{n} e^{-\alpha-\beta f_k}(p_k e^{\alpha+\beta f_k} - 1) = 1 - \sum_{k=1}^{n} e^{-\alpha-\beta f_k} = A$$

where τ_k is a positive number for every k, $(1 \leqq k \leqq m)$, the bound A being independent on p, with equality if and only if

$$p_k = e^{-\alpha-\beta f_k}, \quad (k=1,\dots,n) \qquad (35)$$

which implies $A = 0$. By introducing (35) into (31) we get

$$p_k = \frac{1}{\Phi(\beta)} e^{-\beta f_k}, \quad (k=1,\dots,n) \qquad (36)$$

where

$$\Phi(\beta) = \sum_{k=1}^{n} e^{-\beta f_k} \qquad (37)$$

By introducing (36) into(33) we obtain

$$\frac{d \ln \Phi(\beta)}{d\beta} = - E(f) \qquad (38)$$

which has a unique solution β if f is a nondegenerated random variable (i.e. f is not equal almost everywhere to the constant E(f)) because we can write the previous equation(38) under the form

$$G(\beta) = 0 \tag{39}$$

where the function

$$G(\beta) = \sum_{k=1}^{n} (f_k - E(f))e^{-\beta(f_k - E(f))} \tag{40}$$

is strictly monotonous tending to $+\infty$ when β tends to $+\infty$ and to $-\infty$ when β tends to $-\infty$. Obviously, when f is a nondegenerated random variable it takes on, with positive probability at least one value smaller than E(f) and at least one value greater than E(f).

For the probability distribution (36) where β is the unique solution of the equation (38) we get from (34)

$$H_{max} = \alpha+\beta E(f) = \ln\Phi(\beta) + \beta E(f) \tag{41}$$

We can apply the same technique if instead of the constraint (33) we put several mean values of several random variables. The solution of PEM will give the most uncertain probability distribution which does not ignore any possibility compatible with these new constraints.

4. THE PRINCIPLE OF MINIMUM INTERDEPENDENCE.

For the construction of the product probability distribution when we know the one-dimensional probability distributions and some mixed moments between some random variables (like some covariances or correlations between some random variables for instance) we can adopt the same strategy, based on prudence, by determining the probability distribution which is the nearest one to the independence between the one-dimensional probability distributions given a priori, compatible with the constraints we have on the dependence expressed by the mixed moments. This is the Principle of Minimum Interdependence (or PIM) developed by Guiasu(1978) and Guiasu, Leblanc, Reischer (1982) as a minimization of Watanabe's measure slightly generalized (the measure W*) and which is in fact a special case of Kullback's principle of minimum divergence. For simplifying the writting let us take the case m=2. According to PIM we determine the product probability distribution

$$\pi = (\pi_{k_1k_2})_{1\leq k_1\leq n_1, 1\leq k_2\leq n_2} \tag{42}$$

$$\pi_{k_1k_2} \geqq 0 \quad , \quad \sum_{k_1=1}^{n_1}\sum_{k_2=1}^{n_2}\pi_{k_1k_2} = 1 \tag{43}$$

for which

$$W^* = W^*(\pi;p^*_{(1)},p^*_{(2)}) = \sum_{k_1=1}^{n_1}\sum_{k_2=1}^{n_2}\pi_{k_1k_2}\ln\frac{\pi_{k_1k_2}}{p^*_{k_1}p^*_{k_2}} \tag{44}$$

is minimum, compatible with (in the simplest case) the covariance

$$C(f_1,f_2) = \sum_{k_1=1}^{n_1}\sum_{k_2=1}^{n_2}(f_{k_1}-E(f_1))(f_{k_2}-E(f_2))\pi_{k_1k_2} \tag{45}$$

between two nondegenerate random variables f_1 and f_2. By introducing the Lagrange's multipliers α,γ and by utilizing the equality (25) we obtain

$$W^*+\alpha.1+\gamma C(f_1,f_2) = \sum_{k_1=1}^{n_1}\sum_{k_2=1}^{n_2}\pi_{k_1k_2}\ln\left(\frac{\pi_{k_1k_2}}{p^*_{k_1}p^*_{k_2}}e^{\alpha+\gamma\tilde{f}_{k_1}\tilde{f}_{k_2}}\right) =$$

$$= \sum_{k_1=1}^{n_1}\sum_{k_2=1}^{n_2}p^*_{k_1}p^*_{k_2}e^{-\alpha-\gamma\tilde{f}_{k_1}\tilde{f}_{k_2}}\left(\frac{\pi_{k_1k_2}}{p^*_{k_1}p^*_{k_2}}e^{\alpha+\gamma\tilde{f}_{k_1}\tilde{f}_{k_2}}-1\right) + \tag{46}$$

$$+ \sum_{k_1=1}^{n_1}\sum_{k_2=1}^{n_2}p^*_{k_1}p^*_{k_2}e^{-\alpha-\gamma\tilde{f}_{k_1}\tilde{f}_{k_2}}\frac{1}{2\tau_{k_1k_2}}\left(\frac{\pi_{k_1k_2}}{p^*_{k_1}p^*_{k_2}}e^{\alpha+\gamma\tilde{f}_{k_1}\tilde{f}_{k_2}}-1\right)^2 \geqq$$

$$\geqq 1 - \sum_{k_1=1}^{n_1}\sum_{k_2=1}^{n_2}p^*_{k_1}p^*_{k_2}e^{-\alpha-\gamma\tilde{f}_{k_1}\tilde{f}_{k_2}} = B$$

where $\tilde{f}_{k_j} = f_{k_j} - E(f_j)$, $(j=1,2)$ and $\tau_{k_1k_2}$ is a positive number for each pair

$$1 \leq k_j \leq n_j, \quad (j=1,2)$$

and where the bound B is independent on π, with equality if and only if

$$\pi_{k_1k_2} = p^*_{k_1} p^*_{k_2} e^{-\alpha-\gamma(f_{k_1} - E(f_1))(f_{k_2} - E(f_2))} \tag{47}$$

$$(k_1=1,\ldots,n_1;k_2=1,\ldots,n_2)$$

By introducing (47) into the constraints (43) and (45) we get

$$\pi_{k_1k_2} = \frac{1}{\Psi(\gamma)} p^*_{k_1} p^*_{k_2} e^{-\gamma(f_{k_1} - E(f_1))(f_{k_2} - E(f_2))} \tag{48}$$

$$(k_1=1,\ldots,n_1;k_2=1,\ldots,n_2)$$

where

$$\Psi(\gamma) = \sum_{k_1=1}^{n_1} \sum_{k_2=1}^{n_2} p^*_{k_1} p^*_{k_2} e^{-\gamma(f_{k_1} - E(f_1))(f_{k_2} - E(f_2))} \tag{49}$$

being the solution of the equation

$$\frac{d\ln\Psi(\gamma)}{d\gamma} = -C(f_1,f_2) \tag{50}$$

REMARQUES: 1) When $C(f_1,f_2)=0$, then $\gamma=0$, $\Psi(\gamma)=1$ and (48) becomes

$$\pi_{k_1k_2} = p^*_{k_1} p^*_{k_2} \tag{51}$$

in which case

$$p_{k_1} = \sum_{k_2=1}^{n_2} \pi_{k_1k_2} = p^*_{k_1}, \quad (k_1=1,\ldots,n_1)$$

$$p_{k_2} = \sum_{k_1=1}^{n_1} \pi_{k_1k_2} = p^*_{k_2}, \quad (k_2=1,\ldots,n_2)$$

2) When $p^*_{(1)}$ and $p^*_{(2)}$ are the marginal probability distributions of the solution π, i.e. if we have the equalities

$$p_{(1)} = p^*_{(1)}, \; p_{(2)} = p^*_{(2)}$$

then

$$W^*(\pi;p^*_{(1)},p^*_{(2)}) = W(\pi;p_{(1)},p_{(2)}) = \\ = H(p_{(1)}) + H(p_{(2)}) - H(\pi) \tag{52}$$

In such a case the probability distribution minimizing the interdependence W is just the probability distribution maximizing the entropy $H(\pi)$ because $H(p_{(1)})$ and $H(p_{(2)})$ are constant in this case. Thus in such a case PIM is equivalent with PEM.

5. THE PIM-PEM MODEL FOR EVOLUTION.

Let S be a system and let $t_1 < t_2 < \ldots < t_m$ be an increasing succession of time-moments. We want to determine the probabilities of all possible trajectories of the system when we know:

a) the mean value $E(f(t_j))$ of a state function $f(t_j)$ at the time-moment t_j and its possible values

$$f_{k_j}(t_j), \; k_j=1,\ldots,n_j$$

corresponding to the n_j states of the system at the moment t_j.

b) the covariances

$$C(f(t_j),f(t_{j+1})), \; j=1,\ldots,m-1$$

between the state functions $f(t_j)$ and $f(t_{j+1})$ corresponding to successive time-moments.

We shall use PEM for determining the best probability distribution $p^*(t_j)$ at each moment t_j and thereafter we shall apply PIM for determining the best product probability distribution on the trajectories of the system with respect to the time-moments $t_1,\ldots,t_m$. The adjective "best" is used here in the sense discussed already in the previous paragraphs: we determine, at each time-moment the most uniform probability distribution compatible with the mean value of a state function and finally the product probability distribution, on the successions of system's states at the moments $t_1,\ldots,t_m$ which is the nearest one to the most independent product probability distribution which is just the product of the one-dimensional probability distributions, determined using

PEM , at the time-moments $t_1,\ldots,t_m$. compatible with the covariances between successive state functions. Therefore the largest stochastic model for evolution, which is compatible with the given constraints is obtained by combining PEM and PIM. Both are necessary.

By applying the results obtained in the paragraphs 3 and 4, PEM implies that, at each time-moment t_j, the probability distribution on the n_j states of the system at this time-moment is

$$p^*_{k_j}(t_j) = \frac{1}{\Phi_j(\beta_j)} e^{-\beta_j f_{k_j}(t_j)}, \qquad (k_j=1,\ldots,n_j), \quad (53)$$

where

$$\Phi_j(\beta) = \sum_{k_j=1}^{n_j} e^{-\beta f_{k_j}(t_j)} \quad (54)$$

and β_j is the solution of the equation

$$\frac{d\ln\Phi_j(\beta)}{d\beta} = -E(f(t_j)) \quad (55)$$

For the product probability distribution on the successions of states of the system at the moments $t_1,\ldots,t_m$, PIM gives us, according to the theory developed in the paragraph 4,(in the m-dimensional case)

$$\pi_{k_1\ldots k_m}(t_1,\ldots,t_m) = \frac{\left(\sum_{j=1}^{m}\Phi_j(\beta_j)\right)^{-1}}{\Psi(\gamma_{1,2},\ldots,\gamma_{m-1,m})} \times$$

$$\times e^{-\sum_{j=1}^{m}\beta_k f_{k_j}(t_j) - \sum_{j=1}^{m-1}\gamma_{j,j+1}(f_{k_j}(t_j)-E(f(t_j))(f_{k_{j+1}}(t_{j+1}-E(f(t_{j+1}))} \quad (56)$$

where $\Psi(\gamma_{1,2},\ldots,\gamma_{m-1,m})$ is the normed factor, i.e. the sum of the numerator in the left-hand side of the equality (56), $\gamma_{j,j+1}$ being the solution of the equation

$$\frac{\partial \ln\Psi(\gamma_{1,2},\ldots,\gamma_{m,m+1})}{\partial\gamma j,j+1} = -C(f(t_j),f(t_{j+1}))$$

for $j=1,\ldots,m-1$.

REMARQUES: 1) We can see that in the expression (53) $p^*_{(j)}$,

$j=1,\ldots,m$, are obtained by applying PEM. Therefore $p^*_{(j)}$, $j=1,\ldots,m$, are only an approximation of the reality and, of course, we have no reason at all for imposing them as being the marginal probability distribution of the product probability distribution π. Obviously, the final solution depend on the probability distributions $p^*_{(j)}$, $j=1,\ldots,m$, but we have no justification to impose them among the constraints of the model, as the marginal probability distribution of the solution.

2) If at each moment t_j we know not only the mean value $E(f(t_j))$ of the state function $f(t_j)$ but also its variance $V(f(t_j))$, by applying PEM and taking into account that

$$V(f(t_j)) = \sum_{k_j=1}^{n_j} (f_{k_j}(t_j)-E(f(t_j)))^2 p^*_{k_j}(t_j) \qquad (57)$$

we get for the probability distribution on the system's states at the moment t_j, the expression

$$p^*_{k_j}(t_j) = \frac{1}{\Phi_j(\beta_j,\delta_j)} e^{-\beta_j f_{k_j}(t_j)-\delta_j(f_{k_j}(t_j)-E(f(t_j))^2} \qquad (58)$$

where

$$\Phi_j(\beta_j,\delta_j) = \sum_{j=1}^{m} e^{-\beta_j f_{k_j}(t_j)-\delta_j(f_{k_j}(t_j)-E(f(t_j)))^2}$$

β_j and δ_j respectively satisfying the equations

$$\frac{\partial \ln \Phi_j(\beta,\delta)}{\partial \beta} = -E(f(t_j))$$

$$\frac{\partial \ln \Phi_j(\beta,\delta)}{\partial \delta} = -V(f(t_j))$$

By replacing e^x by $1+x$ we obtain the first order approximation of the solution which has the advantage to be quite simple even if it less accurate

$$\pi_{k_1\ldots k_m}(t_1,\ldots,t_m)=p^*_{k_1}(t_1)\ldots p^*_{k_m}(t_m)A(t_1,\ldots,t_m), \qquad (59)$$

where

$$A(t_1,\ldots,t_m)=1-\sum_{j=1}^{m-1}\rho_{j,j+1}\frac{f_{k_j}(t_j)-E(f(t_j))}{\sigma_j}\,\frac{f_{k_{j+1}}(t_{j+1})-E(f(t_{j+1}))}{\sigma_{j+1}}$$

We have put

$$V(f(t_j)) = \sigma_j^2 \quad (\sigma_j>0),\ j=1,..,m$$

for the variance and

$$\rho_{j,j+1}=\frac{C(f(t_j),f(t_{j+1}))}{(V(f(t_j))\ V(f(t_{j+1})))}$$

for the correlation coefficient between $f(t_j)$ and $f(t_{j+1})$. Of course, the first order approximation of the solution is a probability distribution if and only if

$$\sum_{j=1}^{m-1}\rho_{j,j+1}\frac{f_{k_j}(t_j)-E(f(t_j))}{\sigma_j}\,\frac{f_{k_{j+1}}(t_{j+1})-E(f(t_{j+1}))}{\sigma_{j+1}}\leqq 1$$

for all the values $k_1,\ldots,k_m,(1\leqq k_1\leqq n_1,\ldots,1\leqq k_m\leqq n_m)$. These inequalities are generally satisfied by the random variables weakly correlated.

Looking at the expression (59) we can notice the fact that the "curvature" of the product probability space is determined entirely by the correlation coefficients, i.e. by the interdependences between the state functions of the system. When the state functions $f(t_1),\ldots,f(t_m)$ are non-correlated then

$$\rho_{j,j+1}= 0,\ j=1,\ldots,m-1$$

and the product space becomes "Euclidean":

$$\pi_{k_1\ldots k_m}(t_1,\ldots,t_m)=p^*_{k_1}(t_1)\ldots p^*_{k_m}(t_m)$$

$$(k_j=1,\ldots,n_j;\ j=1,\ldots,m)$$

We want to underline this important fact that the interdependence implies the curvature of the product space.

6. CONCLUSIONS

The study of the behaviour of a gas (Boltzmann, Gibbs) and the attempt to build up a mathematical model for the communication systems (Shannon) determined the introduction in mathematics of a new concept, the entropy,as a measure of uncertainty contained by a probability distribution. Using this entropy both a global measure of interdependence between the marginal probability distributions of a product probability distribution (Watanabe) and an entropic divergence between two arbitrary probability distributions of the same dimensions (Kullback) have been introduced. Two variational principles related to these new measures have been formulated in a quite natural way: 1) The Principle of Maximum Entropy or PEM(Jaynes, Ingarden), giving the most uncertain probability distribution compatible with some constraints represented by some mean values of some random variables; 2) The Principle of Minimum Interdependence, or PIM, (a special but important case of Kullback's principle of minimum divergence, i.e. the minimization of the divergence of an arbitrary product probability distribution from the "independent" direct product of arbitrary one-dimensional probability distributions(Guiasu)),giving the nearest product probability distribution to the independence between components, compatible with some mixed moments between random variables. The aim of the present paper was to show that both these variational principles are necesary for building up a general and prudent model for the evolution of a system compatible with some partial data on this evolution represented by mean values of some random variables (state functions) of the system at some given time-moments and by some covariances (or arbitrary mixed moments) between these random variables.

Department of Mathematics
York University, Downsview,
Toronto, M3J 1P3, C a n a d a

ACKNOWLEDGMENTS: The author is indebted both to Professor Claude P. Bruter for his kind invitation to send a paper to the Colloquium on Mathematical Problems met in the Study of Natural Phenomena and to the Natural Sciences and Engineering Research Council of Canada for my research grant (NSERC Canada A5712).

REFERENCES

Boltzmann, L.:1896, *Vorlesungen über Gastheorie*,J.A.Barth, Leipzig.
Gibbs, J.W.: 1902, *Elementary Principles in Statistical Mechanics Developed with Especial Reference to the Rational Foundation of Thermodynamics*, Yale University Press, New Haven, Conn.
Guiasu, S.: 1977, *Information Theory with Applications*,McGraw-Hill, New York-Düsseldorf-Toronto-London.
Guiasu, S.: 1978, 'On Entropic Measure of Connection and Interdependence between the Subsystems of a Given Large System', in J. Benes and L. Bakule (eds.), *Third FORMATOR Symposium on Mathematical Methods for the Analysis of Large-Scale Systems. Proceedings*, Academia, Prague,pp. 113-124.
Guiasu, S., R. Leblanc and C. Reischer: 1982, 'On the Principle of Minimum Interdependence', *Journal of Information and Optimization Sciences*, 3, 1-24.
Ingarden, R.S.: 1963, 'Information Theory and Variational Principles in Statistical Theories', *Bull. Acad. Polon. Sci. Ser. Sci. Math. Astronom. Phys.*, 11, 541-547.
Jaynes, E.T.: 1957, 'Information Theory and Statistical Mechanics', *Phys. Rev.*, 106, 620-630; 108, 171-190.
Kullback, S.: 1959, *Information Theory and Statistics*,Wiley, New York; Chapman and Hall, London.
Shannon, C.E.: 1948, 'A Mathematical Theory of Communication', *Bell Syst. Techn. J.*, 27, 379-423, 623-656.
Watanabe, S.: 1969, *Knowing and Guessing*, Wiley, New York.

ON SOME VARIATIONAL METHODS

L. NIRENBERG*

When I was invited to participate in this symposium I was asked to present a talk on variational methods. The subject is a large one, with a very old history so in this talk, I will merely take up a few simple variational principles. Some of the topics presented here are contained in the expository article (4).

Let us start with some very simple examples

(a) An elliptic boundary value problem. Let Ω be a bounded domain in R^n with smooth boundary. We wish to solve a semilinear equation

$$\text{(1)} \qquad \Delta u + g(u) = f(x) \text{ in } \Omega$$

$$u = 0 \text{ on } \partial\Omega.$$

Here f is a smooth given function in $\overline{\Omega}$, g is a smooth function and Δ represents the Laplace operator. This is the Euler equation for the variational expression (Lagrangian)

$$\text{(2)} \qquad F(u) = \int_\Omega \{\frac{1}{2} |\text{grad } u|^2 - G(u) + uf\} \, dx$$

which is defined for all functions u in the Sobolev space H_0^1 : the completion of C^∞ functions with compact support in Ω under the norm $\{\int_\Omega |\text{grad } u|^2 \, dx\}^{1/2}$. Here G is the primitive of g :

$$G(u) = \int_0^u g(s) \, ds.$$

* This work was supported by U.S. Army Research Office Grant DAA-G29-81-K-0043

C. P. Bruter et al. (eds.), Bifurcation Theory, Mechanics and Physics, 169–176.

It is easily seen that a function u at which the functional F is stationary, satisfies (1).

(b) Periodic solutions for Hamiltonian systems of ordinary differential equations.

$$\frac{du}{dt} = \frac{\partial H}{\partial q}, \quad \frac{dq}{dt} = \frac{\partial H}{\partial p} \tag{3}$$

Here p(t), q(t) are vector valued (in R^n) functions of time t, ant the Hamiltonian H(p,q) is a real function defined in R^{2n}. One wishes to find a nonstationary solution with given time period - say 2π. If we consider the Lagrangian defined for 2π-periodic vector functions p(t), g(t) :

$$F(p,q) = \int_0^{2\pi} \left[p.\frac{dq}{dt} + H(p(t),q(t))\right] dt \tag{4}$$

we see that (3) is the Euler equation satisfied by a "stationary point" p(t) of F.

(c) Some differential equations don't appear at first sight to come from variational problems, but may, in fact, do so -- from some off-beat Lagrangian. Here is a surprising example due to Brézis and Ekeland (3) for the initial-boundary value problem for the heat equation

$$u_t - \Delta u = f \tag{5}$$

Here u(x,t) is a function of time t and space variables x in a bounded domain Ω in R^n with smooth boundary. We wish to find u in Ω, and for $0 \leq t \leq T$, satisfying the initial and boundary conditions

$$u(x,0) = u_0(x) \tag{6}$$

$$u = 0 \text{ on } \partial\Omega.$$

Here Δ represents the Laplace operator in the space variables only.

They introduced the following unusual variational problem :

Minimize the functional

$$F(v) = \int_0^T \int_\Omega \left[\frac{1}{2} \ |\mathrm{grad}_x v|^2 + \frac{1}{2} \ |\mathrm{grad}_x \Delta^{-1}(f-v_t)| - fv \right] dx\, dt + \frac{1}{2} \int_\Omega v^2(x,T)\, dx.$$

This defined for all functions in the Sobolev space

$v \in H^1(\Omega x\ (0,T))$, with $v(x,0) = u_0$, $v(x,t) = 0$ for $x \in \partial\Omega$.

Here $\Delta^{-1}g$ is the solution $w \in H^1_0(\Omega)$ of $\Delta w = g$, $\forall$ t.

This function F is *convex* and so it is easy to see that a minimum exists. It is then a simple exercise to verify that the minimum point u(x,t) satisfies (5) and (6).

This is not the simplest way to solve this classical problem but in (3) they treat also more general, nonlinear, equations.

We will present some methods for finding stationary points of functionals.

First let us consider problem (a), with $g \equiv 0$. The functional F is, then, convex and so any stationary point will be a minimum. For convex functionals one hase the well known result.

Theorem 1. Let F be a real, lower semicontinous, convex function bounded below, defined in a ball $\|x\| \leq R$ *on a reflexive Banach space X, and such that for some point* x_0 *inside the ball*

$$F(x_0) < \inf_{\|x\|=R} F(x) .$$

Then F attains a minimum inside the ball.

The reason is that one can exploit some compactness that is present : closed bounded convex sets in a reflexive Banach space are compact in the weak topology. In particular one obtains immediately the existence of a minimum point for F given by (2) (with g = 0) in the Sobolev space $H^1_0(\Omega)$

(functions vanishing on $\partial\Omega$ and having square integrable first derivatives).

The result of the theorem does not hold in every Banach space, i.e. F need not achieve a minimum.

For nonconvex functionals F, it is now customary to make use of an associated compactness condition due to R. Palais and S. Smale —— the (PS) condition.

Definition (PS) . A real function F of class C^1 in a Banach space X satisfies the *(PS) condition* if any sequence

$\{x_j\} \in X$ with

$$|F(x_j)| \leq M$$

and $F'(x_j) \to 0$ in norm has a strongly convergent subsequence x_{j_i}.

Here $F'(x)$, the derivative of F at x, is a continuous linear functional on X, i.e. an element of the dual space X^*, and it is assumed that $F'(x_j) \to 0$ strongly in X^*.

The (PS) condition is a very strong one, for example the function $F(x) = e^{-x^2}$ on R^1 does not satisfy it. Indeed one has the following result :

Lemma. Let $F: X \to R^1$ *be of class* C^1, *satisfy (PS) and be bounded from below. Then F achieves its minimum.*

Let us turn to the more general equation (1) and (2). In general F will not be bounded from above or below so a stationary point will not, in general, be a maximum or minimum point —— rather a saddle point.

Saddle points are usually found by min-max procedures. One introduces a class of sets in X, maximizes F over a set Σ in the class and then takes infimum with respect to the set Σ.

To illustrate this idea we describe the, by now familiar, Mountain Pass Lemma (MPL) which we first state for a function F defined on the plane.

Think of F as as representing the height of land above sea level ; we assume $F \in C^1(R^2)$ and satisfies (PS). Suppose a town $u_0 \in R^2$ is surrounded by a mountain range, i.e. there is an open neighbourhood U of u_0 in R^2 such that

$$F(u_0) < c_0 \leq F(u) \qquad \forall\, u \in \partial U.$$

(u_0 might be the lowest point in the valley).
Suppose we are in a town u_1 on the other side of the mountain range, i.e. $u_1 \notin \overline{U}$, with $F(u_1) < c_0$. We wish to walk from u_1 to u_0 in such a way that we climb as little as possible, i.e; we keep F as low as possible.

The answer is, of course, to cross the mountain range via the lowest mountain pass. When we are just at the top of the mountain pass, the ground is horizontal there, and so we are at a stationary point of F. Mathematically this is simply expressed as follows —— even for a function defined in a Banach space X.

Theorem (MPL). *Let* $F: X \to R^1$ *be of class* C^1 *and satisfied* (PS). *Assume that there is an open neighbourhood* $U \subset X$ *and points* $u_0 \in U$, $u_1 \notin U$ *such that*

$$F(u_0),\ F(u_1) < c_0 \leq F(u) \qquad \forall\, u \in \partial U.$$

Then the following number

$$c := \inf_P \max_{u \in P} F(u) \geq c_0$$

is a stationary value of F, i.e. there is a stationary point of F where F has the value c. Here P represents any continuous path going from u_1 to u_0. We first maximize F over any such path P, since it hits ∂U, the max $\geq c_0$.

This intuitively obvious result first appeared in the literature in this form in a paper (1) of Ambrosetti and Rabinowitz in 1973. It might, perhaps, be more appropriately called the lowest mountain pass lemma (LMPL).

To illustrate its use here is a simple example which we learned from K. C. Chang, a special case of (1).

$$(1)' \qquad \begin{aligned} \Delta u + u^2 &= f(x) \quad \text{in } \Omega \subset R^3 \\ u &= 0 \quad \text{on } \partial\Omega \end{aligned}$$

Theorem. Assume f > 0 and smooth. Then (1)' has at least two solutions.

Proof. : We first obtain one solution by constructing "sub and super solutions" $\underline{u} \leq \overline{u}$, i.e. such functions satisfying

$$\Delta\underline{u} + \underline{u}^2 - f \geq 0, \quad \underline{u} \leq 0 \quad \text{on } \partial\Omega$$

$$\Delta\overline{u} + \overline{u}^2 - f \leq 0, \quad \overline{u} \geq 0 \quad \text{on } \partial\Omega.$$

We take, namely,

$$\overline{u} \equiv 0$$

and $\underline{u}$ = the solution v of

$$\Delta v = f, \quad v = 0 \text{ on } \partial\Omega.$$

By the maximum principle,

$$v \leq 0 \text{ in } \Omega,$$

so $\underline{u} \leq 0$ and

$$\Delta\underline{u} + \underline{u}^2 - f = \underline{u}^2 \geq 0 .$$

By a well known iteration procedure, using the maximum principle, one shows the existence of a solution u_0 of (1) satisfying

$$\underline{u} \leq u_0 \leq \overline{u} .$$

Thus $u_0 \leq 0$.

The proof of existence of a second solution is based on the following

Observation. u_0 is a local minimum of the functional (2) :

$$(2)' \qquad F(u) = \int_\Omega \frac{1}{2} |\text{grad } u|^2 - \frac{u^3}{3} + fu$$

defined for $u \in H^1_0(\Omega)$.

This is easily verified. Since u_0 is a solution of (1)' it is a stationary point of F.

Then, for $\|w\| = r$ small, here $\| \ \|$ is the norm

$$\|w\|^2 = \int_\Omega |\text{grad } w|^2,$$

we have

$$F(u_0 + w) - F(u_0) = \int \frac{1}{2} |\text{grad } w|^2 - u_0 w^2 - \frac{w^3}{3}$$

$$\geq \frac{1}{4} \|w\|^2 - \int u_0 w^2$$

by Sobolev's inequality, for w=r small,

$$\geq \frac{1}{4} r^2$$

since $u_0 \leq 0$.

Thus u_0 lies in a valley $U = \{u \mid \|u-u_0\| < r\}$ for $0<r$ small.

Finally it is easy to see that F satisfies (PS) in H_0^1. We may then apply MPL, taking

$$u_1 = tv$$

for $v \in H_0^1$ a fixed function with $\int v^3 > 0$, and t sufficiently large. By the MPL we find that F has a stationary value $> F(u_0)$.

Open problem. It is not known for arbitrary smooth f whether (1)' admits any solutions. A. Bahri (2) has shown (for a more general class of problems) that the set of f in L^2 for which there exists a solution is dense in L^2.

Some further material from (4) was also presented in the lecture : extensions of MPL together with applications to problems of the form (1), as well as illustrations of the fact that a nonlinear equation which is the Euler equation for some variational expression may, in fact, come from some other Lagrangian, perhaps a very unusual looking one. Then it may pay to modify a familiar variational problem to some quite different looking one which may prove more tractable.

References :

(1) A. Ambrosetti, P. H. Rabinowitz, Dual variational problems in critical point theory and applications, J. Funct'l. Anal., 14 (1973), 349-381.

(2) A. Bahri, Résolution générique d'une classe d'équations non linéaires, C. R. Acad. Sci. Paris, to appear.

(3) H. Brézis, I. Ekeland, Un principe variationnel associé à certaines équations paraboliques. Le cas indépendent du temps. Comptes Rendus Acad. Sci. Paris (May 3, 1976) Ser. A, p. 971-974.

(4) L. Nirenberg, Variational and topological methods in nonlinear problems. Bull. (New Ser.) AMS 4 (1981) p. 267-302.

L. Nirenberg*Courant Institute of Mathematical Sciences
New York University
* This work was supported by U.S. Army Research Office Grant DAA-G29-81-K-0043.

H.W. Broer

QUASI PERIODICITY IN LOCAL BIFURCATION THEORY

1. INTRODUCTION

In this paper we present an overview of some results obtained recently in the local theory of bifurcations of dynamical systems, which are concerned with quasi periodic flow.

With this 'local theory' we here mean the qualitative (topological) behaviour of parametrised systems of ordinary differential equations near equilibrium points, especially the changes in this behaviour as the parameters move.

The results of this paper involve some techniques, connected with so called small divisors which come from the Kolmogorov-Arnold-Moser theory into this field of bifurcations. Although these techniques found their origin in Hamiltonian dynamics (celestial mechanics, stability of the solar system), it is by now well known that their scope of applicability is much wider. Compare e.g. Arnol'd [1963] and Moser [1966, 1967].

The results to follow are all centered around some examples, living in $\mathbb{R}^3$ and $\mathbb{R}^4$, of bifurcations of codimension one and two. Such bifurcations will be met in generic one, respectively two parameter families of dynamical systems. Some of the examples are in the conservative category, others occur in the world of dissipative systems. We quote from Broer [1980b], Guckenheimer [1980b], Chenciner [1981a,b] and from Braaksma & Broer [1981].
It should be mentioned here that Scheurle [1980] applies small divisor techniques to bifurcations of equilibrium points in reversible dynamical systems.
Our examples all exhibit the phenomenon of so called 'nearly integrability' which is structurally unstable in the respective sets of parametrised systems. Never the less this phenomenon is persistent for small enough perturbations of such systems. So these examples once more illustrate that for the corresponding dynamical systems with parameters, the property of structural stability is not

C. P. Bruter et al. (eds.), Bifurcation Theory, Mechanics and Physics, 177–208.

generic.

1.1. Outline of the contents

We continue this introduction by giving a brief outline of the contents of this paper:

In §2 we first present a short introduction to bifurcation theory, in order to set a framework for the rest of this paper. We shall describe the so called 'classical' theory of local bifurcations, as can be found in e.g. Arnol'd [1972, 1980]. In this classical theory a central role is played by the well known Hopf-bifurcation, see below.

In §3 the two linear situations are sketched, around which our examples will be centered.

In §4 a normal form will be given for a finite number of terms of the Taylor series which occur in the examples. This normal form possesses a lot of symmetry, which easens our analysis of the bifurcation problems. In normal form coordinates the parametrised system can be expressed as a symmetric truncation plus higher order terms, which constitute a small perturbation of this truncation. The idea of such normalisations goes back to Poincaré [1879].

Finally, in §§5 and 6 we present our overview. We mention here that these sections can be read rather independently from each other.

In §5 we give some results in the volume preserving context, where our analysis involves only one real parameter. These volume preserving, or divergence zero, systems arise naturally in frictionless mechanics. In dimension three they may also serve as models for the velocity field of an incompressible fluid. In that case the real parameter, say μ, may be used to make the velocity field time dependent: just put $\mu = t$, where t denotes the time. For more physical details see below.

In §6 we treat some dissipative examples which will be met in generic two or more parameter families. The analysis of these bifurcation problems is of considerable importance for the study of chaotic motion in systems with infinitely many degrees of freedom. The question then is whether and to which extent the complicated dynamical phenomena in these finite dimensional systems may serve as a mathematical model for turbulent physical systems. Compare Ruelle & Takens [1971], Gollub & Swinney [1975] - who have done some experimental work - and e.g.

Guckenheimer [1980a].

1.2. Acknowledgement

I thank B.L.J. Braaksma, A. Chenciner, F. Takens and G. Vegter for their help during the preparation of this paper. Also I express my gratitude to G.B. Huitema and B. Sytsma who made some of the pictures.

2. AN INTRODUCTION TO BIFURCATION THEORY

In this section we recall a few more or less well known results from the 'classical theory' of local bifurcations. Compare e.g. Arnol'd [1972, 1980]. This summary is meant as a framework for the rest of this paper.

2.1. Topological behaviour near equilibrium points

Consider a system of autonomous differential equations

$$\dot{x} = F(x) \tag{1}$$

where $x \in \mathbb{R}^n$ and $F : \mathbb{R}^n \to \mathbb{R}^n$ is a C^∞-map. In a somewhat more geometric language (1) is called a vector field. Our first interest is with the local behaviour of (1) near points $p \in \mathbb{R}^n$ where $F(p) = 0$: equilibria or stationary solutions of the system, also called singularities, zeros or restpoints of the vector field.

If $F(p) = 0$, write $F(x) = A(x-p) + O(||x-p||^2)$, where A is a $n \times n$-matrix, and consider the linear approximation

$$\dot{y} = Ay \tag{2}$$

of (1) near p.
In the case where A has no purely imaginary eigenvalues (the singularity p is then called hyperbolic) the theorem of Hartman-Grobman [1964] states that the local phase portraits of (1) around p and of (2) around 0, are topologically equivalent.
To be more precise:
There exists a local homeomorphism $h : (\mathbb{R}^n, p) \to (\mathbb{R}^n, 0)$, sending integral curves of (1) to integral curves of (2), not necessarily preserving the time-parametrisation of the solutions, but only their orientation.

In the class of all vector fields (1) now consider the subset, the elements of which only posses hyperbolic

singularities.
In a fairly natural topology on the class of all C^∞ vector fields, this subset is open and dense, and so in a sense very large.

Open and dense sets are very large, but so are also countable intersections of such open and dense sets, since the class of all C^∞-vector fields forms a Baire space. We recall that a set is called residual or of second Baire category if it contains such a countable intersection of open and dense sets. See Oxtoby [1970]. Properties which define such a residual set are called generic. Compare Smale [1967].

So it appears that for vector fields it is a generic property to possess only hyperbolic singularities. Moreover it follows from the above that in this generic situation all singularities are locally structurally stable. This means the following: If F is perturbed slightly to $\tilde{F}$, then the vector field

$$\dot{x} = \tilde{F}(x) \qquad (\tilde{1})$$

possesses a hyperbolic singularity $\tilde{p}$ near p and again a local homeomorphism $h : (\mathbb{R}^n, p) \to (\mathbb{R}^n, \tilde{p})$ exists, putting integral curves of (1) sense preserving onto integral curves of $(\tilde{1})$. We say that h is a (local) topological equivalence between (1) and $(\tilde{1})$.
Therefore 'locally' there exist exactly n+1 equivalence classes for topological equivalence: each class is characterised by the number of eigenvalues the linear approximation in the singularity has in the left half of the complex plane, being exactly the dimension of the basin of attraction of the equilibrium. Moreover each of these equivalence classes possesses a linear 'normal form'. See (2).
Summarising and omitting some details we may rephrase the above as:

(i) For C^∞ vector fields (local) structural stability is a generic property.
(ii) The number of (local) topological equivalence classes is finite, and each equivalence class is nicely characterised by a simple normal form.

2.2. Bifurcations of equilibrium points in one parameter families

The next step in the present local study is the introduction of one real parameter into the system (1), so obtaining

$$\dot{x} = F(x,\mu) \tag{3}$$

where $x \in \mathbb{R}^n$ and $\mu \in \mathbb{R}$. Assume F to be C^∞ in both x and μ. We often write $F_\mu(x)$ in stead of $F(x,\mu)$. Systems like (3) are called one parameter families or arcs of vector fields.

Now let us single out any singularity of (3) and follow it as the parameter μ moves. It is a generic property for systems (3) that in this way in $\mathbb{R}^n \times \mathbb{R}$ a smooth curve is obtained: just use the implicit function theorem for the equation $F(x,\mu) = 0$.

'For such a generic' $F = F_\mu(x)$ then consider the linear approximation into the x-direction, as under 2.1, following this curve of singularities. The eigenvalues of the corresponding $n \times n$-matrix now may occasionally be on the imaginary axis: even for generic families this will be unavoidable.

So in generic one parameter families (3) one meets singularities which are non-hyperbolic. If for $(x_o,\mu_o) \in \mathbb{R}^n$ such a 'degeneracy' happens, we call (x_o,μ_o) a bifurcation point. It is a generic property for $F = F_\mu(x)$ that these degeneracies are as simple as possible: Either

(s) exactly one (real) eigenvalue equals zero, while all other eigenvalues are 'hyperbolic'; or

(h) exactly one complex conjugate pair of eigenvalues becomes purely imaginary (remaining distinct), while all other eigenvalues remain 'hyperbolic'.

Moreover, in both cases the relevant eigenvalues have to trespass the imaginary axis with positive speed as one moves along the curve of singularities in $\mathbb{R}^n \times \mathbb{R}$. Also in both cases some generic conditions are imposed on higher order terms.

Here we speak of codimension one singularities. They occur as bifurcations 'in generic one parameter families' for isolated values of the parameter. (Observe that the conditions (s) and (h) define codimension one subsets of the vector space of $n \times n$ matrices.)

As an example for $n = 1$ consider case (s). The system (3), near such a bifurcation point (x_o,μ_o), appears to be (locally) topologically equivalent to the following normal form

$$\dot{x} = x^2 + \mu$$

around the point $(x,\mu) = (0,0)$. This bifurcation is called the saddle node bifurcation. For a phase portrait see fig. 1.

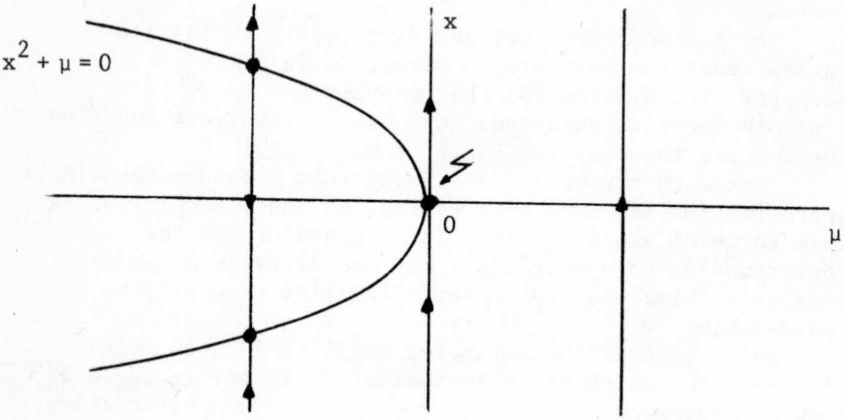

fig.1: phase portrait of the saddle node bifurcation.

As a second example for n = 2 we consider case (h). Now, modulo some changes of direction, the local phase portrait in $\mathbb{R}^2 \times \mathbb{R}$ is topologically equivalent to the normal form

$$\begin{cases} \dot{\varphi} = 1 \\ \dot{r} = r(\mu - r^2) \end{cases} \tag{5}$$

(where r and φ are polar coordinates in $\mathbb{R}^2$) around the point $r = 0$, $\mu = 0$ in $\mathbb{R}^2 \times \mathbb{R}$. This bifurcation is known as the Hopf-bifurcation. Phase portraits are depicted in fig. 2.

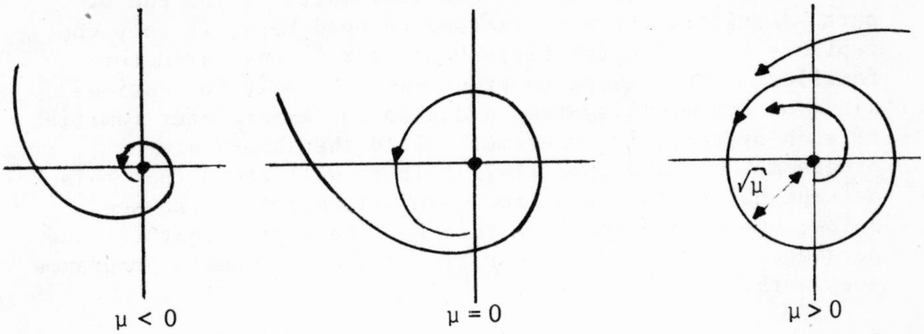

fig.2: phase portrait of the Hopf-bifurcation.

Compare Arnol'd [1972,1980].

Note that we did not precisely define what is to be understood by a picture of a phase portrait in the plane - or on the line -. The idea, which may be clear now, is to draw (the images of) some 'characteristic' integral curves $t \in \mathbb{R} \to \gamma(t) \in \mathbb{R}^2$, such as singularities (dots) and closed integral curves (circles), etc., putting arrows in each curve to indicate the direction of the evolution. See above.

We conclude this section with two remarks.

The first remark is that the normal forms (4) and (5) represent equivalence classes for a kind of topological equivalence quite similar to that introduced under 2.1. Now the homeomorphism sends a neigbourhood of (x_o, μ_o) in $\mathbb{R}^n \times \mathbb{R}$ to a neigbourhood in $\mathbb{R}^n \times \mathbb{R}$ of $(0,0)$. These equivalences may be seen as parameter dependent equivalences in the sense of 2.1. The statements made in the above examples then may be rephrased as:
The saddle node and Hopf bifurcation are (locally) structurally stable, seen as one parameter families on $\mathbb{R}$ and $\mathbb{R}^2$ respectively.

The second remark is that both examples can be extended to higher dimensions by just adjoining a decoupled system of linear equations for the other coordinates. If this adjoined system is hyperbolic then the whole one parameter family is (locally) structurally stable. This is proved by Palis & Takens [1977] using center manifold theory.

Evidently in this way we find all the 'generic' bifurcations as described in (s) and (h), and they apparently are structurally stable.

Now compare the conclusion formulated at the end of part 2.1. Exactly the same statements hold here, if only one replaces "for C^∞ vector fields" by "for C^∞ one parameter families". The attempt to prove such a result for various kinds of dynamical systems and also for k-parameter families of such systems, is sometimes called the Thom-Smale programme. Compare Thom [1972]. It is well known that this attempt has failed in a great many situations, also see below, but we may conclude this section saying that for one parameter families of vector fields the Thom-Smale programme does work.

3. THE EXAMPLES LINEARISED

Our examples will show a stronger non-hyperbolicity than the ones from §2: in the bifurcation the linear approximation of the system exhibits a double degeneracy. In this section we study these linear approximations for their own sake: we shall describe the linear systems in $\mathbb{R}^3$ and $\mathbb{R}^4$ around which the examples lateron are build.

3.1. <u>The three dimensional case</u>

In $\mathbb{R}^3$ consider cartesian coordinates x,y,z or cylindrical coordinates r,φ,z. Consider a linear system (so with the origin as a singularity) having eigenvalues $\pm i\alpha$ and 0, for some positive α. In Jordan normal form we may write

$$\begin{cases} \dot{x} = -\alpha y \\ \dot{y} = \alpha x \\ \dot{z} = 0 \end{cases} \tag{6}$$

or equivalently in cylindrical coordinates

$$\begin{cases} \dot{\varphi} = \alpha \\ \dot{r} = 0 \\ \dot{z} = 0. \end{cases} \tag{7}$$

It is easily seen that in this linear context such a singularity has codimension two: it will be met only in generic families of linear systems if they have at least

two parameters.
Observe that this linear system has divergence zero, meaning that the corresponding matrix has trace zero. If one restricts to divergence free linear systems, then the above singularity has codimension one: it may already occur in generic one parameter families.

3.2. The four dimensional case

In $\mathbb{R}^4$ there exists a similar singularity, now the eigenvalues are $\pm i\alpha$ and $\pm i\beta$ with α and β positive. In cartesian coordinates x_1, x_2, x_3, x_4 the Jordan form for $\alpha \neq \beta$ is

$$\begin{cases} \dot{x}_1 = -\alpha x_2 \\ \dot{x}_2 = \alpha x_1 \\ \dot{x}_3 = -\beta x_4 \\ \dot{x}_4 = \beta x_3 \end{cases} \tag{8}$$

or, in toroidal (double polar) coordinates $r_1, \varphi_1, r_2, \varphi_2$:

$$\begin{cases} \dot{\varphi}_1 = \alpha \\ \dot{\varphi}_2 = \beta \\ \dot{r}_1 = \dot{r}_2 = 0. \end{cases} \tag{9}$$

Again in the general (dissipative) situation this singularity has codimension two, while in the divergence free (conservative) context its codimension equals one.

4. NORMAL FORMS FOR HIGHER ORDER TERMS

We now adjust the Taylor series, up to a certain order, of parametrised systems which begin with linear terms as in (6) or (9). As we explained already in the introduction (§1), the reason to do this is the following: The 'normalised' part of the Taylor series is a polynomial which in our examples displays a lot of symmetry, which in this situation may be analysed rather easily. As soon as this is done we add the higher order terms again, which near the bifurcation point constitute only a small perturbation, and try to evaluate the effects of this perturbation.

4.1. The three dimensional case

We start with the three dimensional case. The idea is as follows: The linear system (6) generates rotations around the z-axis, and by carrying out coordinate transformations, the higher order terms successively can be made invariant under these rotations.
Also compare Takens [1974]. In the divergence zero (or volume preserving) context the needed coordinate transformations can be kept volume preserving. Cf. Broer [1979, 1980a]. We recall the fact that a vector field $\dot{\xi} = F(\xi)$, $\xi \in \mathbb{R}^n$, has divergence zero if and only if for each t its time t-evolution $\xi \in \mathbb{R}^n \to \Phi_t(\xi) \in \mathbb{R}^n$, as far as defined, preserves the standard volume on $\mathbb{R}^n$.
We now formulate more precisely:

Theorem 1: Let $\dot{\xi} = F_\mu(\xi)$ be a family of vector fields, where $\xi = (x,y,z) \in \mathbb{R}^3$ and where μ is an r-dimensional parameter. Suppose that for $\mu = 0$ the vector field in $\xi = 0$ has a singularity with first order term (6). Also let $n \in \mathbb{N}$ be given.
Then, up to a μ-dependent change of coordinates, the right-hand side $F = F_\mu(\xi)$ has the form $F = \widetilde{F} + p$, where:
(i) In cylindrical coordinates $\dot{\xi} = \widetilde{F}_\mu(\xi)$ has the normal form

$$\begin{cases} \dot{\phi} = f(r^2,z,\mu) \\ \dot{r} = rg(r^2,z,\mu) \\ \dot{z} = h(r^2,z,\mu), \end{cases} \tag{10}$$

expressing rotational symmetry;

(ii) $p_\mu(x,y,z) = O((|\mu|^2+x^2+y^2+z^2)^{\frac{n+1}{2}})$ as $|\mu|^2 + x^2 + y^2 + z^2 \to 0$;
(iii) In the case where $\dot{\xi} = F_\mu(\xi)$ has divergence zero the change of coordinates may be chosen volume preserving, such that both $\dot{\xi} = \widetilde{F}_\mu(\xi)$ en $\dot{\xi} = p_\mu(\xi)$ have divergence zero.

Of course the first terms of $\widetilde{F}$ are just (7). See §3. Note that we did not specify the degree of differentiability of the various objects in the above theorem. We here mention that if $F = F_\mu(\xi)$ is C^∞, then so are 'all' other

objects is the theorem.
From now we maintain this convention: 'everything' is C^∞, unless specified otherwise.

In our context of topological equivalence it is not an essential restriction to assume that $f(r^2,z,\mu) \equiv 1$. (In the divergence free case this is not completely trivial.)
So from now on the vector fields $\dot{\xi} = \widetilde{F}_\mu(\xi)$ rotate with angular velocity one.

In general the higher order terms p cannot be made symmetric! Compare Takens [1973] and the results below.

4.2. The four dimensional case

The four dimensional case can be treated completely analogous, we only need to exclude strong resonances: For all k, $\ell \in \mathbb{Z}$ with $0 < |k| + |\ell| \leq n+1$ it is required that $k\alpha + \ell\beta \neq 0$.
Then, as in theorem 1, a normal form decomposition $F = \widetilde{F} + p$ exists, where in toroïdal coordinates $\dot{\xi} = \widetilde{F}(\xi)$, whith $\xi \in \mathbb{R}^4$, has the form

$$\begin{cases} \dot{\varphi}_i = f_i(r_1^2, r_2^2, \mu) \\ \dot{r}_i = r_i g_i(r_1^2, r_2^2, \mu), \quad i = 1,2 \end{cases} \tag{11}$$

which is compatible with (9). Also here the term p, of order $n+1$, in general is not symmetric. In the divergence free context we again may assume that both $\dot{\xi} = \widetilde{F}_\mu(\xi)$ and $\dot{\xi} = p_\mu(\xi)$ have divergence zero.

4.3. A remark on the strategy to follow

In all cases the analysis of the symmetric normal form $\dot{\xi} = \widetilde{F}_\mu(\xi)$ is carried out by first omitting the angular contributions of (10) and (11), so reducing the system to a plane. In the three dimensional cases this is the (r,z)-plane, in the four dimensional cases the (r_1,r_2)-plane. These reduced, planar vector fields have a rather easy structure.

It should be emphasised here that this kind of study in higher dimensions has certain limitations. The reason for this is the fact that if one 'couples many oscillators' even the symmetric normal forms are too difficult to be analysed: The reduced system may for example be a Volterra equation in $r_1, r_2, \ldots, r_k$, which can have 'strange attractors

5. THE DIVERGENCE FREE EXAMPLES

We start with the bifurcations in the divergence zero cases, since they are of codimension one and so involve only one parameter.

The two examples below are studied by different methods. The three-dimensional example, using a Poincaré first return map, is reduced to a perturbation of the twist map, cf. Moser [1962]. In this three-dimensional case we discuss the mathematics - and in particular the geometry - to some detail. We do this because on one hand in this low dimensional situation these details may be easily understood, while on the other hand the following examples have yet a quite similar geometric structure.
In the analysis of the other examples we will often refer to this more elaborate description, stressing both the correspondence and some differences with this first example.
The four dimensional example uses Moser [1967] and deals directly with the parametrised system of differential equations.
This last method has wider applications and it confirms that both examples of this section generalise to higher dimensions, if only more parameters are added. Cf. Braaksma & Broer [1981].

5.1. <u>The dimension three example (Broer [1980b])</u>

5.1.1. Analysis of the normal form

Let $F = \widetilde{F} + p$ be a normal form decomposition as in theorem 1, see (10), where μ is a real parameter.
It is generic to require that $\frac{\partial h}{\partial \mu}(0,0,0) \neq 0$, and rescaling of the parameter then leads to $\frac{\partial h}{\partial \mu}(0,0,0) = 1$.
Here h is the vertical velocity component in $\dot{\xi} = \widetilde{F}_\mu(\xi)$. See (10). As announced in 4.3 we study the normal form $\dot{\xi} = \widetilde{F}_\mu(\xi)$ by omitting the $\dot{\varphi}$-component, so reducing to a system

$$\begin{cases} \dot{r} = rg(r^2, z, \mu) \\ \dot{z} = h(r^2, z, \mu) \end{cases}$$

defined on the (r,z)-plane. In this plane we blow up, or

rescale, with $\sqrt{|\mu|}$: i.e. for $\mu \neq 0$ we introduce new variables $\bar{r}$ and $\bar{z}$, defined by

$$r = \bar{r}\sqrt{|\mu|} \text{ and } z = \bar{z}\sqrt{|\mu|}.$$

Transforming the reduced system to the new variables, rescaling the time with $\sqrt{|\mu|}$ and truncating at order two now yields for $\mu \neq 0$

$$\begin{cases} \dot{\bar{r}} = a\bar{r}\bar{z} \\ \dot{\bar{z}} = b\bar{r}^2 - a\bar{z}^2 - \operatorname{sgn}\{\mu\} \end{cases} + O(\sqrt{|\mu|}) \qquad (12)$$

as $\mu \to 0$, uniformly on compact sets.
Here a and b are real constants. From now on, for simplicity, we restrict to the case where $a < 0$ and $b > 0$. Note that this condition defines an open set in the C^2-topology in the class of all our one parameter families. Also we again replace $\bar{r}$ and $\bar{z}$ by r and z respectively.

This technique of rescaling, leading to 'convenient' lower order terms, is well known.

Now in (12) take the limit for $\mu \downarrow 0$ and observe that the thus obtained family for $\mu \geq 0$ is smoothly parametrised by $\sqrt{\mu}$. According to Broer [1979] the limit for $\mu \downarrow 0$

$$\begin{cases} \dot{r} = arz \\ \dot{z} = br^2 - az^2 - 1 \end{cases}$$

is C^∞-stable within the class of all such blown-up reductions of symmetric, divergence free vector fields on $\mathbb{R}^3$. Fig. 3 depicts the phase portrait of this limit system.

This stability for the symmetric vector field $\dot{\xi} = \tilde{F}_\mu(\xi)$ on $\mathbb{R}^3$ means the following: There exists a positive constant μ_o such that for all $0 < \mu < \mu_o$ the system $\dot{\xi} = \tilde{F}_\mu(\xi)$ possesses an invariant 'ellipsoid'. The inner region of this ellipsoid consists of a 'nested' family of invariant two-dimensional tori, shrinking towards a closed orbit (i.e. a periodic solution). The characteristic distance in the phase portrait of $\dot{\xi} = \tilde{F}_\mu(\xi)$ is of order $\sqrt{\mu}$. In the rescaled coordinates, however, the invariant 'ellipsoid' is of order 1.

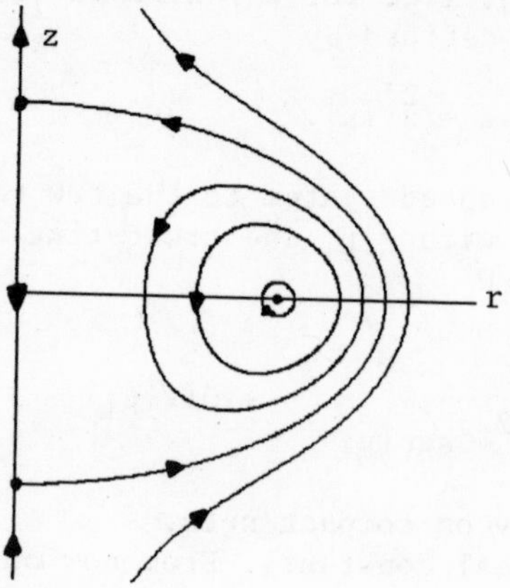

fig.3: phase portrait of the normal form reduced to the (r,z)-plane.

5.1.2. (Re)introduction of the higher order terms; towards a Poincaré map

We now analyse the perturbed system $F_\mu = \tilde{F}_\mu + p_\mu$ (for small, positive μ) by means of its so-called Poincaré map corresponding to the halfplane $\{\varphi = 0\}$. This means the following (see fig. 4):

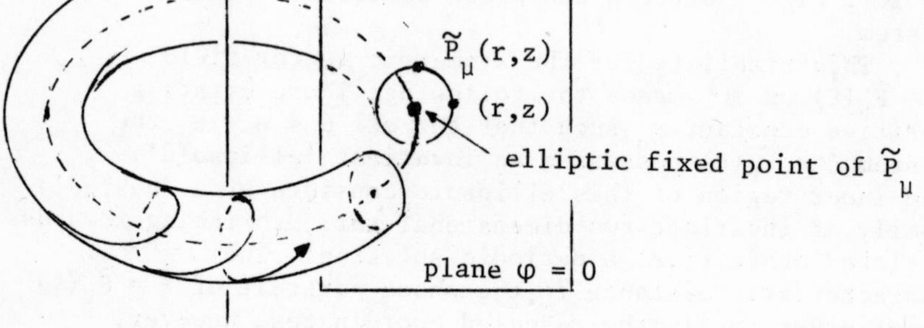

fig.4: description of the Poincaré map.

First consider the symmetric system $\dot{\xi} = \tilde{F}_\mu(\xi)$, which rotates with angular velocity one around the z-axis. We define the map $\tilde{P}_\mu$ from the halfplane $\{\varphi = 0\}$ onto itself by assigning to a point with coordinates (r,z) its time 2π-evolution under the flow of $\dot{\xi} = \tilde{F}_\mu(\xi)$. So $\tilde{P}_\mu(r,z)$ is the point where the integral curve starting in the point (r,z), with $\varphi = 0$, returns the first time to this halfplane. The map $\tilde{P}_\mu$, which clearly is a diffeomorphism, is sometimes called a first return Poincaré map of the system $\dot{\xi} = \tilde{F}_\mu(\xi)$. The definition of a similar first return map P_μ of the perturbed system $\dot{\xi} = F_\mu(\xi)$ involves some minor technicalities, since this system is not necessarily symmetric near the z-axis. Cf. Broer [1979].

Observe that P_μ is a small perturbation of $\tilde{P}_\mu$, where the size of this perturbation is controlled by taking μ close enough to zero.

It easily follows from the fact that $\dot{\xi} = \tilde{F}_\mu(\xi)$ has divergence zero, that the map $\tilde{P}_\mu$ preserves the 'area' $r dr dz$. Similarly the map P_μ preserves a slightly different 'area', but some minor change of coordinates makes P_μ preserve the same 'area' as $\tilde{P}_\mu$.

For a similar construction see e.g. Arnol'd & Avez [1967], §19.

Let us point out some correspondences between the geometric behaviour of the flow of the vector fields $\dot{\xi} = \tilde{F}_\mu(\xi)$ and $\dot{\xi} = F_\mu(\xi)$ on one hand and the respective Poincaré maps $\tilde{P}_\mu$ and P_μ on the other hand.

Firstly note that a closed orbit (periodic solution) of a vector field corresponds to a periodic orbit of the Poincaré map.

So for example $\tilde{P}_\mu$ possesses a fixed point (so of period) corresponding to the closed orbit which is the 'centre of the 'nested' tori. (This fixed point and the corresponding periodic solution are called elliptic, since the eigenvalues of the derivative of $\tilde{P}_\mu$ in this fixed point are on the complex unit circle.) See fig. 4. Secondly observe that an invariant 2-torus of a vector field corresponds to an invariant circle of the corresponding Poincaré map.

5.1.3. The symmetric Poincaré map as a twist map

We now scrutinise the geometry of the symmetric Poincaré map $\tilde{P}_\mu$. The phase portrait of this map is exactly fig. 3: $\tilde{P}_\mu$ is also the time 2π-evolution of the reduced system (12).

Since both this reduced vector field and $\tilde{P}_\mu$ respect the 'area' rdrdz, it can be shown that this reduced system possesses a first integral

$$H_\mu(r,z) = -\frac{1}{4}r^2(br^2-2az^2-2) + O(\sqrt{|\mu|}),$$

uniformly as $\mu \downarrow 0$.
This is the reason why people call the symmetric situation (completely) integrable. For obvious reasons then the original, 'perturbed' situation is called nearly integrable.

The existence of this first integral (a Hamiltonian function) enables us to pass to so called action angle variables (I,θ) in the halfplane $\{\varphi = 0\}$. Compare Arnol'd & Avez [1967], appendix 26 or Arnol'd [1976]. We define

$$I(h,\mu) = \frac{1}{2\pi} \oint_{H_\mu(r,z)=h} \frac{1}{2} r^2 dz$$

being the $\frac{1}{2\pi}$ th part of the 'area' enclosed by the level curve $H_\mu(r,z)=h$ in the region of the half-plane which is of relevance to us now. The other coordinate θ, defined modulo 2π, is the corresponding 'phase angle'.

In the variables (I,θ,φ) the symmetric system $\dot{\xi} = \tilde{F}_\mu(\xi)$ has the form

$$\begin{cases} \dot{I} = 0 \\ \dot{\theta} = \omega(I,\mu) \\ \dot{\varphi} = 1 \end{cases} \tag{13}$$

which is 'linear' on each fixed 2-torus. It can be shown that the frequency $\omega(I,\lambda)$ is of order $\sqrt{\mu}$.

Correspondingly in the variables (I,θ) the symmetric Poincaré map $\tilde{P}_\mu$ has the form

$$\tilde{P}_\mu(I,\theta) = (I,\theta+\lambda(I,\mu)) \tag{14}$$

where $\lambda(I,\mu) = 2\pi\omega(I,\mu)$ is called the rotation number of the invariant circle labelled by I and μ.

If for an invariant circle, labelled by I and μ, one has that $\omega(I,\mu)$ is a rational number, say p/q with $(p,q) = 1$, then this invariant circle is full of periodic points of period q. The corresponding invariant 2-torus is completely

filled with closed integral curves of the period $2\pi q$.
If, on the other hand, $\omega(I,\mu)$ is not rational, then the invariant circle is densely filled up by each of its orbits. Similarly the corresponding invariant 2-torus is the closure of each integral curve that it contains. Here we speak of quasi periodic motions of the system $\dot{\xi} = \tilde{F}_\mu(\xi)$.
The map (14) is called a twist map as soon as

$$\frac{\partial \lambda}{\partial I}(I,\mu) \neq 0, \tag{15}$$

a condition which is verified in Broer [1980b] by estimating some elliptic integrals.

5.1.4. The original Poincaré map as a perturbed twist map

At last we turn to the 'perturbed' twist map P_μ: the Poincaré map of our original vector field $\dot{\xi} = F_\mu(\xi)$.

The elliptic fixed point $I = 0$ of $\tilde{P}_\mu$, mentioned in 5.1.2, is persistent for the perturbation to P_μ, for small enough μ. This means that close to $I = 0$ the map P_μ has a unique fixed point, which again is elliptic. See Broer [1980b]. So the elliptic closed orbit of $\dot{\xi} = \tilde{F}_\mu(\xi)$ survives the perturbation.

The rational tori, mentioned in 5.1.3 , may not be expected to survive the perturbation, since they are completely filled up with closed orbits of the same period. This is a consequence of a so called Kupka-Smale theorem due to Robinson [1970] which states among other things the following:
For divergence zero vector fields in dimension three it is a generic property that closed orbits of bounded period are isolated.

Remain the irrational tori, carrying quasi periodic flow.
We study the corresponding situation for the Poincaré map: invariant circles with a rotationnumber not commensurable with 2π. It appears that many of these invariant circles survive the perturbation from $\tilde{P}_\mu$ to P_μ. They are found back in the perturbed system: slightly deformed, but with the same rotation number as before the perturbation.
One needs an extra condition on these rotationnumbers, meaning that they are even badly commensurable with 2π.
To be more precise:
If λ is the rotationnumber, then we require that there exists real numbers $\tau > 2$ and $\gamma > 0$ such that for all rationals

p/q the following inequality holds:

$$\left|\frac{\lambda}{2\pi} - \frac{p}{q}\right| \geq \frac{\gamma}{q^{\tau}}. \tag{16}$$

Compare e.g. Moser [1962]. The above condition has to do with the presence of so called small divisors.

It is easily proven that the numbers λ which belong to a fixed closed interval, that satisfy the infinitely many estimations of type (16), constitute a Cantor set: a compact, perfect and totally disconnected set. Cantor sets are uncountable, but nowhere dense and so very small in the topological sense. Nevertheless the measure of this specific Cantor set is positive: relative to the closed interval which it was constructed from, its measure tends to full measure as $\gamma \downarrow 0$. For more details see Oxtoby [1970] or Herman [1977].

Due to the twist condition (15) the union of the 'unperturbed' $\tilde{P}_\mu$-invariant circles that are able to survive the perturbation, is the product of a Cantor set and a circle: this union having positive measure in the above sense. For simplicity we here forego the problem of 'small twist', related to the fact that the size of the perturbation is controlled by taking μ small. We just mention that in (16) one has to choose $\gamma = \gamma(\mu)$ in a suitable way, where $\lim_{\mu\to 0} \gamma(\mu) = 0$. See Moser [1962].

Since by (15) all rotationnumbers of these invariant circles are mutually different and since for each circle this number is kept constant 'during the perturbation', it follows that the perturbed twistmap P_μ also possesses an uncountable number of invariant circles with 'strongly' irrational $\lambda/2\pi$.
A question now is whether this union of P_μ-invariant circles, and so whether the union of invariant 2-tori for the system $\dot{\xi} = F_\mu(\xi)$, also has positive measure.

In Pöschel [1981] it is proved that under slightly different circumstances (the systems are Hamiltonian) such a perturbed collection of invariant tori in a nearly integrable system forms a differentiable family parametrised over a Cantor set. Differentiable in the sense of Whitney. Also Pöschel [1981] estimates the measure of the complement of this collection of tori in terms of the perturbation.

Compare the integrable situation, i.e. for $\tilde{P}_\mu$ and $\dot{\xi} = \tilde{F}_\mu(\xi)$, as skeched above: with respect to these quasi periodic motions there seems to be a kind of similarity

between nearly integrable systems and their integrable approximations.
Also compare Arnol'd [1963] and Moser [1969].
For more general information on this subject also see Moser [1973] or Arnol'd & Avez [1967].

If we here assume that Pöschel [1981] generalises to our situation, then we may formulate

Theorem 2: Take any generic one parameter family $\dot{\xi} = F_\mu(\xi)$ of divergence zero systems, as specified above. Then, a constant $\mu_1 > 0$ exists, such that for all $\mu \in (0,\mu_1)$ in $\mathbb{R}^3$ there exists a Cantor set of positive measure, 'relatively tending to full measure as $\mu \downarrow 0$', which is the union of 2-tori which are invariant under the flow of $\dot{\xi} = F_\mu(\xi)$. Moreover this flow on each of these 2-tori is quasi periodic.

5.1.5. Conclusive remarks

(i) The flow of a one parameter family as in theorem 2 is not ergodic, as it is easy to find two disjoint invariant sets of positive measure.

(ii) Near the elliptic fixed point $I = 0$ we can expand P_μ as a powerseries in I. See (14). Thus we obtain the so called Birkhoff normal form

$$P_\mu(I,\theta) = (I, \theta + \beta_1(\mu) + \beta_2(\mu)I) + O(I^2).$$

It can be proven, see Broer [1980b] that for $i = 1,2$

$$\frac{\beta_i(\mu)}{\sqrt{\mu}} \to c_i \quad \text{as } \mu \downarrow 0,$$

where $c_1 \neq 0 \neq c_2$.

The arguments and references given before the formulation of theorem 2 now 'directly' yield a part of the Cantor set mentioned in theorem 2: clearly this argument only works near the elliptic orbit.

(iii) The phase portrait of the nearly integrable system is strikingly different from its integrable approximation, in spite of the similarity with respect to the quasi periodic motions as mentioned above. This can already be guessed from the Kupka-Smale argument given before: there is an important difference on the level of periodic motions. We do not go into this subject here, but refer to Moser [1973] and to Arnol'd & Avez [1967], §20 where an attempt has been made to give a picture of the phases portrait of a perturbed twist map. Also in Abraham & Marsden [1978], p. 585, one may find a picture of the present situation. According to their terminology our system $\dot{\xi} = F_\mu(\xi)$, for small enough $\mu > 0$, contains a vague attractor of Kolmogorov.

(iv) This difference between integrable and nearly integrable implies that our one parameter families are not structurally stable. Which shows that for such families structural stability is not a generic property: the described phenomena have an open occurence. So here the Thom-Smale programme does not work.

(v) The geometry of the bifurcation at hand has some importance in fluid mechanics, compare §1, especially in plasma physics in connection with so called 'compact tori' and 'tokamaks'. This I learned from private communications with A.N. Kaufman (Lawrence Berkely Laborary). Also the geometry of the bifurcation bears some resemblance to smoke rings...

5.2. The dimension four example (Braaksma & Broer [1981]

5.2.1. Analysis of the normal form and its perturbation

We start in the same way as in 5.1. Compare §§3.2 and 4.2. Firstly we require that there be no resonances up to order 12 in the eigenvalues α and β of the linear part (8). So, according to §4.2, we obtain a normal form $F_\mu = \tilde{F}_\mu + p_\mu$, where the Taylor series of p vanishes up to order 11 and where $\tilde{F}_\mu$ possesses toroidal symmetry: according to (11) $\dot{\xi} = \tilde{F}_\mu(\xi)$ in toroidal coodinates has the form

$$\begin{cases} \dot{\varphi}_i = f_i(r_1^2, r_2^2, \mu) \\ \dot{r}_i = r_i g_i(r_1^2, r_2^2, \mu), \quad i = 1,2. \end{cases}$$

As before μ is the real bifurcation parameter. Also as before we have that both $\dot{\xi} = \tilde{F}_\mu(\xi)$ and $\dot{\xi} = p_\mu(\xi)$ are divergence free. Again omitting the $\dot{\varphi}_i$-components, we now reduce to the (r_1,r_2)-plane and follow the same strategy as before:
We rescale the variables r_1 and r_2 with $\sqrt{|\mu|}$, and we also

rescale time with a factor $|\mu|$. Truncation at order 3 now leads to

$$\begin{cases} \dot{r} = r_1(c_1 r_1^2 + 2c_2 r_2^2 - c_3 \operatorname{sgn}\{\mu\}) \\ \dot{r} = -r_2(2c_1 r_1^2 + c_2 r_2^2 - c_3 \operatorname{sgn}\{\mu\}) \end{cases} + O(|\mu|), \qquad (17)$$

as $\mu \to 0$, uniformly on compact sets.
This is completely similar to (12). The real constants c_1, c_2 and c_3 may be arbitrary. From now on, however, we restrict to the open case where $c_3 > 0$ and $c_1 > c_2 > 0$. This restriction is not essential.
As in the three-dimensional case we find a C^∞-stable phase portrait for the limit

$$\begin{cases} \dot{r}_1 = r_1(c_1 r_1^2 + 2c_2 r_2^2 - c_3) \\ \dot{r}_2 = -r_2(2c_1 r_1^2 + c_2 r_2^2 - c_3) \end{cases}$$

obtained by letting $\mu \downarrow 0$. See Broer [1979, 1980a].

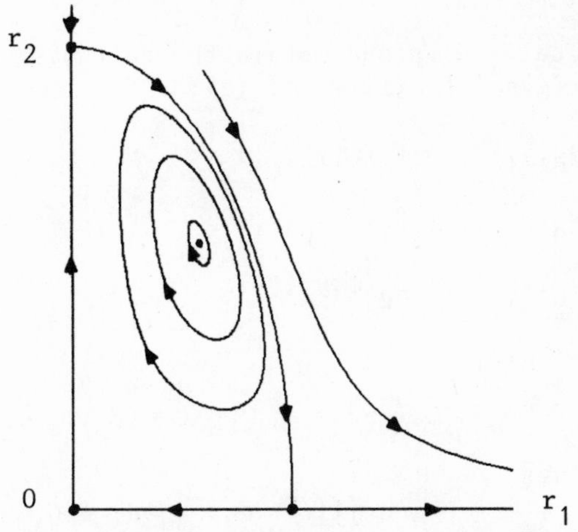

Fig. 5: phase portrait of the normal form reduced to the (r_1, r_2)-plane.

For a phase portrait cf. fig. 5.
In this case the stability means that the symmetric system

$\dot{\xi} = \widetilde{F}_\mu(\xi)$ ($\mu > 0$ and small) possesses an invariant 3-sphere which encloses an invariant 4-disk. This disk is foliated by a one parameter family of invariant 3-tori, which collapses into an invariant 2-torus. (Also the foliation contains two transversal 3-disks.) As before the characteristic distance in the phase portrait of $\dot{\xi} = \widetilde{F}_\mu(\xi)$ is of order $\sqrt{\mu}$. In the rescaled coordinates, however, the 4-disk under consideration is of order 1.

Now, since $\dot{\xi} = \widetilde{F}_\mu(\xi)$ has divergence zero the system (17) respects the 'area' $r_1 r_2 dr_1 dr_2$, which yields a first integral

$$H_\mu(r_1,r_2) = \frac{1}{4}r_1^2 r_2^2(c_1 r_1^2 + c_2 r_2^2 - c_3) + O(\mu)$$

uniformly as $\mu \downarrow 0$.
Proceeding as under 5.1. we define the action integral

$$y_o(h,\mu) = \frac{1}{2\pi}\oint_{H_\mu(r_1,r_2)=h} \frac{1}{2}r_1^2 r_2 dr_2$$

and let φ_o be the corresponding phase angle, defined modulo 2π.
We now substitute $y_1 = \sqrt{\mu}$ and obtain the following normal form for our original family $\dot{\xi} = F_\mu(\xi)$:

$$\begin{cases} \dot{\varphi}_o = a_o(y_o,y_1) + \bar{f}_o(\varphi,y_o,y_1) \\ \dot{\varphi}_1 = a_1(y_o,y_1) + \bar{f}_1(\varphi,y_o,y_1) \\ \dot{\varphi}_2 = a_2(y_o,y_1) + \bar{f}_2(\varphi,y_o,y_1) \\ \dot{y}_o = \bar{g}_o(\varphi,y_o,y_1) \\ \dot{y}_1 = 0 \qquad\qquad (\varphi = (\varphi_o,\varphi_1,\varphi_2)) \end{cases} \tag{18}$$

where $\dfrac{\partial\bar{g}_o}{\partial y_o} + \sum_{j=0}^{2} \dfrac{\partial\bar{f}_j}{\partial\varphi_j} = 0$ and $\iiint \bar{g}_o(\varphi,y_o y_1)d\varphi = 0$, expressing the preservation of volume.
Here the terms $\bar{f}_j$ and $\bar{g}_o$ are small perturbations, which makes the form (18) nearly integrable. Observe that the frequency $a_o(y_o,y_1)$ is slower than a_1 and a_2:
$a_o(y_o,y_1) = O(y_1^2)$, uniformly as $y_1 \downarrow 0$. Compare 5.1.
From now on $\dot{\xi} = \widetilde{F}_\mu(\xi)$ shall denote the 'integrable' part of

(18), so without the perturbation terms $\bar{f}_j$ and $\bar{g}_o$.

Moser [1967] makes a general study of normal forms like (18). He carries out a persistence proof for quasi periodic solutions of $\dot{\xi} = \tilde{F}_\mu(\xi)$ when sufficiently small perturbations are added. His procedure moreover is very general, since it applies to various contexts: to volume preserving, Hamiltonian or to dissipative systems etc. Summarily: to any Lie algebra of vector fields. Also compare Moser [1966].

We now may largely repeat the disccusion held under 5.1.4. Instead of doing that we just formulate a result which would be a consequence of Braaksma & Broer [1981] and an analogue of Pöschel [1981] for this situation. This analogue has not yet been proved however.

Theorem 3: Let $\dot{\xi} = F_\mu(\xi)$ be a generic one parameter family of divergence free vector fields, as specified above. Then there exists a positive constant μ_1, such that the family $\{\dot{\xi} = F_\mu(\xi)\}_{0<\mu<\mu_1}$ possesses a collection of invariant 3-tori, the union of which is a Cantor set of positive measure. The flow in these 3-tori is quasi periodic with three independent frequencies.

5.2.2. Conclusive remarks

(i) As in the former case the three frequencies have to be 'very' independent because of the presence of small divisors. For such frequencies $\omega_0, \omega_1, \omega_2$ one needs a condition like:
There exists constants $\gamma > 0$ and $\tau > 2$ such that for all tri-indices $\nu \in \mathbb{Z}^3 \setminus \{0\}$ one has $|\langle \nu\ \omega \rangle| > \gamma|\nu|^{-\tau}$.
Here $\langle \nu, \omega \rangle = \sum_{j=0}^{2} \nu_j \omega_j$ and $|\nu| = \sum_{j=0}^{2} |\nu_j|$.

(ii) Any of the F-invariant 3-tori has been obtained as a perturbation of an $\tilde{F}$-invariant 3-torus, where the frequency ratios of the quasi periodic motions in these two respective tori are the same.
Compare 5.1. Moreover during this perturbation the parameter value $(\mu = y_1^2)$ may have shifted: i.e. an 'unperturbed' $\tilde{F}_\mu$-invariant torus may be deformed into an $F_{\mu'}$-invariant torus, where μ' near μ must be chosen appropriately.

Apparently there is some difference with the former dimension three case. If here one picks out an arbitrary $\mu \in (0,\mu_1)$, then one does not know whether the vector field $\dot{\xi} = F_\mu(\xi)$ has any invariant 3-tori or not. On the other hand, by Fubini's theorem it follows that in $(0,\mu_1)$ there exists a set C_{μ_1} of positive measure, such that for all $\mu \in C_{\mu_1}$ the system $\dot{\xi} = F_\mu(\xi)$ possesses a number of invariant 3-tori constituting a set of positive measure in $\mathbb{R}^4$.

Remarks on non-ergodicity and structural instability can be made, analogous to the three dimensional case.

6. SOME DISSIPATIVE CASES

From now on we abandon the restriction to volume preservingness. Again on $\mathbb{R}^3$ and $\mathbb{R}^4$ consider the degenerate singularities with linear parts (6) and (8) respectively. Cf. §3. We shall sketch a partly analysis of generic two parameter families of vector fields which unfold these singularities.

6.1. Subordinate Hopf bifurcations in the reduced normal form

The mathematical idea is as before:
At first the lower order terms in the Taylor series are normalised as in §4. For our two parameter family $\dot{\xi} = F_{\mu,\nu}(\xi)$ we write, as before, $F = \widetilde{F} + p$, where $\widetilde{F}$ is symmetric and p contains the higher order terms.
(The real parameters are denoted by μ and ν.)
Secondly $\dot{\xi} = \widetilde{F}_{\mu,\nu}(\xi)$ is reduced to a plane: the (r,z)-plane in the dimension three case and the (r_1,r_2)-plane in the dimension four case.
Let us name this reduction $\dot{\eta} = \overline{F}_{\mu,\nu}(\eta)$ $(\eta \in \mathbb{R}^2)$.
This planar family $\dot{\eta} = \overline{F}_{\mu,\nu}(\eta)$ can be investigated rather easily. It appears that in both the three and the four dimensional case there exists an open class of these planar systems, with two parameters, which exhibit Hopf bifurcations. The meaning of this can be illustrated by fig. 6. For such a family $\dot{\eta} = \overline{F}_{\mu,\nu}(\eta)$ in the (μ,ν)-plane there exists a curve L, starting from the origin, with the following property: If in this parameter plane one moves transversally through L, then the reduced system $\dot{\eta} = \overline{F}_{\mu,\nu}(\eta)$ undergoes Hopf bifurcation. This implies that there exists a wedge $\widetilde{W}$ in the parameter plane, with $L \subset \partial\widetilde{W}$, such that for $(\mu,\nu) \in \widetilde{W}$ the system $\dot{\eta} = \overline{F}_{\mu,\nu}(\eta)$ possesses a closed orbit near the origin in $\mathbb{R}^2$. Cf. fig. 2.
For the corresponding symmetric vector field $\dot{\xi} = \widetilde{F}_{\mu,\nu}(\xi)$ this yields an invariant torus: of dimension two in the

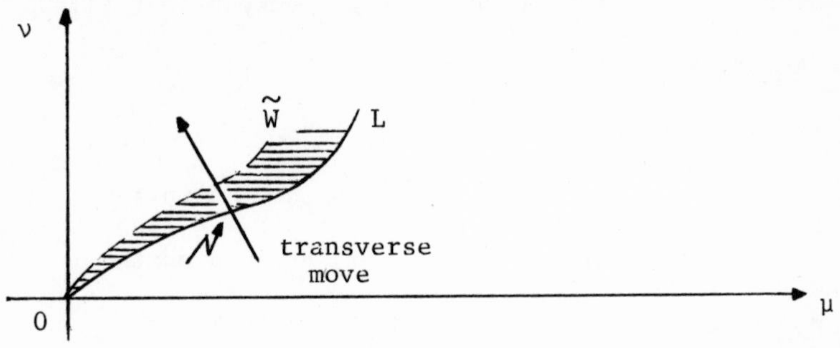

Fig. 6: a line of subordinate Hopf birfurcations in the parameter plane.

$\mathbb{R}^3$-case and of dimension three in the $\mathbb{R}^4$-case. So in both cases the symmetric family $\dot{\xi} = \widetilde{F}_{\mu,\nu}(\xi)$ has a two parameter family of invariant tori, the tori being parametrised by the set $\widetilde{W}$.
Observe that generically there will be many of these tori which carry a quasi periodic $\widetilde{F}$-flow.
As in the previous cases one may apply perturbation theory in order to prove persistence of such tori for the original family $\dot{\xi} = F_{\mu,\nu}(\xi)$, where $F = \widetilde{F} + p$.

In Guckenheimer [1980b] a first attempt is made to investigate quasi periodic flow with three independent frequencies, which occurs in our four dimensional example. It appears that experiments by Fenstermacher, Swinney and Gollub [1979] on fluid flow between rotating cylinders, exhibit quasi periodicity with two and with three independent frequencies. These quasi periodic modi occur as transient stages between laminar flow and full turbulence if one varies some experimental parameter (such as Reynolds or Rayleigh number). It is to be noted that the experimental design, which contains circle-symmetry, is highly responsable for the appearance of the third frequency.
The argument that Guckenheimer uses for the physical relevance of his study is that although the phenomenon of quasi periodicity is not structurally stable (see above), it nevertheless occurs for a set of parameter values having positive measure: so the phenomenon should yet be observable. Again a controversy between topology and measure theory. See above.

For more references on the physical aspects of our

codimension two bifurcation see e.g. Guckenheimer [1980a,b].

6.2. A more specific description

6.2.1. In dimension three

The details here are pretty much the same as in §5. Starting with a family $\dot{\xi} = F_{\mu,\nu}(\xi)$, $F = \tilde{F} + p$, see above, we find a reduced system $\dot{\eta} = \bar{\bar{F}}_{\mu,\nu}(\eta)$ defined on the (r,z)-plane.
For $\mu = \nu = 0$ we truncate at order two:

$$\begin{cases} \dot{r} = crz \\ \dot{z} = ar^2 + bz^2 \end{cases}$$

where a,b and c are real constants. Compare Takens [1974]. For open subcases $a > 0$, $b \neq 0$, $b > c$ and $a > 0$, $b < c < 0$, in both cases we require $c + 2b \neq 0$, one finds the Hopf bifurcations we spoke of in the introduction of this section. Addition of generic third order terms yields that these Hopf bifurcations are 'stable', implying that the closed orbit is hyperbolic (attracting or repelling in a non-degenerate way, compare (5) and fig. 2). Cf. Guckenheimer [1980a].

Now let $\tilde{M}$ be the two parameter family of 2-tori, parametrised by $\tilde{W}$. See above, especially fig. 6. Since $\tilde{M}$ is an invariant manifold for the family $\dot{\xi} = \tilde{F}_{\mu,\nu}(\xi)$, and since $\tilde{M}$ is normally hyperbolic - in 'normal directions there is non-degenerate attraction or repelling, see above - one may wish to apply invariant manifold theory (Hirsch, Pugh & Shub [1977] or rather Palis & Takens [1977]) to prove persistence of $\tilde{M}$ for the higher order terms p. The problem here is the dependence on the parameters μ and ν: both the normal hyperbolicity and the size of p strongly depend on it.

Our present concern is with the fact whether such a perturbed manifold, invariant for $\dot{\xi} = F_{\mu,\nu}(\xi)$, contains quasi periodic flow.
To solve this problem we may apply perturbation theory following both of the methods displayed in §5: the method of the first return Poincaré map or the one which uses Moser [1967]. Compare Braaksma & Broer [1981]. Tacitly assuming that Pöschel [1981] can be extended to these circumstances, we now formulate the result:

Theorem 4: Let $\dot{\xi} = F_{\mu,\nu}(\xi)$ be a generic two parameter family of vector fields on $\mathbb{R}^3$ in the above sense. Then in the 5-dimensional (x,y,z,μ,ν)-space there exists a Cantor set M near $\tilde{M}$, which is the union of F-invariant 2-tori carrying quasi periodic flow. M foliates over a Cantor set W in the (μ,ν)-plane. The measure of W is positive.

This Cantor set W measure-theoretically covers a large part of the wedge $\tilde{W}$. Again the F-invariant tori are obtained as perturbations of $\tilde{F}$-invariant tori, where shifts in both parameter values must be allowed for, while the frequency ratios are kept constant 'during the perturbation'
As before theorem 4 once again illustrates that structural stability is not generic for such families.

Added in proof:
Results similar to theorem 4 were obtained independently by Scheurle and Marsden [1982]. Their methods - independent of Pöschel's work - give somewhat more information on the structure of the set W.

6.2.2. In dimension four (Guckenheimer [1980b], Braaksma & Broer [1981])

We omit all details here, since everything is completely analogous to the above case 6.2.1.
We remark that Guckenheimer [1980b] on a formal level (Fourier series) attacks the problems, using a Poincaré return map. Braaksma & Broer [1981] indicate a proof in the line of Moser [1967]. Compare section 5.2.
We now formulate the result:

Theorem 5: For a generic two parameter family of vector fields $\dot{\xi} = F_{\mu,\nu}(\xi)$ on $\mathbb{R}^4$ there exists a Cantor set M in the 6-dimensional $(x_1,x_2,x_3,x_4,\mu,\nu)$-space, being union of F-invariant 3-tori with quasi periodic flow. Moreover M foliates over a set W in the (μ,ν)-plane, where W has positive measure.

Also here M is close to a 5-dimensional manifold $\tilde{M}$ which is invariant for an 'integrable' approximation' $\tilde{F}$ of F. Generically $\tilde{M}$ is normally hyperbolic. It seems that recently Iooss and Langford have proved persistence of this manifold, using invariant manifold theory. See 6.2.1.

6.2.3. A non normally hyperbolic analogue (Chenciner [1981a,b])

If one allows for more parameters the dynamical complexity of the appearing bifurcations increases. In the above examples 6.2.1 and 6.2.2 the Hopf bifurcations are normally hyperbolic, compare the normal form (5). If one introduces one extra parameter, generically one may also meet non normally hyperbolic Hopf bifurcations.

The situation studied by Chenciner may be obtained by e.g. reducing the three dimensional example to a Poincaré map of the plane. See above.

In fact Chenciner right away starts from a Hopf bifurcation for planar diffeomorphisms, which is non normally hyperbolic, but where the degree of degeneration further is as low as possible. This planar bifurcation has codimension two, so we consider generic two parameter unfoldings of it. Techniques, similar to §4, yield that such a family can be put into a normal form

$$P_{\mu,a}(z) = z[1 + f(\mu,a|z|^2) + O(|z|^{2n+2})]e^{2\pi i[\omega(\mu,a) +}$$

$$+ g(\mu,a,|z|^2) + O(|z|^{2n+2})]$$

where $\mathbb{R}^2$ is identified with $\mathbb{C}$ and the parameters are denoted by μ and a. Here it is needed that $\omega_o = \omega(0,0)$ has no strong resonances: $\omega_o \neq p/q$ for $q \leq 2n+3$.
If one writes $f(\mu,a,s) = \mu + as + a_2(\mu,a)s^2 + .. + a_n(\mu,a)s^n$ and $g(\mu,a,s) = b_1(a,\mu)s + ... + b_n(\mu,a)s^n$, then assume generically that $a_2(0,0) \neq 0 \neq b_1(0,0)$.

If $\tilde{P}_{\mu,a}$ denotes the normalised, symmetric part of $P_{\mu,a}$, then by the implicit function theorem we find in the (μ,a)-plane a curve $\tilde{\Gamma}$, such that for $(\mu,a) \in \tilde{\Gamma}$ the map $\tilde{P}_{\mu,a}$ near the origin of $\mathbb{R}^2$ possesses a unique invariant circle which is not normally hyperbolic. See fig. 7.
The search now is for invariant circles of the 'perturbed' map $P_{\mu,a}$ where the pair (μ,a) varies over a certain wedge shaped neighbourhood V of $\tilde{\Gamma}$.

Chenciner applies Rüssmann [1970] to this situation. As a further generic condition it is needed that the rotation number of $P_{\mu,a}$, restricted to its unique invariant circle 'varies enough' along the curve $\tilde{\Gamma}$. This is what the 'twist condition', see (15), boils down to in the present case. It follows:

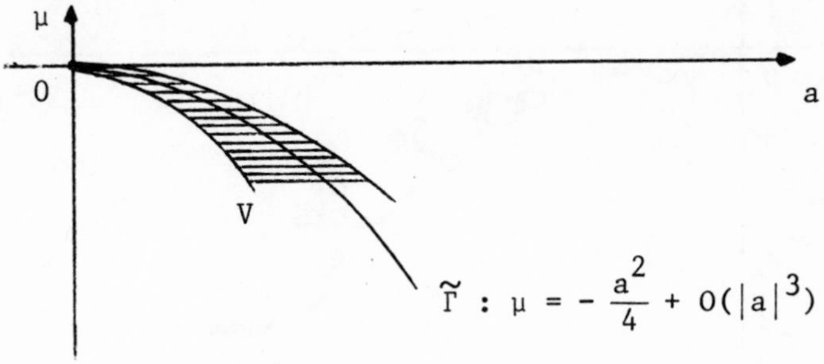

fig. 7: a line of subordinate degenerate Hopf bifurcations in the parameter plane occurring for the normal form.

Theorem 6: For a generic family $P_{\mu,a}$ of planar diffeomorphisms as specified above there exists a Cantor set Γ near $\tilde{\Gamma}$ (with $(\mu,a) = (0,0)$ as density point) such that:

(i) For $(\mu,a) \in \Gamma$ the map $P_{\mu,a}$ possesses a non-normally hyperbolic invariant circle $C_{\mu,a}$. The restriction of $P_{\mu,a}$ to $C_{\mu,a}$ is C^∞-conjugate to a rigid rotation over an angle $2\pi\omega(\mu,a)$.

(ii) Every point $(\mu_0,a_0) \in \Gamma$ is centre of a piece of curve γ_{μ_0,a_0}, with its extremities outside the wedge V, such that for $(\mu,a) \in \gamma_{\mu_0,a_0}$, $(\mu,a) \neq (\mu_0,a_0)$ the map $P_{\mu,a}$ possesses a normally hyperbolic invariant circle on which P_μ, is C^∞-conjugate to a rigid rotation over an angle $2\pi\omega(\mu_0,a_0)$. These normally hyperbolic circles are attractors on one side of the centre (μ_0,a_0) and repellors on the other side.

(iii) In fact the curve $\gamma_{\mu_0,a}$ can be extended to a small double cone with (μ_0,a_0) as its centre: for (μ,a) in this cone the diffeomorphism $P_{\mu,a}$ has a normally hyperbolic invariant circle. These circles are attractors in one half of the cone and repellors in the other half.

(iv) If (μ,a) is in the intersection of two of such open half cones then $P_{\mu,a}$ possesses two invariant circles, where one is attracting and the other repelling.

See fig. 8. For more details compare Chenciner [1981a,b]. Clearly also such generic two parameter families of diffeomorphisms are not structurally stable.

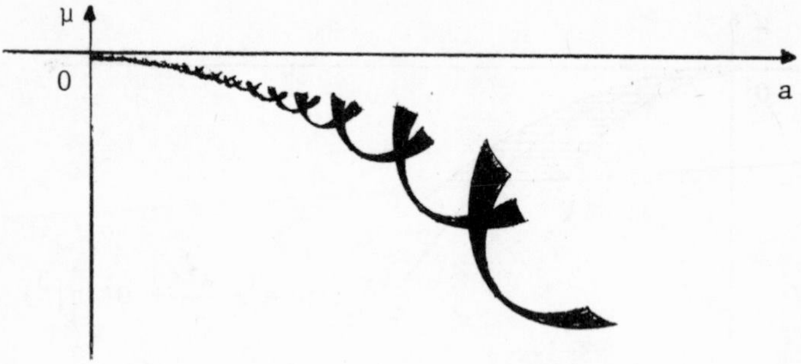

fig. 8: a bundle of cones over a Cantor set near $\tilde{\Gamma}$ in the parameter plane occurring for the perturbed normal form.

REFERENCES

Abraham, R. and J.E. Marsden: 1978, Foundations of Mechanics, Benjamin.

Arnol'd, V.I.: 1963, 'Proof of A.N. Kolmogorov's Theorem on the Preservation of Quasiperiodic Motions under Small Perturbations of the Hamiltonian', Usp. Mat. Nauk. SSR 18, no. 5, 13-40.

Arnol'd, V.I.: 1972, 'Lectures on Bifurcations and Versal Families', Russ. Math. Surveys 27, 54-123.

Arnol'd, V.I.: 1976, Méthodes Mathématiques de la Mécanique Classique, Éditions MIR, Moscow.

Arnol'd, V.I.: 1978, Chapitres Supplémentaires de la Théorie des Équations Différentielles Ordinaires, Éditions MIR, Moscow.

Arnol'd, V.I. and A. Avez: 1967, Problèmes Ergodiques de la Mécanique Classique, Gauthier-Villars, Paris.

Braaksma, B.L.J. and H.W. Broer: 1981, Quasi Periodic Flow near a Codimension One Singularity of a Divergence Free Vector Field in Dimension Four, to appear in Revue Astérisque.

Broer, H.W.: 1979, Bifurcations of Singularities in Volume Preserving Vector Fields, Ph.D.Thesis, Groningen.

Broer, H.W.: 1980a, 'Formal Normal Form Theorems for Vector Fields and some Consequences for Bifurcations in the Volume Preserving Case, in D.A.Rand and L.-S.Young (eds.), Dynamical Systems and Turbulence, Warwick 1980,

Springer-Verlag, pp. 54-74.
Broer, H.W.: 1980b, 'Quasi Periodic Flow Near a Codimension One Singularity of a Divergence Free Vector Field in Dimension Three', in D.A.Rand and L.-S.Young (eds), Dynamical Systems and Turbulence, Warwick 1980, Springer-Verlag, pp. 75-89.
Chenciner, A.: 1981a, Courbes Fermées Invariantes Non Normalement Hyperboliques au Voisinage d'une Bifurcation de Hopf Dégénéré de Diffémorphismes de ($\mathbb{R}^2$,0), Preprint Paris VII.
Chenciner, A.: 1981b, 'Bifurcations de Difféomorphismes de $\mathbb{R}^2$ au Voisinage d'un Point Fixe Elliptique', Cours à l'école d'été des Houches.
Fenstermacher, P.R., H.L. Swinney and J.P. Gollub: 1979, 'Dynamical Instabilities and the Transition to Chaotic Taylor Vortex Flow', J. Fluid Mech. 94, 103- .
Gollub, J.P. and H.L. Swinney: 1975, 'Onset of Turbulence in a Rotating Fluid', Physical Review Letters 35, 927-930.
Guckenheimer, J.: 1980a, 'On a Codimension Two Bifurcation', in P.A.Rand and L.-S.Young (eds.), Dynamical Systems and Turbulence, Warwick 1980, Springer-Verlag, pp. 99-142.
Guckenheimer, J.: 1980b, On Quasi Periodic Flow with Three Independent Frequencies, Preprint Santa Cruz.
Hartman, P.: 1964, Ordinary Differential Equations, Wiley and Sons.
Herman, M.R.: 1977, Mesure de Lebesgue et Nombre de Rotation, in J. Palis and M. do Carmo (eds.), Geometry and Topology, Rio de Janeiro, July 1976, Springer-Verlag, pp. 271-293.
Hirsch, M.W., C.C. Pugh and M.Shub: 1977, Invariant Manifolds, Springer-Verlag.
Moser, J.K.: 1962, 'On Invariant Curves of Area-Preserving Mappings of an Annulus', Nachr. Akad. Wiss. Göttingen, Math. Phys. Kl. II, pp. 1-20.
Moser, J.K.: 1966, 'On the Theory of Quasi Periodic Motions', Siam Review 8, no. II, pp. 145-172.
Moser, J.K.: 1967, 'Convergent Series Expansions for Quasi Periodic Motions,' Math. Ann. 169, pp. 136-176.
Moser, J.K.: 1969, 'On the Construction of Almost Periodic Solutions for Ordinary Differential Equations', Proc. Int. Conf. on Functional Analysis and Related Topics, pp. 60-67. Tokyo.
Moser, J.K.: 1973, 'Stable and Random Motions in Dynamical Systems', Annals of Mathematical Studies no. 77, Princeton.
Oxtoby, J.: 1970, Measure and Category, Springer-Verlag.

Palis, J. and F. Takens: 1977, 'Topological Equivalence in Normally Hyperbolic Dynamical Systems', Topology 16, pp. 335-345.

Poincaré, H.: 1897, Thèse, in Oeuvres 1, Gauthier-Villars 1928, pp. LIX-CXXIX.

Pöschel, J.: 1981, Integrability of Hamiltonian Systems on Cantor Sets, Preprint ETH-Zürich.

Robinson, R.C.: 1970, 'Generic Properties of Conservative Systems I, II', Amer. J. Math. 92, pp. 562-603, 879-906.

Ruelle, D. and F. Takens: 1971, 'On the Nature of Turbulence', Comm. Math. Phys. 20, pp. 167-192 and 23, pp. 343-344.

Rüssmann, H.: 1970, 'Über Invariante Kurven Differentierbarer Abbildungen eines Kreisringes', Nachr. Akad. Wiss. Göttingen, Math. Phys. Kl. II, pp. 67-105.

Scheurle, J.: 1980, Bifurcations of Quasi-periodic Solutions from Equilibrium Points of Reversible Dynamical Systems, Preprint University of Stuttgart.

Scheurle, J. and J.E. Marsden: 1982, Bifurcation to Quasi-periodic Tori in the Interaction of Steady State and Hopf Bifurcations, Preprint University of California, Berkely.

Smale, S.: 1967, 'Differentiable Dynamical Systems', Bull. Am. Math. Soc. 73, pp. 747-817.

Takens, F.: 1973, 'A Nonstabilisable Jet of a Singularity of a Vector Field, in M.M. Peixoto (ed.), Dynamical Systems, Acad. Press, pp. 583-598.

Takens, F.: 1974, 'Singularities of Vector Fields', Publ. Math. IHES, 43, pp. 48-100.

Thom, R.: 1972, Stabilité Structurelle et Morphogénèse Benjamin.

This paper appears by kind permission of the Editors of the Nieuw Archief voor Wiskunde.

J. Rappaz

NUMERICAL ANALYSIS OF BIFURCATION PROBLEMS FOR PARTIAL DIFFERENTIAL EQUATIONS.

1. INTRODUCTION.

The aim of this paper is to present a survey of some results recently obtained by the author and by different co-workers on the numerical approximation of bifurcation problems.

In order to justify the abstract theory which will be developped in the following, we begin to consider a model problem. Let $\Omega \subset \mathbb{R}^2$ be a bounded domain and $W_0^{1,p}(\Omega)$ with $2<p<\infty$, be the usual Sobolev space of functions in $L^p(\Omega)$, the first derivatives of which belong to $L^p(\Omega)$, and zero on the boundary $\partial\Omega$ of Ω; we are interested to find pairs $(\lambda,u) \in \mathbb{R}\times W_0^{1,p}(\Omega)$ such that

$$-\Delta u = \lambda e^u \text{ in } \Omega \tag{1.1}$$

Concerning a numerical approximation of Problem (1.1), we consider a conforming finite element method by defining a family of finite dimensional subspaces V_h of $W_0^{1,p}(\Omega)$, h beeing a positive parameter which tends to zero (see [9] for instance), and by solving the nonlinear problems: find pairs $(\lambda,u) \in \mathbb{R}\times V_h$ such that

$$\int_\Omega \text{grad}u.\text{grad}v \, dx = \lambda \int_\Omega e^u.v \, dx , \quad \forall v \in V_h . \tag{1.2}$$

Let now $T: L^2(\Omega) \to W_0^{1,2}(\Omega)$ and $T_h: L^2(\Omega) \to V_h$ be the continuous linear operators defined, for $f \in L^2(\Omega)$, by the relations:

$$\int_\Omega \text{grad}(Tf).\text{grad}v \, dx = \int_\Omega f.v \, dx , \quad \forall v \in W_0^{1,2}(\Omega), \tag{1.3}$$

$$\int_\Omega \text{grad}(T_hf).\text{grad}v \, dx = \int_\Omega f.v \, dx , \quad \forall v \in V_h ; \tag{1.4}$$

T is the inverse of the operator $(-\Delta)$ with zero boundary condition and, by assuming some regularity conditions on $\partial\Omega$,

C. P. Bruter et al. (eds.), Bifurcation Theory, Mechanics and Physics, 209–223.

T is a continuous linear operator from $L^2(\Omega)$ into $W^{2,2}(\Omega)$ and consequently from $L^2(\Omega)$ into $W_o^{1,p}(\Omega)$. It is easy to show that Problems (1.1) and (1.2) are quite equivalent to find $(\lambda,u) \in \mathbb{R}\times W_o^{1,p}(\Omega)$, respectively $(\lambda,u) \in \mathbb{R}\times V_h$, which are solutions of

$$F(\lambda,u) \equiv u - \lambda T e^u = o\,, \tag{1.5}$$

respectively

$$F_h(\lambda,u) \equiv u - \lambda T_h e^u = o. \tag{1.6}$$

Let us notice that if a pair $(\lambda,u) \in \mathbb{R}\times W_o^{1,p}(\Omega)$ satisfies $F_h(\lambda,u) = o$, then u belongs to the range of T_h which is V_h so that we can equivalently solve the approximate problem (1.6) in $\mathbb{R}\times W_o^{1,p}(\Omega)$.

The general context which is induced by this model example is the following.

Let V and W be two real Banach spaces with norms $||\cdot||_V$ and $||\cdot||_W$ respectively, G: $\mathbb{R}\times V \to W$ be a $C^r (r\geq 2)$ nonlinear mapping such that the r-th Frechet derivative $D^r G$ is bounded on all bounded subsets of $\mathbb{R}\times V$, $T \in L(W,V)$ be a continuous linear operator from W into V; we are interested by the problem: find $(\lambda,u) \in \mathbb{R}\times V$ such that

$$F(\lambda,u) \equiv u + TG(\lambda,u) = o. \tag{1.7}$$

If $T_h \in L(W,V)$, $o < h \leq 1$, is a family of continuous linear operators (h will tend to zero) satisfying:

$$\lim_{h\to o} ||T - T_h||_{L(W,V)} = o\,, \tag{1.8}$$

we can consider the approximate problems: find $(\lambda,u) \in \mathbb{R}\times V$ solutions of the equation

$$F_h(\lambda,u) \equiv u + T_h G(\lambda,u) = o. \tag{1.9}$$

With regard to the numerical analysis, the range V_h of the operator T_h is a finite dimensional subspace of V and we can exhibit a basis $\chi_1,\chi_2,\ldots,\chi_N$ of V_h ($N = \dim V_h$) in a such way that the approximate problem (1.9) is equivalent to a finite dimensional problem

$$\mathcal{F}_h(\lambda,y) = o \tag{1.10}$$

where $F_h: \mathbb{R}^{N+1} \to \mathbb{R}^N$ is the C^r mapping of components F_{ih}, $1 \leq i \leq N$, satisfying the relation:

$$F_h(\lambda, \sum_{i=1}^{N} y_i \chi_i) = \sum_{i=1}^{N} F_{ih}(\lambda,y)\chi_i \qquad (1.11)$$

with y_i, $1 \leq i \leq N$, beeing the components of $y \in \mathbb{R}^N$.

We first remark that this general context is justified by our model example above in which we set $V = W_0^{1,p}(\Omega)$, $2<p<\infty$, $W = L^2(\Omega)$ and $G(\lambda,u) = -\lambda e^u$. However the form of problems (1.7),(1.9),(1.10) is sufficiently large in order to include a variety of classical approximation schemes such as conforming finite element methods, mixed finite element methods, spectral methods,... for nonlinear partial differential equations.

Setting the two problems (1.7) and (1.9); we now can ask the following questions:

(i) If $\Gamma = \{(\lambda(\alpha),u(\alpha)): |\alpha| \leq \alpha_0\}$ is a part of a solution branch of Problem (1.7) and if we assume Hypothesis (1.8), is it true that, for h small enough, there exists a part $\Gamma_h = \{(\lambda_h(\alpha),u_h(\alpha)): |\alpha| \leq \alpha_0\}$ of a solution branch of Problem (1.9) which is near from Γ, even if Γ passes through a turning point or a bifurcation point of (1.7)? If it is the case, is it possible to check some optimal error estimates for $|\lambda(\alpha) - \lambda_h(\alpha)|$ and $\|u(\alpha) - u_h(\alpha)\|$ where $\|\cdot\|$ is the norm in V or an other norm?

(ii) Do all solutions of the approximate problem (1.9) belong to a "small" neighborhood of the solutions of the exact problem (1.7) when h is sufficiently small?

(iii) Are we able to give "good" algorithms in order to solve the finite dimensional problem (1.10)?

In this lecture we don't talk about questions (ii) and (iii); for a work in connection with the question (ii) which is negative in general, see [2,3] for instance; the reader interested by question (iii) is refered to the paper of H.B. Keller [17] and the references therein.

In Section 2 of this paper, we give some results of [6,7,8,

13,22] in order to treat question (i). By applying our general results to our model example (1.1),(1.2) in the case where V_h is a finite element space constructed with piecewise polynomial functions of degree 1, we obtain the following result. If $\Gamma = \{(\lambda(\alpha),u(\alpha)) : |\alpha| \leq \alpha_0\}$ is a part of a solution branch of (1.1) such that for $\alpha \neq o$ $(\lambda(\alpha),u(\alpha))$ is a regular solution of (1.1) and $(\lambda(o),u(o))$ is a turning point of (1.1) (see [21]), then in all neighborhood of Γ in $\mathbb{R}\times W_o^{1,p}(\Omega)$, there exists, for $h \leq h_o$ small enough, a part $\Gamma_h = \{(\lambda_h(\alpha),u_h(\alpha)) : |\alpha| \leq \alpha_0\}$ of a solution branch of (1.2); moreover we have the error estimates for $|\alpha| \leq \alpha_o$:

$$|\lambda^{(\ell)}(\alpha) - \lambda_h^{(\ell)}(\alpha)| \leq ch^2 , \tag{1.12}$$

$$\| u^{(\ell)}(\alpha) - u_h^{(\ell)}(\alpha)\|_{W^{1,2}(\Omega)} \leq ch , \tag{1.13}$$

$$\| u^{(\ell)}(\alpha) - u_h^{(\ell)}(\alpha)\|_{L^2(\Omega)} \leq ch^2 , \tag{1.14}$$

$$\| u^{(\ell)}(\alpha) - u_h^{(\ell)}(\alpha)\|_{L^\infty(\Omega)} \leq ch^2|\ell nh| , \tag{1.15}$$

where $\lambda^{(\ell)}, u^{(\ell)}$ is the ℓ-th derivative of λ and u with respect to α, $\ell = o,1,2,..,c = c(\ell)$ is a constant independent of h and α.

In section 3, we give some remarks on possible extensions of our theory and we comment some open problems.

Throughout the paper, we shall constantly use the following notations. Given a smooth function Φ from $X\times Y$ into Z for some Banach spaces X,Y,Z, we denote by $D^m\Phi(x,y) \in L_m(X\times Y,Z)$ the m-th total derivative of Φ at the point $(x,y) \in X\times Y$, where $L_m(X\times Y,Z)$ is the space of all continuous m-linear mappings from $X\times Y$ into Z. We denote by $D_x\Phi(x,y)$, $D_y\Phi(x,y)$, $D^2_{xx}\Phi(x,y)$, $D^2_{xy}\Phi(x,y)$.. the corresponding partial derivatives.

We shall also use the classical Sobolev spaces $W^{m,p}(\Omega)$ and $H^m(\Omega) = W^{m,2}(\Omega)$ with the norms $\|\cdot\|_{m,p} = \|\cdot\|_{W^{m,p}(\Omega)}$ and $\|\cdot\|_m = \|\cdot\|_{H^m(\Omega)}$.

2. ABSTRACT RESULTS.

In this section we consider the problems (1.7) and (1.9) by assuming the hypothesis (1.8).

Moreover, we introduce a real Banach space $H \supset V$ of norm $\|\cdot\|_H$; we suppose that the injection from V into H is continuous and that the C^r mapping $G:\mathbb{R}\times V \to W$ satisfies the following hypothesis:

For all $1 \leqslant m \leqslant \ell \leqslant r$, *for all* $\lambda \in \mathbb{R}$ *and for all* $u, u_1, u_2, \ldots, u_{m-1} \in V$, *the linear continuous operator*

$$A(\lambda,u,u_1,\ldots,u_{m-1}) \equiv D^{\ell}_{\underbrace{\lambda\lambda\ldots\lambda}_{(\ell-m)\,\text{times}}\ \underbrace{u\ldots u}_{m\,\text{times}}} G(\lambda,u)(u_1,u_2,\ldots,u_{m-1},\cdot)$$

from V *into* W *is extensible to a continuous linear operator from* H *into* W *and, for all bounded subset* B *of* $\mathbb{R}\times V^m$, *there exists a constant* c *such that*

$$\sup_{(\lambda,u,u_1,\ldots,u_{m-1})\in B} \|A(\lambda,u,u_1,\ldots,u_{m-1})\|_{L(H,W)} \leqslant c. \qquad (2.1)$$

Remark 2.1. If we set H = V, this hypothesis is naturally satisfied because we have supposed that the C^r mapping $G:\mathbb{R}\times V \to W$ is such that D^rG is bounded on all bounded subsets of $\mathbb{R}\times V$.

Remark 2.2. In our model example of Section 1 where $V = \overset{\circ}{W}{}^{1,p}(\Omega)$, $2 < p < \infty$, $W = L^2(\Omega)$ and $G(\lambda,u) = -\lambda e^u$, we can easily verify, by using the Sobolev imbedding theorem $(\overset{\circ}{W}{}^{1,p}(\Omega) \subset C^0(\bar{\Omega}))$, that Hypothesis (2.1) holds when we choose $H = \overset{\circ}{H}{}^1(\Omega)$ or $H = L^2(\Omega)$ or $H = L^{\infty}(\Omega)$.

Now we are able to answer at question (i) of Section 1 and to establish some error estimates in $\mathbb{R}\times H$ between the solutions of the exact problem (1.7) and those of the approximate problem (1.9).

2.1. Regular solution branches.

In this subsection we are concerned with the approximation of branches $\{(\lambda,u(\lambda)) : \lambda \in \Lambda\}$ of nonsingular solutions of (1.7), (here Λ is a compact interval of $\mathbb{R}$), i.e. which satisfy the following properties:

$\lambda \in \Lambda \to u(\lambda) \in V$ *is a continuous function from* Λ *into* V, (2.2)

$$F(\lambda,u(\lambda)) = o , \quad \forall\lambda \in \Lambda , \tag{2.3}$$

$$D_uF(\lambda,u(\lambda)) \text{ is an isomorphism of } V,\ \forall\lambda \in \Lambda \tag{2.4}$$

Remark that, by using the implicit function theorem, the function $\lambda \to u(\lambda)$ is of class C^r.

Theorem 1. *(see [6] and [7, Theorem 1]). We suppose that the hypotheses* (1.8), (2.2) - (2.4) *hold. Then there exist positive constants* h_o, ε *and for* $h \leq h_o$ *a unique function* $\lambda \in \Lambda \to u_h(\lambda) \in V$ *of class* C^r *such that:*

$$F_h(\lambda,u_h(\lambda)) = o , \quad \forall\lambda \in \Lambda , \tag{2.5}$$

$$\| u(\lambda) - u_h(\lambda)\|_V \leq \varepsilon , \quad \forall\lambda \in \Lambda ; \tag{2.6}$$

moreover we have

$$\lim_{h\to o}\ \sup_{\lambda\in\Lambda} \| u^{(\ell)}(\lambda) - u_h^{(\ell)}(\lambda)\|_V = o , \tag{2.7}$$

for all $o \leq \ell \leq r-1$, *where* $u^{(\ell)}, u_h^{(\ell)}$ *are the* ℓ*-th derivatives of* u *and* u_h *with respect to* λ.

Theorem 1 is obtained by first proving a generalization of the implicit function theorem and then using continuation arguments; these results may be viewed as a generalization of pointwise results of [16].

Theorem 2. *(see [6] and [7, Theorem 1] in the case where* $V = H$; *see [13] and [22] for the general case). In addition of hypotheses of Theorem 1 we assume that* (2.1) *holds. Then for all* $o \leq \ell \leq r-1$ *there exists a constant* c *such that for all* $\lambda \in \Lambda$ *and all* $h \leq h_o$ *we have the error estimates:*

$$\|u^{(\ell)}(\lambda) - u_h^{(\ell)}(\lambda)\|_H \leq c \sum_{j=0}^{\ell} \|(T-T_h)\frac{d^j}{d\lambda^j}G(\lambda,u(\lambda))\|_H . \tag{2.8}$$

Error estimates (2.8) are obtained (see [13,22]) by setting two other problems, i.e. find $(\lambda,w) \in \mathbb{R}\times W$ such that $L(\lambda,w) \equiv w + G(\lambda,Tw) = o$ or $L_h(\lambda,w) \equiv w + G(\lambda,T_hw) = o$, and by establishing some specific connections between these two problems and Problems (1.7) and (1.9).

Remark 2.3. In a large number of practical problems (conforming approximations by Galerkin methods for nonlinear elliptic equations), the operators T_h are given by $T_h = \Pi_h T$ where $\Pi_h: V \to V_h$ are projectors onto the finite dimensional subspaces V_h of V. In this case, the error estimates (2.8) become

$$\|u^{(\ell)}(\lambda) - u_h^{(\ell)}(\lambda)\|_H \leqslant c \sum_{j=0}^{\ell} \|(I - \Pi_h)u^{(j)}(\lambda)\|_H \tag{2.9}$$

since $u(\lambda) = - TG(\lambda,u(\lambda))$, (I is the identity in V); under smoothness conditions for $u^{(j)}$, it is classical to obtain orders of convergence for the right member of (2.9) when we use a finite element method (see [9] for instance).

Remark 2.4. We go back to our model example (1.1),(1.2) in which we choose V_h as a finite element space constructed with piecewise polynomial functions of degree 1. Assume that $\lambda \in \Lambda \to u(\lambda) \in \overset{o}{W}{}^{1,p}(\Omega)$, $2 < p < \infty$, is a regular solution branch of Problem (1.1). By applying Theorem 1, there exist $h_o, \varepsilon > o$ and for $h \leqslant h_o$, there exists a unique function $\lambda \in \Lambda \to u_h(\lambda) \in V_h$ which is of class C^∞ and which is a solution branch of the approximate problem (1.2) with $\|u(\lambda) - u_h(\lambda)\|_{1,p} \leqslant \varepsilon$; moreover we have $\lim_{h \to o} \sup_{\lambda \in \Lambda} \|u(\lambda) - u_h(\lambda)\|_{1,p} = o$. If Ω is a convex polygonal domain, we have $u(\lambda) \in H^2(\Omega)$, $\forall \lambda \in \Lambda$ and by using (2.8) successively with $H = H^1(\Omega)$, $H = L^2(\Omega)$ and $H = L^\infty(\Omega)$ together with Remarks 2.2, 2.3 and classical results on the approximation of linear elliptic problems, we obtain the following error estimates for $\ell = o,1,2,\ldots$:

$$\|u^{(\ell)}(\lambda) - u_h^{(\ell)}(\lambda)\|_{H^1(\Omega)} \leqslant ch, \tag{2.10}$$

$$\|u^{(\ell)}(\lambda) - u_h^{(\ell)}(\lambda)\|_{L^2(\Omega)} \leqslant ch^2, \tag{2.11}$$

$$\|u^{(\ell)}(\lambda) - u_h^{(\ell)}(\lambda)\|_{L^\infty(\Omega)} \leqslant ch^2|\ell n h| \quad \text{if } u^{(\ell)} \in W^{2,\infty}(\Omega), \tag{2.12}$$

where $c = c(\ell)$ is independent of λ and h.

2.2 Simple limit points.

In this subsection, we assume that $(\lambda_0,u_0) \in \mathbb{R}\times V$ is a simple limit point of Problem (1.7), i.e. (λ_0,u_0) satisfies the following relations:

$$F(\lambda_0,u_0) = o, \tag{2.13}$$

$$D_uF(\lambda_o,u_o) \text{ is a Fredholm operator of index zero with } \dim\operatorname{Ker}(D_uF(\lambda_o,u_o)) = 1, \tag{2.14}$$

$$D_\lambda F(\lambda_0,u_0) \notin \operatorname{Range}(D_uF(\lambda_o,u_o)). \tag{2.15}$$

In this case, if $o \neq \varphi^* \in V'$ (=dual space of V) is such that $D_uF(\lambda_o,u_o)^*\varphi^* = o$, where $D_uF(\lambda_o,u_o)^* \in L(V',V')$ is the adjoint operator of $D_uF(\lambda_o,u_o) \in L(V,V)$, we can show (see [7] for instance) there exist a positive number α_0 and two C^r mappings $\lambda: \alpha\in [-\alpha_0,\alpha_0] \to \lambda(\alpha) \in \mathbb{R}$ and $u: \alpha\in [-\alpha_0,\alpha_0] \to u(\alpha) \in V$ such that

$$F(\lambda(\alpha),u(\alpha)) = o\ , \quad \forall|\alpha| \leqslant \alpha_0\ , \tag{2.16}$$

$$\lambda(o) = \lambda_0 \text{ and } u(o) = u_0\ , \tag{2.17}$$

$$\langle u(\alpha) - u_0\ , \varphi^*\rangle = \alpha\ , \quad \forall|\alpha| \leqslant \alpha_0\ , \tag{2.18}$$

where $\langle .,.\rangle$ denotes the duality pairing between V and V'. Note here that the functions $\lambda(\alpha)$ and $u(\alpha)$ are univoquely determined by the relations (2.16) to (2.18).

Theorem 3. *(see [7]). We suppose that the hypotheses* (1.8) (2.13) - (2.15) *hold. Then there exist positive numbers* h_0 *and* ε *and for* $h \leqslant h_0$ *two unique* C^r *mappings* $\lambda_h: \alpha\in[-\alpha_0,\alpha_0]\to \lambda_h(\alpha) \in \mathbb{R}$ *and* $u_h: \alpha \in [-\alpha_0,\alpha_0] \to u_h(\alpha) \in V$ *such that*

$$F_h(\lambda_h(\alpha),u_h(\alpha)) = o\ , \quad \forall|\alpha| \leqslant \alpha_0\ , \tag{2.19}$$

$$\langle u_h(\alpha) - u_0\ , \varphi^*\rangle = \alpha\ , \quad \forall|\alpha| \leqslant \alpha_0\ , \tag{2.20}$$

$$|\lambda(\alpha) - \lambda_h(\alpha)| + \| u(\alpha) - u_h(\alpha)\|_V \leqslant \varepsilon\ , \ \forall|\alpha| \leqslant \alpha_0\ ; \tag{2.21}$$

moreover we have

$$\lim_{h\to o} \sup_{\alpha\in[-\alpha_0,\alpha_0]} |\lambda^{(\ell)}(\alpha)-\lambda_h^{(\ell)}(\alpha)|+\| u^{(\ell)}(\alpha)-u_h^{(\ell)}(\alpha)\|_V = o \tag{2.22}$$

for all $0 \leqslant \ell \leqslant r-1$.

Mention here that for proving this theorem, we use the Lyapunov-Schmidt procedure and the same techniques as in Theorem 1.

Theorem 4. (*see* [7] *for the case where* $H = V$ *and see* [13,22] *for the general case*). *In addition of hypotheses of Theorem* 3 *we assume that* (2.1) *holds. Then for all* $0 \leqslant \ell \leqslant r-1$ *there exists a constant* c *such that for all* $|\alpha| \leqslant \alpha_0$ *and all* $h \leqslant h_0$ *we have the error estimates:*

$$\begin{aligned} &|\lambda^{(\ell)}(\alpha) - \lambda_h^{(\ell)}(\alpha)| + \| u^{(\ell)}(\alpha) - u_h^{(\ell)}(\alpha) \|_H \leqslant \\ &\leqslant c \sum_{j=0}^{\ell} \|(T - T_h) \frac{d^j}{d\alpha^j} G(\lambda(\alpha), u(\alpha))\|_H . \end{aligned} \qquad (2.23)$$

For proving Theorem 4, we use the same technique as in the proof of Theorem 2.

Remark 2.5. If we suppose that (λ_0, u_o) is a nondegenerate turning point, we prove, by using (2.22), that the approximate problem (1.9) has indeed a unique turning point (λ_h^o, u_h^o) in a neighborhood of (λ_0, u_0) in $\mathbb{R} \times V$ and we can derive estimates of $|\lambda_0 - \lambda_h^o|$ and $\| u_0 - u_h^o \|_H$ which are of the same form as (2.23). We still mention that, in many practical cases, the error bound for $|\lambda_o - \lambda_h^o|$ can be improved (see [7, Theorem 5]).

Remark 2.6. If we apply our Theorem 4 to our model example (1.1) and (1.2) in a same way as in Remark 2.4, we prove the error estimates (1.12) to (1.15).

2.3. Simple bifurcation points.

Now we assume that $(\lambda_0, u_o) \in \mathbb{R} \times V$ satisfies the hypotheses (2.13) and (2.14) and we replace (2.15) by

$$D_\lambda F(\lambda_0, u_0) \in \text{Range}(D_u F(\lambda_0, u_0)); \qquad (2.24)$$

we recall very briefly the Lyapunov-Schmidt procedure in order to reduce Problem (1.7) to the bifurcation equation.

Let $V_1 = \mathrm{Ker}(D_uF(\lambda_o,u_o))$, ($\dim V_1 = 1$), $V_2 = \mathrm{Range}(D_uF(\lambda_o,u_o))$, $\tilde{V}_1$ and $\tilde{V}_2$ be closed subspaces of V such that $V = V_1 \oplus \tilde{V}_2 = \tilde{V}_1 \oplus V_2$; let $Q: V \to V_2$ be the projector associated to the decomposition $V = \tilde{V}_1 \oplus V_2$; Equation (1.7) is equivalent to the system

$$QF(\lambda,u) = o\,, \qquad (I-Q)F(\lambda,u) = o\,. \tag{2.25}$$

Since $QD_uF(\lambda_o,u_o)$ defines an isomorphism from $\tilde{V}_2$ onto V_2, we can use the implicit function theorem to solve the first equation of (2.25); we obtain the existence of the C^r mapping $v\colon \mathbb{R}\times V_1 \to \tilde{V}_2$ defined in a neighborhood of $(\lambda_0,o) \in \mathbb{R}\times V_1$ and such that $QF(\lambda,u_0+\sigma+v(\lambda,\sigma)) = o$ with $v(\lambda_0,o) = o$. In order to solve Problem (1.7) in a neighborhood of (λ_0,u_0), it remains to consider the bifurcation equation

$$(I-Q)F(\lambda,u_0+\sigma+v(\lambda,\sigma)) = o$$

which is equivalent to

$$f(\lambda,x) \equiv \langle F(\lambda,u_0+x\varphi+v(\lambda,x\varphi)),\varphi^*\rangle = o, \tag{2.26}$$

where $o \neq \varphi \in V_1$ and φ^* is defined as in Subsection 2.2. We can easily verify that $f(\lambda_0,o) = \frac{\partial}{\partial\lambda} f(\lambda_0,o) = \frac{\partial}{\partial x} f(\lambda_0,o) = o$ and we assume in all the following that the determinant of the hessian matrix of f at the point (λ_0,o) is negative, i.e.

$$\frac{\partial^2}{\partial\lambda^2} f(\lambda_0,o)\cdot\frac{\partial^2}{\partial x^2} f(\lambda_0,o) < \left(\frac{\partial^2}{\partial x\partial\lambda} f(\lambda_0,o)\right)^2. \tag{2.27}$$

In this situation, it is well known (see the Morse Lemma) that Equation (2.26) possesses two C^{r-2} branches of solutions in a neighborhood of (λ_0,o) of $\mathbb{R}^2$; these branches can be parametrized by $t\in[-t_0,t_0] \to (\lambda_i(t),x_i(t)) \in \mathbb{R}^2$, $i=1,2$, i.e. $f(\lambda_i(t),x_i(t)) = o$, $\lambda_i(o) = \lambda_0$, $x_i(o) = o$, $|x_i'(o)|+|\lambda_i'(o)| \neq o$, $i=1,2$, so that for $|t| \leqslant t_0$:

$$(\lambda_i(t),\ u_i(t) \equiv u_0 + x_i(t)\varphi + v(\lambda_i(t),\ x_i(t)\varphi)),\ i=1,2,$$

are C^{r-2} solution branches of (1.7) in a neighborhood of (λ_0,u_0), i.e. we have for $|t| \leqslant t_0$: $F(\lambda_i(t),u_i(t)) = o$, $\lambda_i(o) = \lambda_0$, $u_i(o) = u_0$, $i=1,2$.

Assume now that $r \geqslant 4$; by using Hypothesis (1.8) and the

Lyapunov-Schmidt procedure on Problem (1.9) in a same way as for the exact problem, we can reduce (see [7,8] for instance) the equation $F_h(\lambda,u) = o$ to the equation

$$f_h(\lambda,x) \equiv \langle F_h(\lambda,u_o + x\varphi + v_h(\lambda,x\varphi)),\varphi^*\rangle = o \qquad (2.28)$$

where $v_h(\lambda,x\varphi) \in \tilde{V}_2$ satisfies $QF_h(\lambda,u_o + x\varphi + v_h(\lambda,x\varphi)) = o$ for (λ,x) in a neighborhood of (λ_o,o). Now, by using Hypothesis (1.8), it is possible to show (see [8]) the existence of a point $(\lambda_{oh},x_{oh}) \in \mathbb{R}^2$ such that $\lim_{h\to o} \lambda_{oh} = \lambda_o$, $\lim_{h\to o} x_{oh} = o$ and $\frac{\partial}{\partial\lambda} f_h(\lambda_{oh},x_{oh}) = \frac{\partial}{\partial x} f_h(\lambda_{oh},x_{oh}) = o$ for all $h \leqslant h_o$ sufficiently small. In general (λ_{oh},x_{oh}) is not a zero of f_h and we must distinguish two possibilities: $f_h(\lambda_{oh},x_{oh}) = o$ and $f_h(\lambda_{oh},x_{oh}) \neq o$. The second possibility is discussed in Remark 2.7. In the following we consider the case where $f_h(\lambda_{oh},x_{oh}) = o$; sufficient practical conditions (symmetry-breaking bifurcation and bifurcation from the trivial branch) can be given for this situation occurs (see [8] for instance).

Theorem 5. *(see [8]). We suppose that* $r \geqslant 4$ *and that the hypotheses* (1.8), (2.13), (2.14), (2.24) *and* (2.27) *hold; in addition we assume that* $f_h(\lambda_{oh},x_{oh}) = o$. *Then there exist positive numbers* h_o *and* ε *and for* $h \leqslant h_o$ *two* C^{r-2} *mappings* $t \in [-t_o,t_o] \to (\lambda_{ih}(t),x_{ih}(t)) \in \mathbb{R}^2$, $i = 1,2$, *such that* $\lambda_{ih}(o) = \lambda_{oh}$, $x_{ih}(o) = x_{oh}$ *for* $i = 1,2$ *and:*

$$f_h(\lambda_{ih}(t),\, x_{ih}(t)) = o, \quad \forall |t| \leqslant t_o\,, \quad i = 1,2, \qquad (2.29)$$

$$|\lambda_{ih}(t)-\lambda_i(t)|+|x_{ih}(t)-x_i(t)| \leqslant \varepsilon, \ \forall |t| \leqslant t_o, i=1,2. \qquad (2.30)$$

Moreover, by setting

$$u_{ih}(t) = u_o + x_{ih}(t)\varphi + v_h(\lambda_{ih}(t),x_{ih}(t)\varphi) \quad i=1,2,$$

we have

$$F_h(\lambda_{ih}(t),u_{ih}(t)) = o, \quad \forall |t| \leqslant t_o\,, \quad i = 1,2, \qquad (2.31)$$

and for all $o \leqslant \ell \leqslant r-3$:

$$\lim_{h\to o}\ \max_{i=1,2}\ \sup_{|t|\leqslant t_o} |\lambda_{ih}^{(\ell)}(t)-\lambda_i^{(\ell)}(t)|+\| u_{ih}^{(\ell)}(t) - u_i^{(\ell)}(t)\|_V = o\,. \qquad (2.32)$$

Theorem 6. *(see [8] for the case where* V = H *and see* [13,22] *for the general case). In addition of hypotheses of Theorem* 5 *we assume that* (2.1) *holds. Then for all* $o \leq \ell \leq r-3$ *there exists a constant* c *such that for all* $h \leq h_o$ *we have the error estimates:*

$$\max_{i=1,2} \sup_{|t|\leq t_o} \{|\lambda_{ih}^{(\ell)}(t) - \lambda_i^{(\ell)}(t)| + \| u_{ih}^{(\ell)}(t) - u_i^{(\ell)}(t)\|_H\} \leq$$
$$\leq \max_{i=1,2} \sup_{|t|\leq t_o} \sum_{j=o}^{\ell+1} \|(T - T_h) \frac{d^j}{dt^j} G(\lambda_i(t), u_i(t))\|_H \tag{2.33}$$

Remark 2.7. In the general case where $f_h(\lambda_{oh}, x_{oh}) \neq o$, the approximate problem (1.9) does not have some bifurcation point in a "small" neighborhood of (λ_o, u_o). In this situation, one can prove (see [8]) that there exist a neighborhood $\mathcal{O}$ of the point (λ_o, u_o) in $\mathbb{R}\times V$ and a positive constant h_o such that, for $h \leq h_o$, the set S_h of the solutions of (1.9) contained in $\mathcal{O}$ consists of two C^{r-2} branches which are diffeormorphic to (a part of) a nondegenerate hyperbola. In this case we shall say that we have an imperfect numerical bifurcation and it is possible to estimate the distance between S_h and the set S of the solutions of (1.7) contained in $\mathcal{O}$ (see [8]).

3. REMARKS AND COMMENTS.

By applying the results of Section 2 to the Galerkin approximations of nonlinear variational problems, we can improve and generalize some results of [14,18,19]. Moreover, in [6,7,8, 26] we can see several applications of above results (in the case V = H) to mixed finite element method schemes for the von Karman equations or the Navier-Stokes equations.

The results of Section 2 have been extended so as to cover the case of Hopf bifurcation [1,10] and have been generalized to the case of multiple bifurcation [23,24,25].

An other important generalization of these results can be found in [11], where a general theory for the approximation of regular and bifurcating branches of solutions of nonlinear equations is presented. This theory can be applied to numerous problems, including differential equations on unbounded domains (see [12]; T is not compact and Hypothesis (1.8) does not hold), in connection with various numerical algorithms (for example: Galerkin methods with numerical integration).

In [4,5] we can find an analysis of the relationships between the imperfections introduced by the discretization of nonlinear problems and the theory of perturbations of singularities, in particular from the point of view of Golubitsky-Schaeffer [15]. (see also [14] for some particular cases).

Finally we mention some open problems which are the numerical approximation of bifurcating solutions from continuous spectrum, the approximation of sheets of solutions of nonlinear problems, the approximation of nondifferentiable problems...

For a complete bibliography on numerical methods for bifurcation problems, see [20].

Jacques Rappaz
Département de Mathématiques
Ecole Polytechnique Fédérale Lausanne
CH 1015 Lausanne-Dorigny(Switzerland)

REFERENCES

[1] Bernardi, C.: 1981,'Approximation numérique d'une bifurcation de Hopf', C.R. Acad.Sc. Paris 292 Série I, 103-106.

[2] Beyn,W.-J. and Lorenz, J.: 'Spurious solutions for discrete superlinear boundary value problems', to appear in Computing.

[3] Bohl, E.: 1979 'On the bifurcation diagram of discrete analogues of ordinary bifurcation problems', Math. Meth. Appl. Sci.1, 566-571.

[4] Brezzi, F.,Descloux, J., Rappaz, J., Zwahlen, B.:'Numerical approximation of an imperfect bifurcation problem', to appear.

[5] Brezzi, F., Fujii, H.: 'Numerical imperfections and perturbations in the approximation of nonlinear problems', to appear.

[6] Brezzi, F., Rappaz, J., Raviart, P.A.: 1980, 'Finite dimensional approximation of nonlinear problems; Part I: branches of nonsingular solutions', Numer. Math. 36, 1-25.

[7] Brezzi, F., Rappaz, J., Raviart, P.A.: 1981, 'Finite dimensional approximation of nonlinear problems; Part II: limit points', Numer. Math. 37, 1-28.

[8] Brezzi, F., Rappaz, J., Raviart, P.A.: 1981, 'Finite dimensional approximation of nonlinear problems; Part III: bifurcation points', Numer. Math. 38, 1-30.

[9] Ciarlet, P.G.: 1978, 'The finite element method for elliptic problems', North-Holland, Studies in Mathematics and its applications.

[10] Descloux, J.: 'Numerical approximation of Hopf bifurcation', to appear.

[11] Descloux, J., Rappaz, J.: 'Approximation of solution branches of nonlinear equations', to appear in RAIRO Numer. Anal.

[12] Descloux, J., Rappaz, J.: 'Numerical approximation of a nonlinear Sturm-Liouville problem on an infinite interval', to appear in Teubner, Proceedings of Equadiff V.

[13] Descloux, J., Rappaz, J., Scholz, 'On error estimates for the approximation of nonlinear problems', to appear.

[14] Fujii, H. and Yamaguti, M.: 1980, 'Structure of singularities and its numerical realization in nonlinear elasticity', J. Math. Kyoto Univ. 20-3, 489-590.

[15] Golubitsky, M., Schaeffer, D.: 1979, 'A theory for imperfect bifurcation via singularity theory', Comm. Pure Appl. Math. 32, 21-98.

[16] Keller, H.B.: 1975, 'Approximation methods for nonlinear problems with applications to two-point boundary value problems', Math. Comp. 29, 464-474.

[17] Keller, H.B.: 1977, 'Numerical solution of bifurcation and nonlinear eigenvalue problems', in P.H. Rabinowitz (ed), Applications of bifurcation theory, Academic Press, New-York, pp.359-384.

[18] Kesavan, S.: 1979, 'La méthode de Kikuchi appliquée aux équations de von'Kármán', Numer. Math. 32, 209-232.

[19] Kikuchi, F.: 1976, 'An iterative finite element scheme for bifurcation analysis of semilinear elliptic equations', Report Inst. Space. Aero. Sc. 542, Tokyo Univ.

[20] Mittelmann, H.D., Weber, H.: 1981, 'A bibliography on numerical methods for bifurcation problems', Report Dpt. Math., Dortmund University.

[21] Paumier, J.C.: 1981, 'Une méthode numérique pour le calcul des points de retournement. Application à un problème aux limites non-linéaire', Numer. Math. 37, 433-462.

[22] Rappaz, J.: 'Estimations d'erreur dans différentes normes pour l'approximation de problèmes de bifurcation', to appear.

[23] Rappaz, J., Raugel, G.: 1981, 'Finite dimensional approximation of bifurcation problems at multiple eigenvalues', Report No 71, Centre Math. Appl., Ecole Polytechnique Paris.

[24] Rappaz, J., Raugel, G.: 'Approximation of double bifurcation points for nonlinear eigenvalue problems', to appear in J.R. Whiteman (ed), MAFELAP 81, Academic Press.

[25] Raugel, G.: 1982, 'Finite dimensional approximation of bifurcation problems in presence of symmetries', Report No 81, Centre Math. Appl., Ecole Polytechnique Paris.

[26] Reinhart, L.: 'On the numerical analysis of the von Kármán equations: mixed finite element approximation and continuation techniques', to appear in Numer.Math.

Martin Golubitsky*

THE BÉNARD PROBLEM, SYMMETRY AND THE LATTICE OF ISOTROPY SUBGROUPS

INTRODUCTION

In this lecture I would like to describe some of the effects that are forced on steady state bifurcation problems by the existence of a group of symmetries. I shall discuss this relationship between symmetry and bifurcation by describing several mathematical problems which are motivated by the Bénard problem.

The Bénard problem in its simplest form is the study of the transition from pure conduction to convective motion in a contained fluid heated on (part of) its boundary. The model equations which lie behind the analysis are the Navier-Stokes equations in the Boussinesq approximation. The purpose of this exposition is to indicate the type of information that can be obtained through the use of singularity theory and group theory. For this reason the exact form of the Boussinesq equations is not needed and they will not be presented. The interested reader can consult the paper by Fauvre and Libchaber [1983] in this volume or the extensive and very interesting work of Busse [1962, 1975, 1978].

The specific results outlined below rely for their proofs on the machinery of group theory and singularity theory plus very extensive calculations. What is remarkable is that after this effort has been expended the final answer has a delightful and compelling organization based on the lattice of isotropy subgroups of the given group representation. It is our intention to emphasize this relationship throughout.

The paper is divided into four sections. The first three sections concern specific realizations of the Bénard problem while the last section presents certain general results concerning bifurcation with symmetry.

C. P. Bruter et al. (eds.), Bifurcation Theory, Mechanics and Physics, 225–256.

A. The spherical Bénard problem:

The fluid is contained between two fixed concentric spheres and is heated along the inner sphere. This is a model for convection of the molten layer of the Earth contained between the solid inner core and the mantle. Since the Earth rotates this is a simplified model. The symmetry group for this problem is O(3).

B. The planar Bénard problem with non-symmetric boundary conditions:

The fluid is contained between two parallel planes and is heated from below. Moreover, the boundary conditions on the upper plane are assumed to be different from those on the lower plane. This is the situation found in Bénard's original experiment as there one finds a free boundary on top and a rigid boundary below. One should note that we consider an infinite plane while all experiments are - of course - performed on a finite plane. So the relationship between the mathematics presented here and any given experiment is, at best, arguable.

In addition, the fact that this form of the Bénard problem is posed on the infinite plane makes the mathematical analysis - even a local one near the pure conduction solution - extremely difficult. A popular restriction (Busse [1962], Sattinger [1978]) which alleviates some of the technical difficulties is to look only for solutions which are doubly periodic in the plane. There are, however, several different types of double periodicity possible and it is here that Bénard's experiment serves as a useful guide. In his experiment, Bénard found that convection patterns in the shape of hexagons occur and, moreover, that these hexagons are arranged on the hexagonal lattice - at least away from the boundary. Given this fact, it seems reasonable to look first for solutions which are doubly periodic with respect to the hexagonal lattice. We note, however, that among the many patterns of convection found by experiment there are some which are not doubly periodic with respect to the hexagonal lattice.

We now describe the symmetry group for this problem. The Boussinesq equations have the symmetry of the Euclidean group in the plane generated by rotations, reflections and translations. The assumption of double periodicity implies that the translations act periodically; that is, the 2-torus

T^2 is part of the group of symmetries. The assumption of double periodicity with respect to the hexagonal lattice implies that the Boussinesq equations are left invariant only by those rotations and reflections which preserve the hexagonal lattice. Let D_6 denote the dihedral group of symmetries of the regular hexagon in the plane. Then the group of symmetries for this form of the Bénard problem is $T^2 + D_6$.

C. <u>The planar Bénard problem with symmetric boundary conditions:</u>

This problem is posed like the preceding one with a single exception. We assume that the boundary conditions on the top plane are identical to those on the bottom. This is the case found in experiments where the fluid is contained between two bounding planes of the same type and, moreover, such experiments are performed frequently.

For this form of the Bénard problem the symmetries include $T^2 + D_6$ as above and, in addition, a reflectional symmetry obtained by reflecting the fluid layer about it's midplane. This reflection interchanges top and bottom and - with the assumption on the boundary conditions - leaves the Boussinesq equations invariant. Thus the symmetry group is $T^2 + D_6 + Z_2$.

Each of the first three sections is devoted, in order, to a description of the bifurcation behavior found in the above problems. The exposition involves, in each case, describing how one reduces these problems to the problem of finding the zeroes of a mapping g: $\mathbf{R}^n \to \mathbf{R}^n$ depending on a parameter where g commutes with (a given representation of) the groups indicated above. Specifically we describe how group theory restricts the form of g and how singularity theory allows one to find the zero set of g. We shall describe, in each case, an a postiori relationship between the bifurcation structure found and the lattice of isotropy subgroups of the group.

In the last section we present several general results - some new - which indicate that it may be possible in the future to obtain much of the structure of bifurcation problems directly from a knowledge of the group of symmetries of the given problem. Such a result would give a good beginning to the resolution of the problem of spontaneous symmetry breaking - which is the ultimate mathematical goal of this line of research.

The specific results described in the sections below have been obtained in collaboration with the following individuals: David Schaeffer, Ernesto Buzano, Jim Swift and Edgar Knobloch, and Ian Stewart.

1. THE BÉNARD PROBLEM IN SPHERICAL GEOMETRY

The details of the results outlined in this section are contained in Busse [1975], Chossat [1979], Golubitsky and Schaeffer [1982] and the Thèse D'État of Chossat [1982].

Given a fluid contained between two concentric spheres, let τ denote the temperature difference between the inner and outer spheres. If τ is increased there is a first τ_0 at which point the pure conduction solution looses stability and convective motion begins. In the mathematical formulation of this problem τ_0 is the first value of τ where the linearization of the Boussinesq equations about the pure conduction solution, $L(\tau)$, has a zero eigenvalue. Let $V = \ker L(\tau_0)$. It is known that V is the space of spherical harmonics of order p and that p depends on the aspect ratio η where η is the ratio of the radius of the inner sphere to the radius of the outer sphere. See Chossat [1979,1982].

In particular, if one views the spherical Bénard problem as a model for convection in the Earth's molten inner core then $.25 < \eta < .5$ and V is either the spherical harmonics of order 2 or 4. We consider here the case $\eta \sim .3$ and $V = Y_2$ the spherical harmonics of order 2. So dim V = 5. Moreover SO(3) acts irreducibly on Y_2.

Let $\lambda = \tau - \tau_0$. The Liapunov-Schmidt procedure (or the implicit function theorem) shows that finding all solutions to the Boussinesq equations near the pure conduction solution and $\lambda = 0$ is equivalent to finding the zeroes of a C^∞ mapping

$$g: V \times \mathbf{R} \to V; \; g(A,\lambda) = 0$$

which is defined on some neighborhood of $(0,0) \in V \times \mathbf{R}$ and satisfies

(i) $g(0,\lambda) = 0$ where $0 \in V$ corresponds to the pure conduction solution

(ii) $(d_A g)_{0,0} \equiv 0$ (1.1)

(iii) $g(\gamma \cdot A,\lambda) = \gamma \cdot g(A,\lambda)$ for all $\gamma \in O(3)$.

The important point here is that one is looking for the zeroes of an equivariant mapping near a singular point. We analyse such mappings first by group theory and then using singularity theory.

Another realization of the five dimensional irreducible representation of O(3) is as follows: Let V be the vector space of all 3 × 3 symmetric trace 0 matrices. Note that dim V = 5. Let $\gamma \in O(3)$ act on $A \in V$ by $\gamma A = \gamma A \gamma^t$. Via this linear action O(3) acts irreducibly on V. We prefer using this presentation of the representation as opposed to the presentation on spherical harmonics.

Let **E** be the space of C^∞ equivariant (germs of) mappings $g: V \times \mathbf{R} \to V$; i.e. g satisfies (1.1)(iii) above. Let **I** denote the ring of (germs of) invariant C^∞ functions $f: V \times \mathbf{R} \to \mathbf{R}$; i.e., $f(\gamma \cdot A, \lambda) = f(A, \lambda)$ for all $\gamma \in O(3)$. Then **E** is a module over **I** and one can describe this module structure explicitly.

Proposition 1.1: (a) Let f be in **I**. Then there is a smooth function $p: \mathbf{R}^3 \to \mathbf{R}$ such that

$$f(A,\lambda) = p(u,v,\lambda)$$

where $u = tr(A^2)$ and $v = \det A$.
(b) Let g be in **E**. Then there exist invariant functions p and q such that

$$g(A,\lambda) = p(u,v,\lambda)A + q(u,v,\lambda)(A^2 - \frac{1}{3}tr(A^2)I).$$

For details see Golubitsky and Schaeffer [1982].

Remarks: (a) If g comes from a Liapunov-Schmidt reduction then (1.1)(i) implies that $p(0,0,0) = 0$.
(b) In the Bénard problem the pure conduction solution looses stability at τ_0 (which we have identified in the Liapunov-Schmidt reduction with $\lambda = 0$). Observe that the chain rule applied to (1.1)(iii) implies

$$(dg)_{\gamma\cdot A,\lambda}\gamma = \gamma(dg)_{A,\lambda}.$$

In particular $(dg)_{0,A}\gamma = \gamma(dg)_{0,A}$ for all γ. Now this representation of O(3) on V is absolutely irreducible, the only linear maps commuting with O(3) are scalar multiples of the identity. Thus

$$(dg)_{0,\lambda} = p(0,0,\lambda)I.$$

Since the stability of the pure conduction solution is assumed to change at τ_0 as τ is varied, it is reasonable to assume - and may be computed for the Boussinesq equations - that $f_\lambda(0,0,0) \neq 0$.

(c) In general one might assume that $q(0,0,0) \neq 0$. Chossat [1979] has shown that in one important case $q(0,0,0) = 0$ and this is the basis for his analysis. More precisely, let β_i and β_0 be the thermal conductivities of the materials in the inner sphere and the outer shell respectively. Let ρ_i and ρ_0 denote the respective densities of these materials. One finds that the linearization $L(\lambda_0)$ of the Boussinesq equations is self-adjoint if the constants β and ρ satisfy

$$\beta_i/\beta_0 = \rho_i/\rho_0 \tag{1.2}$$

The identity (1.2) defines the self-adjoint case. Chossat shows that in the self-adjoint case $q(0,0,0) = 0$.

Using singularity theory one can analyse two questions.

(1) (Recognition Problem) When does one have enough information about the Taylor expansion of g to ignore higher order terms?

(2) (Imperfect Bifurcation) Determine all possible perturbations of the given bifurcation problem by finding the universal unfolding.

We give the answers to these questions in two cases.

Proposition 1.2: Let $g \in \mathbf{E}$ satisfy $p(0,0,0) = 0$ and $p_\lambda(0,0,0) \neq 0$. Then

(a) if $g(0,0,0) \neq 0$ then g is O(3)-equivalent to

$$h(A,\lambda) = \varepsilon_1 \lambda A + \varepsilon_2 (A^2 - \frac{1}{3}\mathrm{tr}(A^2)I)$$

where $\varepsilon_1 = \mathrm{sgn}(p_\lambda(0,0,0))$ and $\varepsilon_2 = \mathrm{sgn}(q(0,0,0))$.

(b) if $q(0,0,0) = 0$ and

(i) $p_u(0,0,0) \neq 0$

(ii) $C \equiv p_\lambda q_u - p_u q_\lambda \neq 0$ at $(0,0,0)$

(iii) $D \equiv p_u q_v - p_v q_u \neq 0$ at $(0,0,0)$

then g is O(3)-equivalent to

$$h(A,\lambda) = (\varepsilon_1 u + \varepsilon_2\lambda)A + (\varepsilon_3 u + Dv)(A^2 - \frac{1}{3}\,tr(A^2)I).$$

where $\varepsilon_1 = \text{sgn}\, p_u(0,0,0)$, $\varepsilon_2 = \text{sgn}\, p_\lambda(0,0,0)$ and $\varepsilon_3 = \varepsilon_1\, \text{sgn}\, C$.

Moreover the universal unfolding of h is

$$H(A,\lambda,\alpha) = (\varepsilon_1 u+\varepsilon_2\lambda)A + (\varepsilon_3 u+Dv+\alpha)(A^2 - \frac{1}{3}\,tr(A^2)I).$$

Remark: Chossat [1979] has computed analytically ε_1 and ε_2 showing, in particular, that $\varepsilon_1\varepsilon_2 = -1$. Chossat [1982] has computed numerically the sign ε_3 about which we shall say more later.

For a discussion of O(3)-equivalence see Golubitsky and Schaeffer [1979,1982]. The main attribute we use here is the observation that O(3)-equivalence does not change, in a precise qualitative way, the zeroes of g.

We now present a schematic rendering of the bifurcation diagrams, $h(A,\lambda) = 0$, occurring in the normal forms of Proposition 1.2. Note that the equivariance properties imply that if $h(A,\lambda) = 0$ then $h(\gamma\cdot A,\lambda) = \gamma h(A,\lambda) = 0$. So h is zero on orbits of the action of O(3). We wish to identify all solutions which are on the same orbit as one. See Figure 1.1 for the unperturbed bifurcation diagrams.

We note that two pieces of information have been added to the diagrams in Figure 1.1. The first is the stability assignments. We have used the notation "s" for stable and "u" for unstable. These stability assignments refer to linearized orbital stability. To determine linearized stability we ask, "what are the signs of the real parts of the eigenvalues of dh along a given solution branch". Note, however, that equivariance implies that at least one eigenvalue of dh evaluated at a non-trivial solution to h = 0 will be zero. To determine linearized orbital stability we must ask "what are the signs of the real part of those eigenvalues of dh which are not forced by symmetry considerations to be zero?" Thus linearized orbital stability is a kind of conditional stability.

Remarks: (a) In the first bifurcation diagram of Figure 1.1 none of the bifurcating solutions are stable. Thus, in this situation, the non-self-adjoint case in the Bénard problem, no physically reasonable information would be gained by a local analysis.

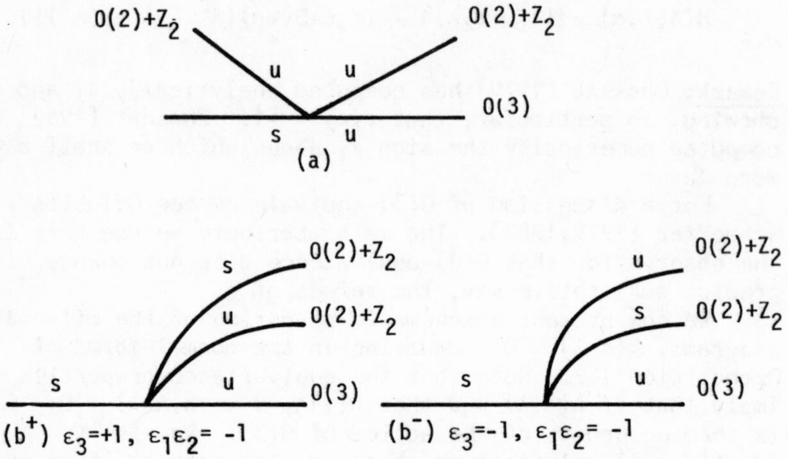

Figure 1.1 Unperturbed bifurcation diagrams.

(b) Note that in the self-adjoint case there is a stable solution which bifurcates from the pure conduction solution; however, which solution branch is stable depends on the information in the higher order term ε_3.

The second piece of extra information which is given on the bifurcation diagrams in Figure 1.1 is the isotropy subgroup of solutions on the given branch. Let A be in V then

$$\Sigma_A = \{\gamma \in O(3) \mid \gamma \cdot A = A\}$$

is the <u>isotropy subgroup</u> corresponding to A. For a solution A to $h(A,\lambda) = 0$, Σ_A is the set of symmetries that the solution A has and these symmetries are usually observable. For example if $\Sigma_A \cong O(2) \oplus Z_2$ then A has an axis of rotation given by the rotation group $SO(2) \subset O(2)$. Such solutions are called <u>axisymmetric</u>.

Remarks: (a) The only solutions which appear in the unperturbed normal forms of Proposition 1.2 are axisymmetric. The isotropy subgroups are preserved up to isomorphism (in fact, inner automorphism) by O(3) equivalences.
(b) One of the main points in Chossat's analysis is that there are two families of axisymmetric solutions which bifurcate supercritically in the self-adjoint case. One can ask how these two families of solutions differ physically. Both represent flows with two cells as shown in Figure 1.2. The difference is that in one case the flow has upwelling at the poles (defined by the axis of symmetry) and downwelling at the equator with the reverse flow the situation for solutions in the other family. (These descriptions are obtained using spherical harmonics.)

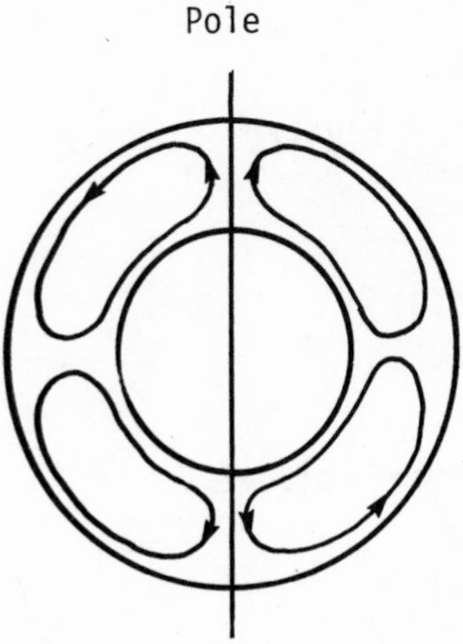

(b^+) upwelling at poles

(b^-) downwelling at poles

Figure 1.2. Two cell axisymmetric solutions

(c) Both families of axisymmetric solutions appear in the non-self-adjoint case. One appears supercritically and one subcritically, the choice depending on the sign of $\varepsilon_1\varepsilon_2$.

Next we discuss the perturbed bifurcation diagrams associated with Figure 1.1(b^+), the case (b^-) is similar. The analytic expression for these zero sets is given by the universal unfolding H in Proposition 1.2 (b). These diagrams are presented in Figure 1.3.

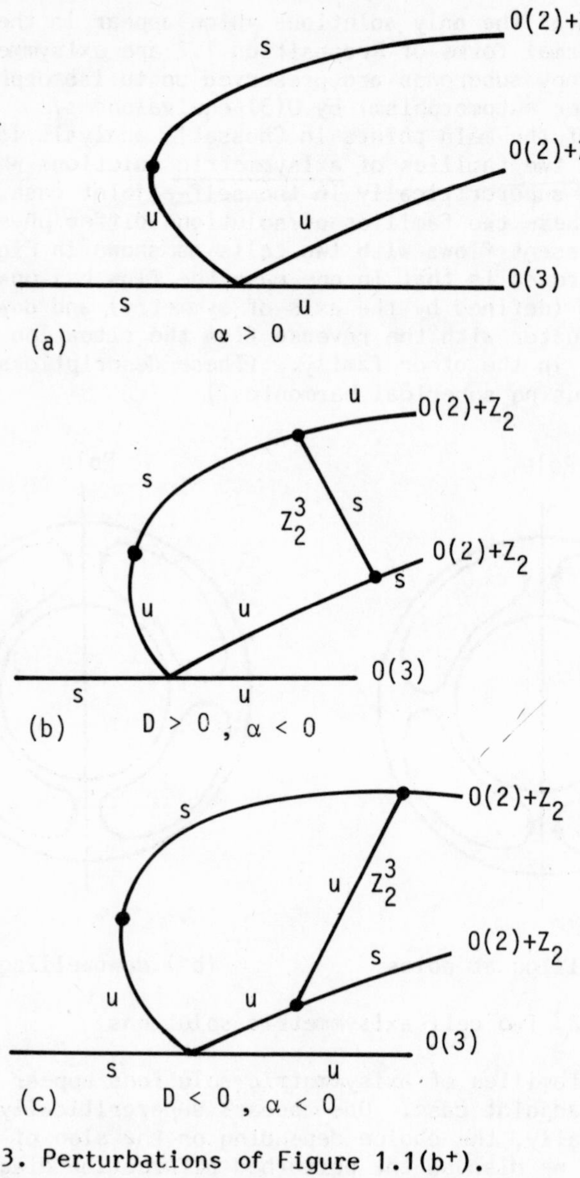

Figure 1.3. Perturbations of Figure 1.1(b^+).

<u>Remarks</u>: (a) When $\alpha < 0$ there exists a new branch of solutions whose isotropy subgroup is the eight element subgroup Z_2^3; these solutions are non-axisymmetric.

(b) Non-axisymmetric solutions may be stable depending on the sign of D.

(c) In the unperturbed problem only axisymmetric solutions exist and only one of the two families of axisymmetric solutions are stable. In the perturbed case both families may be stable, depending on the sign of the perturbation α.

(d) Physically the unfolding parameter α corresponds to making the Bénard problem slightly non-self-adjoint, that is, violating (1.2) by a small amount.

We end this section with a discussion of the lattice of isotropy subgroups. The isotropy subgroups of O(3) - corresponding to the five dimensional irreducible representation - are all isomorphic to $O(3), O(2)+Z_2, Z_2^3$. If Σ_1 and Σ_2 are two isotropy subgroups we define $\Sigma_1 < \Sigma_2$ if some conjugate of Σ_1 is contained in Σ_2. This definition of $<$ makes the set of isotropy subgroups a lattice. In Figure 1.4 we give this lattice structure, the arrows indicating inclusion.

$$
\begin{array}{c}
O(3) \\
\uparrow \\
O(2) + Z_2 \\
\uparrow \\
Z_2^3
\end{array}
$$

Figure 1.4: Lattice of Isotropy Subgroups of O(3).

Remarks: (a) The least degenerate bifurcation problem has non-trivial solution branches whose isotropy subgroups are maximal in the lattice of isotropy subgroups.

(b) The universal unfolding of the next degenerate bifurcation problem has solution branches which have maximal and submaximal isotropy subgroups. Since the lattice has only three subgroups this includes all the isotropy subgroups.

(c) Secondary bifurcation branches, see Figure 1.3, correspond to submaximal isotropy subgroups and these branches connect branches with maximal isotropy subgroups.

2. THE PLANAR BÉNARD PROBLEM WITH NONSYMMETRIC BOUNDARY CONDITIONS

The reader should recall from the Introduction that we study only solutions to the Boussinesq equations which are doubly periodic with respect to the hexagonal lattice in the plane. Moreover, the assumption that the boundary conditions on the upper and lower planes are different implies that the group of symmetries for this problem is $\Gamma = T^2 + D_6$.

Busse [1962] has shown that $\ker L(\tau_0)$, where L is the linearization of the Boussinesq equations about the pure conduction solution and τ_0 is the first eigenvalue, is six-dimensional. The basic idea is that the eigenfunctions of $L(\tau_0)$ are plane waves. Once one has one plane wave then translation gives a second (e.g., sine and cosine). Rotation by 120° and 240° yields four more independent plane waves. Thus $\dim L(\tau_0) \geq 6$ and in the case we consider, it is exactly 6. As above, set $\lambda = \tau - \tau_0$.

The Liapunov-Schmidt procedure implies that finding solutions to the Boussinesq equations (which are doubly periodic in the respect to the hexagonal lattice) reduces to finding the zeroes of a mapping $g: \mathbf{R}^6 \times \mathbf{R} \to \mathbf{R}^6$; i.e., solving $g(x,\lambda) = 0$, where

(1) $(d_x g)_{0,0} \equiv 0$

(2) $g(\gamma x,\lambda) = \gamma g(x,\lambda)$ for all $\gamma \in \Gamma$.

The description of the group theory and singularity theory is much more complicated in this case than in the case of Propositions 1.1 and 1.2. The reader is referred to Buzano and Golubitsky [1983] for details. Our interest centers in the bifurcation diagrams and the lattice of isotropy subgroups which we describe below.

We make several remarks about the structure of g as it relates to the planar Bénard problem.

Remarks: (a) The action of $\Gamma = T^2 + D_6$ on $\mathbf{R}^6$ is absolutely irreducible. Therefore $(d_x g)_{0,\lambda} = p(\lambda)I$ where I is the 6×6 identity matrix. The assumption that the pure conduction solution looses stability at the bifurcation point indicates that $p_\lambda(0) \neq 0$, which we assume.

(b) Symmetry implies that only one quadratic term, the sum of the squares of the coordinates which we denote by Q, can be non-zero. For an idealized Boussinesq fluid (i.e., no surface tension, no temperature dependent viscosity, etc.) Q is, in fact, zero. See Busse [1962].

(c) After symmetry considerations there are two cubic terms which are permitted to be non-zero. We denote the ratio of the coefficients of these cubic terms by a. The value of a will enter our discussion later.

We now describe part of the lattice of isotropy subgroups of Γ acting on R^6. See Figure 2.1. This lattice has two maximal subgroups and two submaximal subgroups. The notation used is S^1 for the rotation group, D_3 for the dihedral group of symmetries of the equilateral triangle and Z_2 for a reflectional symmetry.

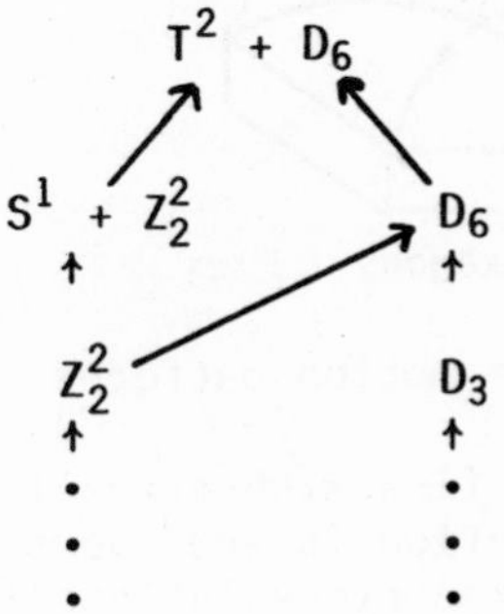

Figure 2.1: The lattice of isotropy subgroups of Γ acting on R^6

Next we describe what solutions with the various isotropy subgroups look like when related to the planar Bénard problem. This should give a better understanding of the effects of the isotropy subgroups. The reader should beware that the results are more complicated than one might think as each isotropy subgroup has several physical realizations. The easiest case is $T^2 + D_6$ which correspond to the pure conduction solution; there is no convective motion. The next simplest is the isotropy subgroup $S^1 + Z_2^2$ which corresponds to <u>rolls</u> as pictured in Figure 2.2(a).

The isotropy subgroup D_6 has two physical realizations. In Figure 2.2(b) one sees a fluid flow which is upwelling at the center and downwelling along the edges of the hexagon.

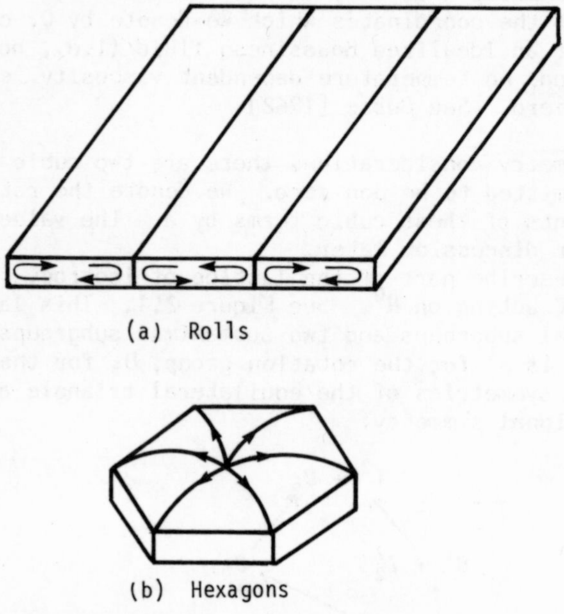

Figure 2.2 Convective motion patterns

Of course the reverse flow with downwelling at the center is possible; this observation is analogous to the existence of two families of axisymmetric solutions in the spherical Bénard problem. Such solutions are called <u>hexagons</u>. The isotropy subgroups for these solutions are easier to visualize if one lets ψ be the vertical velocity component of the (linearized) fluid flow evaluated halfway between the bounding planes and the graphs $\psi = 0$. The results for rolls and hexagons are given in Figure 2.3.

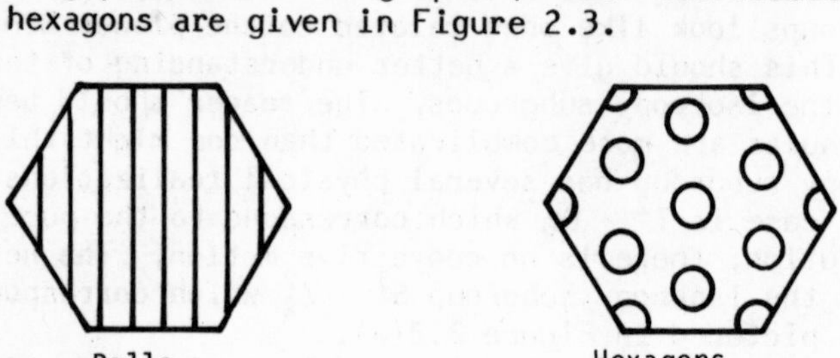

Figure 2.3 $\psi = 0$ for Rolls and Hexagons.

Note that the oval-like closed curve in Figure 2.3(b) is really a smoothed-out hexagon with D_6 symmetry.

For D_3, the triangles, the zero set of ψ comes in two types as shown in Figure 2.4. The flow corresponding to triangles (a) has two realizations either upwelling or

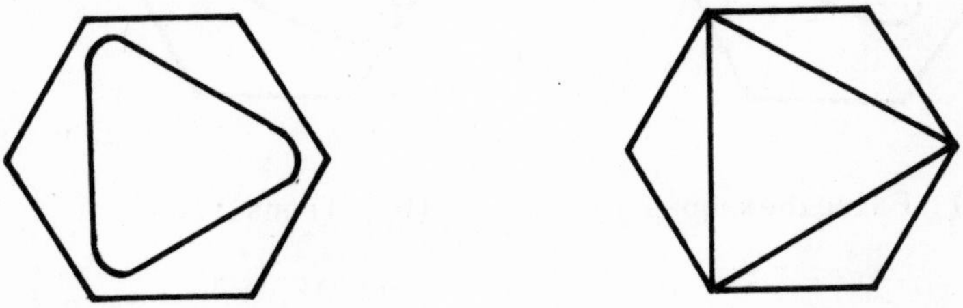

(a) Triangles (b) Regular triangles

Figure 2.4: $\psi = 0$ for D_3 solutions.

downwelling at the center of the triangle-like curve in the figure. This is analogous to the situation for hexagons. The regular triangles have only one realization as the periodicity implies that upwelling at the center of one triangle implies downwelling in the adjacent triangles.

There are four types of zero sets of ψ for solutions with Z_2^2 isotropy subgroups. They are pictured in Figure 2.5. In the first case (a) the zero set of ψ is a smoothed out rectangle-like figure with Z_2^2 symmetry. These solutions we call false hexagons and they come with the two standard physical realizations given by upwelling or downwelling at the center. As the false hexagons could easily be confused with hexagons in an experimental situation. These rectanglar-like figures can grow until they touch the sides of the hexagon (b) and break through the boundary (c). If (c) is continued periodically one gets a zero set for ψ which resembles those of rolls Figure 2.3 (a) except for the periodic behavior along the axis of the rolls; hence the term wavy rolls.

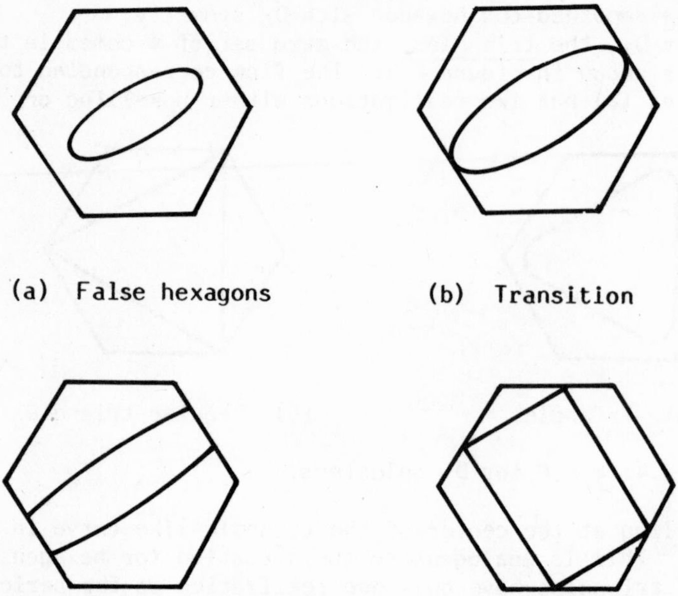

(a) False hexagons (b) Transition

(c) Wavy rolls (d) Patchwork quilt

Figure 2.5 $\psi = 0$ for Z_2^2 solutons.

Finally Patchwork quilt is obtained by letting the rectangle like figure actually approach a rectangle. This case (d) is analogous to the regular triangles described above.

"Theorem 2.1"(a): If the quadratic term $Q \neq 0$ and certain non-degeneracy conditions on the higher order terms hold then the associated bifurcation diagram for g is given in Figure 2.6

(b) If $Q = 0$ and certain non-degeneracy conditions on higher order terms hold then the associated bifurcation diagrams for g depend on the cubic term a. There are four possibilities two of which are given in Figure 2.7.

Typical perturbed bifurcation diagrams for the cases of Figure 2.7 are given in Figure 2.8.

Remarks: (a) Busse [1962] has shown that for an ideal Boussinesq fluid $a < -1$. So the bifurcation diagram for the

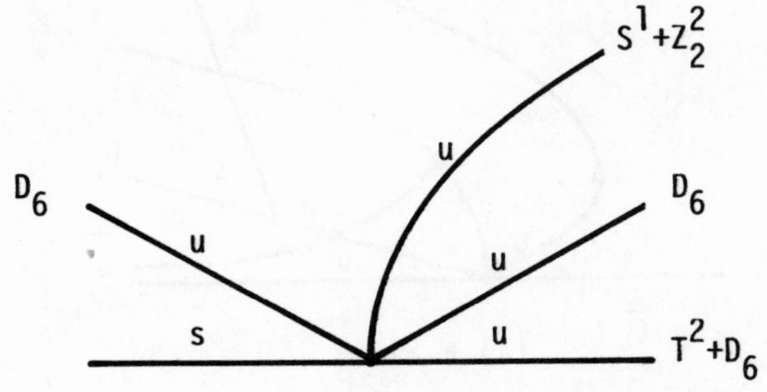

Figure 2.6: Simplest bifurcation diagram with $T^2 + D_6$ symmetry

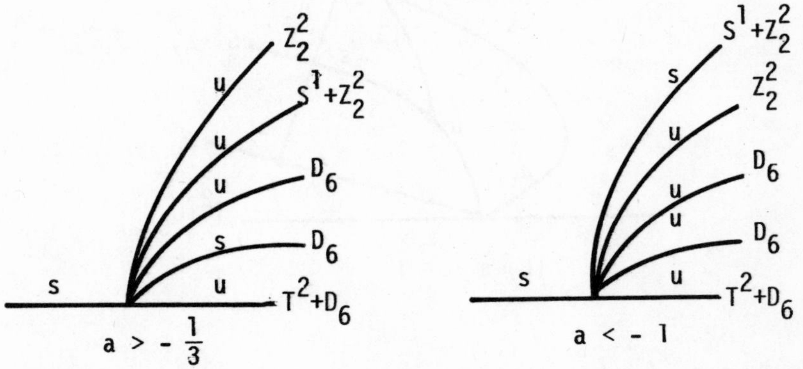

Figure 2.7: Some of the unperturbed bifurcation diagrams for the second least degenerate bifurcation problem with $T^2 + D_6$ symmetry.

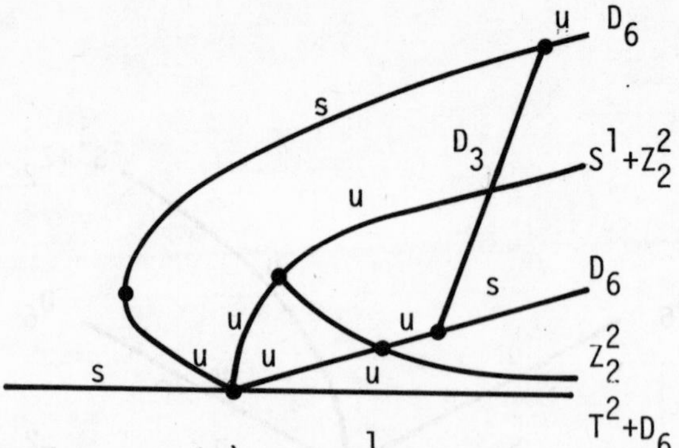

(a) $a > -\frac{1}{3}$

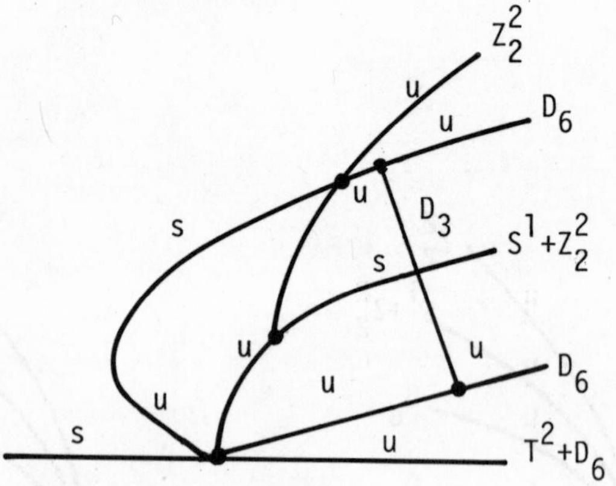

(b) $a < -1$

Figure 2.8: Perturbed bifurcation diagrams: Intersection of branches only occur at black dots.

planar Bénard problem has a stable family of rolls bifurcating supercritically. If the fluid is slightly non-idealized then one expects to see stable hexagon solutions with a jump to rolls as the temperature gradient is increased.

(b) For other convection situations stable hexagons may be the idealized case (as with $a > -1/3$). This is a mathematical possibility consistent with the symmetry. In such a case one may observe stable triangles. The stability of the solutions along the branch of triangle solutions has not been rigorously established; there are indications - using symmetry - that in the case given in Figure 2.8(b) they will be stable.

(c) There are two branches of D_6 solutions in each figure; they correspond to upwelling and downwelling at the center of the hexagon. Which family occurs supercritically and which subcritically depends on higher order terms. In Figure 2.7 (b) either branch could be stable depending on higher order terms.

The most important point in this description is given by comparing the solution branches which actually occur with the lattice of isotropy subgroups given in Figure 2.1. Note that only the maximal isotropy subgroups occur in the least degenerate bifurcation problem while both the maximal and submaximal isotropy subgroups appear in the universal unfolding of the second least degenerate bifurcation problems. Observe that when a submaximal isotropy subgroup appears, the associated solution branch connects branches of solutions corresponding to maximal isotropy subgroups containing that submaximal isotropy subgroup.

3. THE PLANAR BÉNARD PROBLEM WITH SYMMETRIC BOUNDARY CONDITIONS

I shall describe here joint work with Jim Swift and Edgar Knobloch. One changes the formulation of the planar Bénard problem given in the last section by assuming that the boundary conditions on the bounding planes of the fluid are the same. Typically one might assume that the fluid is contained between two identical surfaces - say glass - so that rigid boundary conditions above and below are reasonable. The effect of this change is to add a symmetry to the problem.

We shall still look for solutions to the Boussinesq equations near the pure conduction solution which are doubly periodic with respect to the hexagonal lattice, so that the symmetry group includes $\Gamma = T^2 + D_6$. The added assumption on the boundary conditions implies that reflection about the midplane (in the vertical direction) commutes with the

Boussinesq equations. Thus the symmetry group for this problem is $\Gamma' = \Gamma + Z_2$. Under the same assumption of the six-dimensional kernel of $L(\tau_0)$ as in Section 2, one applies the Liapunov-Schmidt method to obtain $g: \mathbf{R}^6 \times \mathbf{R} \to \mathbf{R}^6$ such that

(a) $(d_x g)_{0,0} \equiv 0$

(b) $g(\gamma x,\lambda) = \gamma g(x,\lambda)$ for all $\gamma \in \Gamma$

(c) $g(-x,\lambda) = -g(x,\lambda)$,

the zeroes of g corresponding to solutions to the Boussinesq equations.

Remark: This new symmetry implies that the reduced bifurcation mapping g is odd in x. See (c) above. It follows that the quadratic term Q of g described in Section 2 must be zero; one might be tempted to conclude that the analysis outlined in "Theorem 2.1(b)" is applicable. One should note however that the hypotheses in that theorem included "certain non-degeneracy conditions on higher order terms". These hypotheses fail when g is odd in x so that the analysis of "Theorem 2.1(b)" is definitely not applicable. As we shall see below the introduction of even a single reflectional symmetry into a bifurcation problem can alter quite substantially the resulting bifurcation pattern.

The beginning of the lattice of isotropy subgroups for the action of Γ' on $\mathbf{R}^6$ is given in Figure 3.1. Note that even though we give only the maximal isotropy subgroups of Γ' that the lattice structure is quite different from that of Γ given in Figure 2.1. This difference will be made more apparent in "Theorem 3.1" below.

$$T^2 + D_6 + Z_2$$

$$\uparrow \qquad \uparrow$$

$$S + Z_2^3 \qquad D_6 \qquad D_3 + Z_2 \qquad Z_2^3$$

$$\uparrow \qquad \uparrow \qquad \uparrow \qquad \uparrow$$

$$\vdots \qquad \vdots \qquad \vdots \qquad \vdots$$

Figure 3.1: The maximal isotropy subgroups of Γ' acting on $\mathbf{R}^6$.

Before describing the bifurcation structure we indicate the physical structures of the convection solutions corresponding to each maximal isotropy subgroup. The first subgroup $S^1 + Z_2^3$ corresponding to rolls. Note that flipping the rolls solution about the midplane and translating by one cell perpendicular to the axis of the rolls gives a symmetry for rolls in Γ' which was not present in Γ. See Figure 2.2(a). This observation explains the extra factor of Z_2 which appears in the isotropy subgroup for rolls in Γ'. Reflection about the midplane for hexagon solutions, D_6, takes a hexagon with upwelling at the center to a hexagon with downwelling at the center. Thus there is only one type of orbit of solutions of hexagons in this formulation, that is, if upwelling occurs as a solution so must downwelling. Note that the new symmetry is not included in the isotropy subgroup for hexagons in Γ'.

The triangle solutions, D_3, intertwine with the new symmetry in a more complicated way. As in the case of hexagons the non-regular triangles have both upwelling and downwelling solutions which are identified by the new symmetry. Moreover the isotropy subgroup for these solutions remains the same D_3 in Γ'. However for the regular triangle solutions a new symmetry is added to the isotropy subgroup. This symmetry is obtained by flipping about the midplane and translating from the center of one triangular cell to the center of an adjacent triangular cell as shown by the arrow in Figure 3.2(a). Thus the isotropy subgroup for regular triangles is $D_3 + Z_2 < D_6$ and we have a new maximal isotropy subgroup.

The situation for the Z_2^2 solutions is similar to the case of triangles. For wavy rolls and false hexagons there is an identification made by upwellings and downwellings so that the isotropy subgroup remains Z_2^2. However, in the single case of patchwork quilt (see Figure 2.5(d)) the new symmetry does add a reflectional symmetry to the isotropy subgroup. This symmetry is obtained by flipping about the midplane and then translating the cells as in the case of regular triangles. See the arrow indicating a relevant translation in Figure 3.2(b). We claim that the isotropy subgroup Z_2^3 obtained this way for patchwork quilt is also a maximal isotropy subgroup. Recall from Figure 2.1 that Z_2^2 is contained (up to conjugacy) in $S^1 + Z_2^2$ and D_6. As with regular triangles $Z_2^3 \not< D_6$ since a flip type symmetry has

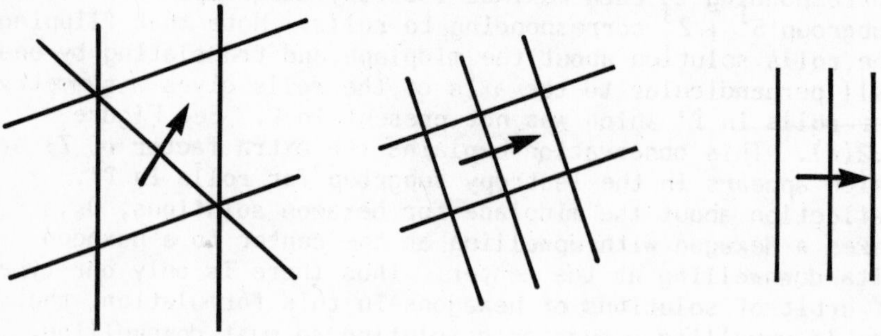

(a) Regular triangles (b) Patchwork quilt (c) Rolls

Figure 3.2. Convection patterns with extra symmetry

been added to the isotropy subgroup. Since the isotropy subgroup for rolls is now $S^1 + Z_2^3$, a flip type symmetry having been added, one might question why Z_2^3 is a maximal isotropy subgroup is Γ'. The answer lies in the direction of the translation which appears in the flip-type symmetry added to the isotropy subgroups of rolls and patchwork quilt. For rolls the translation points from the center of the basic hexagon in the hexagonal lattice to one of the vertices in that hexagon. See Figure 3.2(c). In the case of patchwork quilt it does not. So $Z_2^3 \not\subset S^1 + Z_2^3$ and we have the fourth maximal isotropy subgroup.

We now describe the simplest bifurcation problem commuting in the action of Γ' or R^6.

<u>"Theorem 3.1"</u> Let g be the reduced bifurcation mapping obtained by the Liapunov-Schmidt reduction. Assuming certain non-degeneracy assumptions on g one finds the possibilities for the bifurcation diagrams as shown in Figure 3.3. Recall that a is the ratio of cubic terms. Other possibilities with no stable bifurcating branches exist for $-1 < a < -1/3$.

<u>Remarks</u>: (a) The interesting observation here is the mathematical possibility of the existence of stable regular triangle solutions.

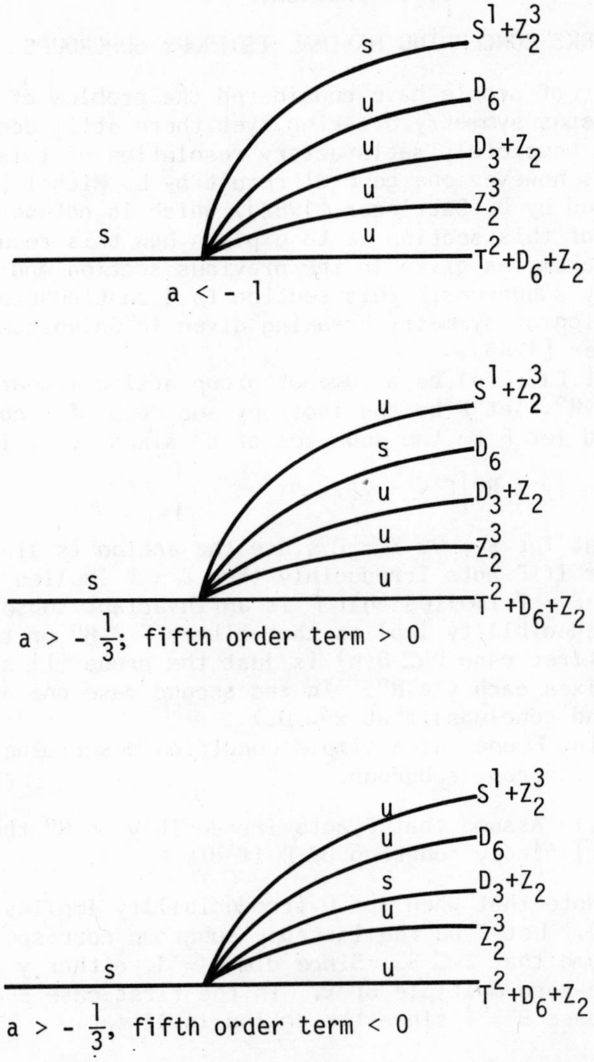

Figure 3.3. Least degenerate bifurcation diagrams with symmetric boundary conditions.

(b) Note once again that there is a one to one correspondence between maximal isotropy subgroups and branches of bifurcating solutions in the least degenerate symmetry preserving bifurcation problem.

4. REMARKS CONCERNING MAXIMAL ISOTROPY SUBGROUPS

A number of people have considered the problem of spontaneous symmetry breaking, yet there still does not exist a completely satisfactory resolution of this problem. There is however one general result by L. Michel [1976] (as described by D. Sattinger [1982]) which is noteworthy. The object of this section is to explain how this result relates to the examples given in the previous section and to maximal isotropy subgroups. This section is a continuation of a discussion of symmetry breaking given in Golubitsky and Schaeffer [1983].

Let $\Gamma \subset O(n)$ be a compact group acting linearly on $\mathbf{R}^n$. For $x \in \mathbf{R}^n$, let Σ be the isotropy subgroup of Γ corresponding to x and let F be the subspace of $\mathbf{R}^n$ fixed by Σ; i.e.,

$$F = \{y \in \mathbf{R}^n \mid \Sigma y = y\}.$$

Note that for $x = 0$, $\Sigma = \Gamma$ since the action is linear. Moreover if Γ acts irreducibly then $\Sigma = \Gamma$ implies $x = 0$. (Proof: $\Sigma = \Gamma$ implies that F is an invariant subspace under Γ. Irreducibility implies that either $F = \mathbf{R}^n$ or that $F = \{0\}$. In the first case $\Gamma \subset O(n)$ is just the group $\{I\}$ since every $\gamma \in \Gamma$ fixes each $y \in \mathbf{R}^n$. In the second case one notes that $x \in F$ and concludes that $x = 0$.)

Using F one has a simple condition describing when Σ is a maximal isotropy subgroup.

<u>Lemma 4.1</u>: Assume that Γ acts irreducibly on $\mathbf{R}^n$ then Σ is a maximal isotropy subgroup of Γ if $\dim F = 1$.

<u>Proof</u>: Note that when $x = 0$ irreducibility implies $\dim F = 0$. So $x \neq 0$. Let Ξ be the isotropy subgroup corresponding to y and assume that $\Sigma \subset \Xi$. Since $\dim F = 1$, either $y = 0$ or y is a non-zero multiple of x. In the first case $\Xi = \Gamma$. In the second case $\Xi = \Sigma$ since the action is linear.

<u>Note</u>: The converse of Lemma 4.1 is not true. As pointed out to us by George Bergman the six-dimensional irreducible representation of the permutation group S_5 has the cycle group Z_5 as a maximal isotropy subgroup. For this example $\dim F = 2$. We shall give other examples below.

Michel and Sattinger have used the condition $\dim F = 1$ to good avail in studying solution branches of bifurcation problems. We describe their result. Let $g: \mathbf{R}^n \times \mathbf{R} \to \mathbf{R}^n$ be a bifurcation problem equivariant with respect to Γ, i.e. $g(\gamma x,\lambda) = \gamma g(x,\lambda)$ for all $\gamma \in \Gamma$. The following lemma summarizes some basic results concerning the relationship between g and isotropy subgroups.

In the following lemma we use the notation $\Sigma(y)$ to indicate the isotropy subgroup corresponding to y. Thus $\Sigma(x) = \Sigma$.

Lemma 4.2: Let $g: \mathbf{R}^n \times \mathbf{R} \to \mathbf{R}^n$ commute with Γ. Then
(a) $\Sigma \subset \Sigma(g(x,\lambda))$

(b) $g:F \times \mathbf{R} \to F$

(c) Let $N(\Sigma)$ be the normalizer of Σ in Γ. Then $\gamma(F) = F$ for all $\gamma \in N(\Sigma)$.

(d) $g|F \times \mathbf{R}$ commutes with the group $D = N(\Sigma)/\Sigma$.

Proof: (a) If $\gamma x = x$ then $g(x,\lambda) = g(\gamma x,\lambda) = \gamma g(x,\lambda)$. Thus $\gamma \in \Sigma(g(x,\lambda))$.

(b) Apply (a) to $y \in F$ to see that $\Sigma(y) \subset \Sigma(g(y,\lambda))$. Since $y \in F$ it follows that $\Sigma \subset \Sigma(y)$. Hence $g(y,\lambda) \in F$.

(c) Let δ be in Γ. Observe that

$$\delta\Sigma\delta^{-1} = \Sigma(\delta x). \tag{4.1}$$

For if $\gamma x = x$, then $\delta\gamma\delta^{-1}(\delta x) = \delta\gamma x = \delta x$. So $\delta\gamma\delta^{-1} \in \Sigma(\delta x)$. Now if $\delta \in N(\Sigma)$ then (4.1) implies that $\Sigma = \Sigma(\delta x)$.

Suppose $y \in F$. Then $\Sigma \subset \Sigma(y)$. For $\delta \in N(\Sigma)$ one has

$$\Sigma = \delta\Sigma\delta^{-1} \subset \delta\Sigma(y)\delta^{-1} = \Sigma(\delta y).$$

Thus $\Sigma \subset \Sigma(\delta y)$ and $\delta y \in F$. (Note that $N(\Sigma)$ is the largest subgroup of Γ which leaves the subspace F invariant.)

(d) Since g commutes with Γ one has that $g|F \times \mathbf{R}$ commutes with $N(\Sigma)$ using (c). But Σ acts as the identity on F so D acts on F and $g|F \times \mathbf{R}$ commutes with D.

The following proposition, due to Michel [1976], is the first general result about the existence of bifurcating branches corresponding to maximal isotropy subgroups. First note that if Γ acts absolutely irreducibly, (i.e., the only matrices on $\mathbf{R}^n$ which commute with Γ are multiples of the identity matrix), then $(dg)_{0,\lambda} = c(\lambda)I$ since the chain rule implies that $(dg)_{0,\lambda}$ commutes with Γ. We shall say that the trivial solution $x = 0$ changes stability non degenerately if $c(0) = 0$ and $c'(0) \neq 0$. Note that if Γ acts irreducibly then $g(0,\lambda) = 0$ for all λ and $x = 0$ is a solution.

Proposition 4.3: Let g commute with Γ. Assume

(i) Γ acts absolutely irreducibly

(ii) $\dim F = 1$, so Σ is a maximal isotropy subgroup

(iii) the trivial solution changes stability non-degerately.

Then there exists a solution branch bifurcating from the origin whose solutions have isotropy subgroup Σ.

Proof: By Lemma 4.2(b) $g: F \times \mathbf{R} \to F$. Let y be the single coordinate in F. Then $g(0,\lambda) \equiv 0$ since the trivial solution persists for all λ. Since g is assumed to change stability at (0,0) one sees that $g_y(0,0) = 0$ and $g_{y\lambda}(0,0) \neq 0$.

Since $g(0,\lambda) \equiv 0$ one may write, using Taylor's theorem, $g(y,\lambda) = yh(y,\lambda)$. The assumptions $g_y(0,0) = 0$ and $g_{y\lambda}(0,0) \neq 0$ imply $h(0,0) = 0$ and $h_y(0,0) \neq 0$. Now one can solve the equation $h(y,\lambda) = 0$ by the implicit function theorem for a unique smooth function $\lambda = \Lambda(y)$ so that $\Lambda(0) = 0$ and $h(y,\Lambda(y)) \equiv 0$. The curve $\lambda = \Lambda(y)$ is the desired branch of solutions.

Remarks: (a) From the point of view of bifurcation theory the information given by Proposition 4.3 is insufficient in several ways. First of all no information is given about the stability of the solutions on the new branch. Michel's interest in this problem came from assuming that $g = \nabla f$ where $f: \mathbf{R}^n \times \mathbf{R} \to \mathbf{R}$ is invariant under Γ. Once one has a potential function the problem of linearized stability is easier. Second, Proposition 4.3 gives no information about how many solutions y exist for each λ. For example, it is possible, though quite improbable, that $\Lambda(y) \equiv 0$. So all the new solutions occur at $\lambda = 0$, a rather unreasonable eventuality. Third, no information is given about the existence or non-existence of other branches.

(b) Sattinger [1982] has used this proposition along with

some standard though sophisticated techniques from group representation theory to make statements about the existence of solution branches for bifurcation problems commuting with the higher (than five) dimensional irreducible representations of O(3).

(c) All of the examples in the first three sections satisfy the assumptions of Proposition 4.3. In particular, all of the maximal isotropy subgrups Σ have $\dim F = 1$. Thus the existence of each of the branches of solutions corresponding to maximal isotropy subgroups is guaranteed by this proposition.

Michel [1976] claimed that a partial converse to this theorem is also true. More precisely, Michel claimed that if a solution exists for every g satisfying (iii), Γ is assumed to act absolutely irreducibly (i), then $\dim F = 1$. Dancer and Sattinger noted that an extra hypothesis is needed to prove this converse. One has:

<u>Proposition 4.4</u>: Assume that Γ acts absolutely irreducibly on $\mathbf{R}^n$. Let Σ be an isotropy subgroup. Assume

(i) The group $D = N(\Sigma)/\Sigma$ is finite.

(ii) For every $g: \mathbf{R}^n \times \mathbf{R} \to \mathbf{R}^n$ commuting with Γ such that the trivial solution changes stability non-degenerately there is a solution branch of $g = 0$ with isotropy subgroup Σ bifurcating from the origin.

Then $\dim F = 1$.

<u>Proof</u>: See Sattinger [1982], §4.

This proposition states, in a sense, that if one wants to find a solution branch corresponding to Σ for <u>every</u> bifurcation problem then one needs to know that $\dim F = 1$. This seems to us to be a misplaced emphasis. Perhaps one really wants to know which conditions on Σ imply that for almost every g there is a solution branch corresponding to Σ. We suggest that the appropriate set of g's to investigate are those with (topological) Γ-codimension equal to zero, that is, those g's whose singularities are the simplest possible consistent with Γ-equivariance. These g's are the equivalent of Morse functions in the Γ-equivariant bifurcation theory context. For these g's one might conjecture that there is a one to one correspondence between maximal isotropy subgroups and solution branches. This conjecture is not true as stated and will be refined in the discussion below.

We now return to the extra hypothesis to Proposition 4.4. As observed in Lemma 4.2 the group D acts on F. We can make a further observation about this action. The following observations were made jointly with Ian Stewart.

Lemma 4.5: Assume that Σ is a maximal isotropy subgroup of Γ and that Γ acts irreducibly on R^n. Then the action of D on F is fixed point free.

Proof: Suppose the action of D on F is not fixed point free. Then there is a $y \neq 0$ in F and $\delta \in N(\Sigma) \sim \Sigma$ which satisfies $\delta y = y$. It follows that $\Sigma(y) \underset{\neq}{\supset} \Sigma$ since $\delta \notin \Sigma$. By the maximality of Σ it follows that $\Sigma(y) = \Gamma$. Since Γ acts irreducibly $y = 0$ contradicting our assumption and the lemma is proved.

Proposition 4.6: Let Γ act irreducibly on R^n and let Σ be a maximal isotropy subgroup of Γ. Let D^0 be the connected component of the identity of $D = N(\Sigma)/\Sigma$. (D is compact since Γ is assumed to be compact.) Then either

(a) $D^0 = \{1\}$,

or (b) $D^0 = S^1$ and F is the direct sum of irreducible subspaces under D^0, $\cong$ **C**, where S^1 is identified with the unit complex numbers and the action of S^1 on **C** is given by complex multiplication

or (c) $D^0 = SU(2)$ and F is the direct sum of irreducible subspaces under D^0, $\cong$ **Q**, where **Q** is the skew field of quaternions, SU(2) is identified with the unit quaternions and the action of D^0 on **Q** is given by quaternionic multiplication.

Definition 4.7: We call a maximal isotropy subgroup of a compact group Γ acting irreducibly on R^n either real, complex, or quaternionic depending on whether D^0 is $\{1\}$, S^1, or SU(2).

Proof: The basic observation is the one given in Lemma 4.5

that D and hence D^0 acts fixed point free. The result then following from Theorem 8.5 in Bredon [1972]. We include a proof here as it is short and it does not appear in the bifurcation theory literature. Assume that $\dim D^0 \geq 1$. Since D^0 is compact it has a maximal torus T^ℓ. We claim that if $\ell \geq 2$ then the action cannot be fixed point free. First note that if S^1 acts on F fixed point free then it acts fixed point free in each irreducible subspace V. Irreducibility implies that $\dim V = 1$ or $\dim V = 2$ and S^1 cannot act in a fixed point free way on **R**. Thus $\dim V = 2$. Moreover the irreducible actions of S^1 on $\mathbf{R}^2 \simeq \mathbf{C}$ are enumerated by $\theta \to \exp(m\theta i)$ for some positive integer m. If $m > 1$ then this action is not fixed point free, take $\theta = 2\pi/m$. So we may assume that $m = 1$.

Now suppose that $T^2 = S^1 \times S^1$ acts on F. Let V be an irreducible subspace of F under this action of T^2. Again irreducibility implies that $\dim V = 1$ or 2 with $\dim V = 1$ and a fixed point free action being incompatible. So we identify V with **C**. The result above states that $(\theta,0)$ acts on **C** by $(\theta,0) \to \exp(i\theta)$ and $(0,\theta)$ acts on C by $(0,\theta) \to \exp(i\theta)$. It follows that the diagonal (θ,θ) of T^2 acts on **C** by $\exp(2i\theta)$. However the diagonal of T^2 is S^1 and such an action of the diagonal is not fixed point free. The claim is proved.

Using the classification theorem for compact, connected Lie groups of positive dimension one sees that there are only three whose maximal torus is one dimensional, namely, S^1, SU(2), and SO(3). Suppose $D^0 = SO(3)$. Then write F as a sum of irreducible subspaces. All of the irreducible actions of SO(3) are odd-dimensional. As a rotation matrix acting on an odd-dimensional space always has an axis of rotation, such actions cannot be fixed point free. So D^0 is either 1, S^1 or SU(2).

As discussed above the irreducible decomposition of S^1 acting on F is as stated in the proposition. Finally one checks that the only fixed point free, irreducible action of SU(2) is given by SU(2) acting as the unit quaternions on the quaternions.

Remarks: (a) In the complex case $\dim F = 0 \bmod 2$ and in the quaternionic case $\dim F = 0 \bmod 4$. Here one can obtain more examples of cases where $\dim F > 1$ and Σ is a maximal isotropy subgroup.

(b) Recall from Lemma 4.2(d) that $g|F \times \mathbf{R}$ commutes with D.

Suppose that $D = S^1$ and F is two dimensional. Identify F with C and note that if g commutes with S^1 then g has the form

$$g(z,\lambda) = p(z\overline{z},\lambda)z + q(z\overline{z},\lambda)iz$$

where p and q are real valued. (Cf. Golubitsky and Langford [1981].) If g has a singularity at the origin then $p(0,0) = 0 = q(0,0)$. One can show easily that $g = 0$ reduces to $z = 0$ or

$$p(z\overline{z},\lambda) = 0 = q(z\overline{z},\lambda)$$

since z and iz are independent if $z \neq 0$. Generally the solution to a system of two equations in two unknowns is a discrete set of points, so generically $z = 0$ is the only solution. No branch of solution bifurcates in $F \times \mathbf{R}$ from the origin. (Aside: if g depends on an extra parameter τ then one can obtain a curve of solutions. This happens in Hopf bifurcation where τ is the perturbed period. See Golubitsky and Langford [1981].)

A similar situation occurs when $D = SU(2)$ and $F = Q$ only there one needs to add three additional parameters in order to find a solution branch. I know of no interesting situation (such as Hopf bifurcation) where this phenomenon occurs. It is an interesting question!

Proposition 4.6 puts the extra assumption in Proposition 4.4 into perspective. The remarks above suggest the following:

Conjecture: Let Γ act absolutely irreducibly on $\mathbf{R}^n$. Let $g: \mathbf{R}^n \times \mathbf{R} \to \mathbf{R}^n$ commute with Γ, have a singularity at (0,0) and have (topological) codimension 0. (In particular, this implies that there is a non-degenerate change of stability along the trivial solution at $\lambda = 0$.) Then each non-trivial branch of solutions to $g = 0$ corresponds to a real, maximal isotropy subgroup. Moreover, each real maximal isotropy subgroup corresponds to a branch of solutions to $g = 0$ for some g with (topological) codimension 0.

I feel confident that this conjecture is true if all of the real, maximal isotropy subgroups also satisfy $\dim F = 1$. This is the case for the examples in the first three sections. In fact, more is true.

Recall that D acts fixed point free on F. If $\dim D = 1$ then either $D = \{1\}$ or $D = Z_2$. If $D = Z_2$ then $g: F \times \mathbf{R} \to F$

is odd in y by Lemma 4.2(d). The simplest such bifurcations for odd functions is the pitchfork bifurcation, $y^3 \pm \lambda y$. See Golubitsky and Schaeffer [1979]. The following fact is true for the examples given in Figure 1.1(a), Figure 2.6 and Figure 3.3. If $D = Z_2$ then the branch of solutions corresponding to Σ is parabola-like as in the pitchfork. If $D = \{1\}$ then the branch of solutions is transcritical, $y^2-\lambda y$, and has two components, one supercritical and one subcritical.

We are in the situation where by abstract techniques one can recover much of the information in the bifurcation diagrams associated to the simplest, least degenerate cases of the examples in the previous sections. Those results were obtained by long, tedious calculations and to be able to replace them by only group theoretic considerations would be a major accomplishment. We are not there yet but the project seems feasible. Finally we note that we havenot yet considered here, in a coherent way, the problem of linearized orbital stability nor have we considered the problem of submaximal isotropy subgroups from an abstract point of view. The examples in Sections 1 and 2, in particular, show that such considerations are absolutely necessary if the abstract theory is to be truely applicable.

Martin Golubitsky, Department of Mathematics, Arizona State University, Tempe, Arizona 85287.

Research supported in part by ARO Contract DAAG-79C-0086 and by NSF Grant MCS-8101580.

REFERENCES

Bredon, G.E., 1972. Introduction to Compact Transformation Groups. Academic Press, New York.

Busse, F.H., 1962. Das Stabilitätsverhalten der Zellarkonvektion bei end licker amplitude, Dissertation, University of Munich. (Engl. Transl. by S.H. Davis, Rand Rep. LT-66-19, Rand Corp. Santa Monica, CA.)

Busse, F.H., 1975. Pattern of convection in spherical shells, J. Fluid Mech. 72, 65-85.

Busse, F.H., 1978. Nonlinear properties of thermal convection. Rep. Prog. in Phys. 41, 1929-1967.

E. Buzano and M. Golubitsky, 1983. Bifurcation on the hexagonal lattice and the planar Bénard problem. Phil. Trans. Roy. Soc. London. To appear.

Chossat, P., 1979. Bifurcation and stability of convective flows in a rotating or not rotating spherical shell, SIAM J. Appl. Math. 37, 624-647.

Chossat, P., 1982. Le probleme de Bénard dans une couche sphérique. Thésc d'État. Université de Nice.

S. Fauve and A. Libchaber, 1983. Rayleigh-Bénard experiments and dynamical systems. This volume.

M. Golubitsky and D. Schaeffer, 1979. Imperfect bifurcation in the presence of symmetry. Commun. Math. Phys. 67, 205-232.

M. Golubitsky and D. Schaeffer, 1982. Bifurcation with O(3) symmetry including applications to the Bénard problem. Commun. Pure and Appl. Math. 35, 81-111.

M. Golubitsky and D. Schaeffer, 1983. A discussion of symmetry and symmetry breaking. Proc. Symposia in Pure Mathematics, 40, A.M.S. To appear.

M. Golubitsky, J. Swift and E. Knobloch, 1983. Boussinesq convection with symmetric boundary conditions. In preparation.

Michel, L., 1976. Simple mathematical models of symmetry breaking, application to particle physics. Mathematical Physics and Physical Mathematics, Polish Scientific Publishers, Warsaw, 251-262.

Sattinger, D.H., 1978. Group representation theory, bifurcation theory, and pattern formation. J. Func. Anal. 28, 58-101.

Sattinger, D.H., 1979. Group Theoretic Methods in Bifurcation Theory. Lec. Notes in Math 762, Springer-Verlag, Berlin.

Sattinger, D.H., 1982. Branching in the presence of symmetry CBMS Lectures, Gainsville, FL. Dec. 1981. To appear.

S. Fauve and A. Libchaber

RAYLEIGH-BÉNARD EXPERIMENTS AND DYNAMICAL SYSTEMS

INTRODUCTION

The problem of explaining the origin of turbulent flows has been recognized for more than a century. However there is not yet a generally accepted definition of turbulence. The motions of the atmosphere and oceans, are to a large extent called turbulent, because they apparently possess complex and random spatial structure, and erratic time dependence. In most of laboratory experiments, the onset of time-dependent turbulence is often preceded by the development of complex stationary spatial patterns. The experiences we are going to describe, and the dynamical system theory, have nothing to say about the interaction between different spatial scales. The convective flows we have observed, have most of their kinetic energy contained in a small wave number range, and the experimental conditions are chosen to get time dependence with the simplest possible convective pattern. In the past years substantial experimental attention has been given to the onset of a temporal chaotic regime of such "simple" physical systems, and to the description of their temporal characteristics by dynamical system theory. However one should not confuse the chaotic regimes we shall describe, (sometimes called weak turbulence), with strongly turbulent flows, involving interactions between different spatial scales.

Until the early seventies, there exists only one qualitative model for the transition to chaos, due to Landau and Hopf, and describing a chaotic regime as a quasiperiodic one, with many independent frequencies. This view was challenged by Ruelle and Takens in 1971 ; they have proved that quasiperiodic attractors are structurally unstable for dynamical systems, (finite dimensional deterministic systems of differential equations), and have portrayed turbulent flows as motion on a strange attractor in the phase space of the system. Their main conjecture is that the phenomena underlying chaos are finite dimensional, although the state space for a fluid system is infinite

C. P. Bruter et al. (eds.), Bifurcation Theory, Mechanics and Physics, 257–276.

dimensional. In 1975 Gollub and Swinney found experimental results that coroborate the ideas of Ruelle and Takens. They observed in a Couette flow experiment, that a few non linearly coupled modes are sufficient to produce a chaotic behavior. It should be noted that the fact that systems of non linear ordinary equations have regions of chaotic motion, has been known since the work of Chirikov (1959) on hamiltonian systems, and the numerical computations of Lorenz (1963) on a three mode truncation of the equations governing finite amplitude convection in a fluid layer heated from below, (the Rayleigh-Bénard problem). The Lorenz system has motivated a variety of numerical experiments on systems of differential equations and on finite difference equations, that have exhibited various possible routes to chaos. The best theoretically and experimentally documented one, is the period-doubling scenario : the system achieves chaotic temporal behavior through the production of successive half subharmonics. This behavior has been understood in the framework of renormalization theory, and qualitative and quantitative experimental agreement has been found. The relevance of dynamical system theory to the transition to temporal chaos of some fluid systems, is therefore an experimental evidence. However this is not yet supported by any precise analysis of the partial differential equations governing fluid motions. For the physicists, the next task is to get a mean to derive ordinary differential equations to predict the time-dependence of a system governed by partial differential equations, when the phenomena underlying chaos are clearly finite dimensional. This has been recently done in the case of competing instabilities, by Arnéodo, Coullet, Spiegel and Tresser, and the example of rotating thermohaline convection has been studied in detail. An experimental study of this problem would be of great interest.

This paper has the following organization. In the first section we describe the Rayleigh-Bénard instability and we give the needed definitions. In the second section, we show that in some parameter range, the observed routes to chaos are in qualitative and quantitative agreement with dynamical system theory, and moreover that all the scenarios might be understood in the framework of one or two-dimensional mappings. The last section contains examples of more complex routes to chaos. Their connection to the physical parameters of the experiment is discussed.

1. THE RAYLEIGH-BÉNARD PROBLEM

1.1. Dimensionless numbers

The Rayleigh-Bénard problem is concerned with convection in a horizontal layer of fluid heated from below. When the temperature difference ΔT across the layer is small enough, there is a trivial solution, representing pure upward heat conduction. Above a critical value of ΔT, the buoyancy force overcomes the viscous force, and convection sets in the layer. The dimensionless number proportional to the ratio of the buoyancy force to the viscous force, is the Rayleigh number

$$R = \frac{g\alpha d^3 \Delta T}{\nu K}$$

where g is the acceleration of gravity, α is the isobaric thermal expansion coefficient, d is the layer thickness, ν is the kinematic viscosity, and K is the heat diffusivity.

The mathematical description of convection is based on the equations of conservation of mass, momentum and energy. The diffusion terms of momentum and energy define the two most significant relaxation time scales of the fluid, d^2/ν and d^2/K, for diffusion of momentum and energy, respectively. The ratio

$$P = \frac{d^2/K}{d^2/\nu} = \frac{\nu}{K}$$

is the Prandtl number of the fluid.

1.2. Oberbeck-Boussinesq equations

The Rayleigh and Prandtl numbers are the two relevant dimensionless numbers, describing the physical conditions of the fluid in the Boussinesq approximation. Indeed, if we use a dimensionless description by introducing d, d^2/K, ΔT as scales for length, time and temperature, respectively, the equations for the velocity vector $\vec{v}$, and for the deviation θ from the temperature distribution in the static case are

$$\vec{\nabla}.\vec{v} = 0 \qquad (1)$$

$$\frac{1}{P}\left(\frac{\partial \vec{v}}{\partial t} + \vec{v}.\vec{\nabla}\vec{v}\right) = - \vec{\nabla}p + \Delta\vec{v} + R\theta\hat{z} \qquad (2)$$

$$\frac{\partial \theta}{\partial t} + \vec{v}.\vec{\nabla}\theta = \vec{v}.\hat{z} + \Delta\theta \qquad (3)$$

where $\hat{z}$ is the vertical ascendent unit vector, and $\vec{\nabla}p$ includes all the terms that can be written in the form of a gradient.

In many laboratory experiments, the top and bottom boundaries consist in layers of material, (often copper), which is a much better heat conductor than the fluid. The boundary conditions are therefore

$$\vec{v} = \vec{0} \quad \text{at} \quad z = \pm\frac{1}{2} \tag{4}$$

$$\theta = 0 \quad \text{at} \quad z = \pm\frac{1}{2} \tag{5}$$

1.3. Linear theory

At a critical Rayleigh number $R_c(k)$, the static solution loses stability. The resulting bifurcation is supercritical. The linear stability analysis shows that the instability of the static layer occurs in the form of a non oscillatory growing disturbance of horizontal wave number k. With the boundary conditions (4) and (5), the most unstable disturbance corresponds to

$$k_c = 3.12$$

and the corresponding Rayleigh number is

$$R_c = 1708$$

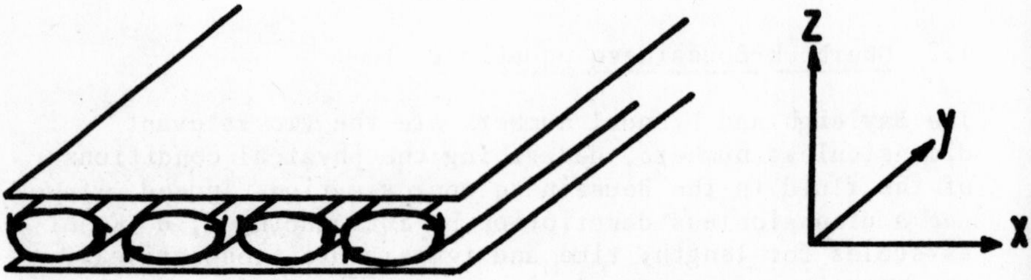

Fig. 1. The convective structure.

The main problem posed by the results of linear theory is the infinite degeneracy of the eigenvalue R_c. Schlüter, Lortz and Busse (1965) have shown that the pattern degeneracy is removed when the non linear terms are taken into account, and that the stable pattern consists in two-dimensional parallel convective rolls (see figure 1). But an orientational degeneracy may still persist, depending on the geometrical configuration of the lateral boundaries. In a layer which is infinite in the horizontal direction, the eigenvalue $R_c(k)$ depends continuously on the wave number k. When lateral boundary conditions are taken into account, $R_c(k)$ splits into a number of discrete values, that decreases with decreasing aspect ratio, (the ratio of the larger horizontal dimension to the height). In large aspect ratio containers, many modes are near instability at the onset of convection. A complicated spatial structure results (Normand 1981, Gollub and Steinman 1981), and a sharp transition to chaos has been experimentally observed (Ahlers 1974). In small aspect ratio containers, only a few modes are unstable near the onset of convection, and the transition to chaos corresponds to the prediction of dynamical system theory (see for instance Eckmann 1981). Consequently, the experiments we shall describe in this paper concern only small aspect ratio parallelepipedic containers ($2 < \Gamma < 6$).

1.4. Instabilities of convection rolls

The analysis of the instabilities of two-dimensional convective rolls has been done numerically by Busse and Clever (1974, 1979), in the case of an infinite aspect ratio layer. They have found that when the Rayleigh number is increased, two-dimensional rolls become unstable to disturbances that lead to a new convective regime, the pattern of which depends essentially on the Prandtl number. The new patterns are generated by modes that are damped at the convection onset, but become coupled with basic two-dimensional rolls as the Rayleigh number is increased. The coupling terms are the non linear terms of advection of heat and momentum $\vec{v}.\vec{\nabla}\theta$ and $\vec{v}.\vec{\nabla}\vec{v}$. The Prandtl number providing a measure of their relative importance, the new growing modes depend essentially on its value. The results of Busse and Clever are summarized on figure 2. For high Prandtl number fluids, (P = 300 : figure 2a), the term $\vec{v}.\vec{\nabla}\theta$ generates an additional roll-like motion at right angles to the basic

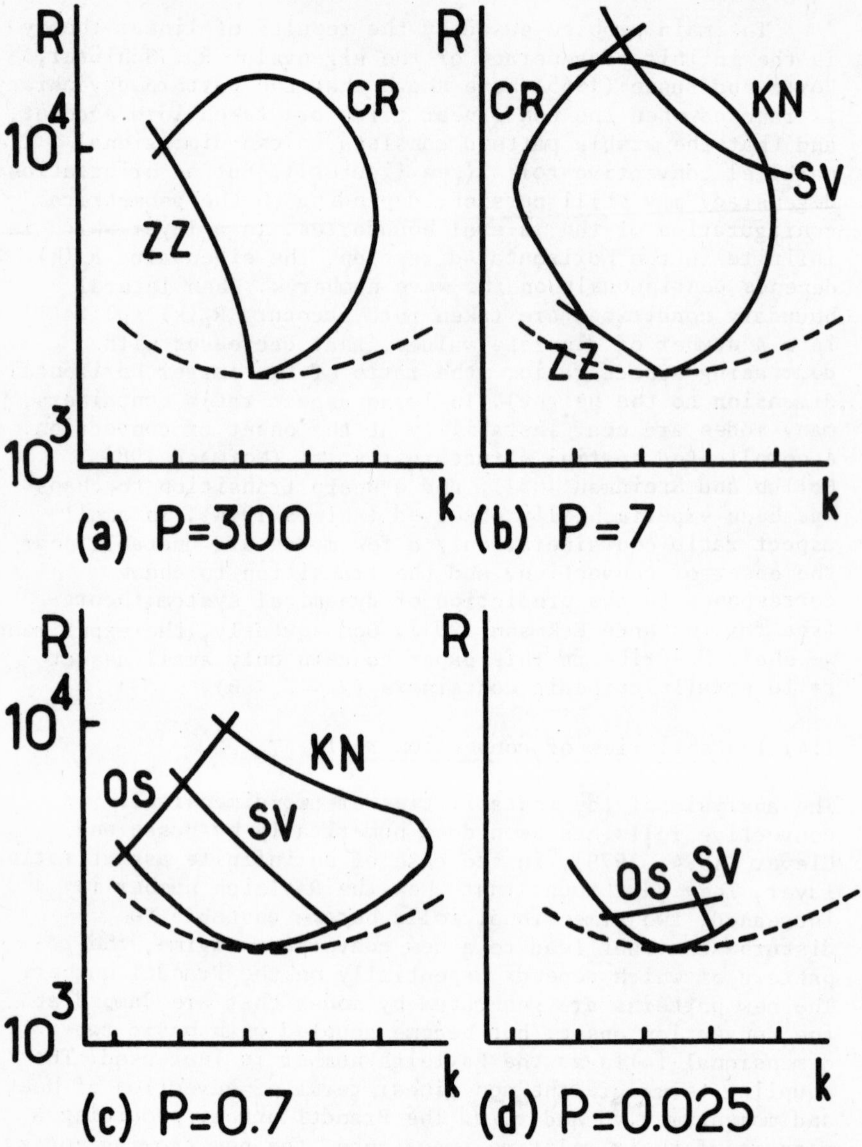

Fig. 2. Stability diagrams for convection rolls (Busse and Clever 1979).

rolls, when the Rayleigh number exceeds a value of order 2.10^4. This new regime, connected with the cross-roll instability (CR), is called bimodal convection. It becomes time-dependent for higher Rayleigh numbers (Busse and Whitehead 1974). For very low Prandtl number fluids, ($P = 0.025$: figure 2d), the intertial term $\vec{v}.\vec{\nabla}\vec{v}$ couples vertical vorticity modes with basic two-dimensional rolls, and generates the oscillatory instability (OS). This instability corresponds to a wave travelling in the direction of the axis of the rolls, and consequently to a transverse time dependent oscillation of the rolls. It occurs for a critical Reynolds number of the basic flow, and therefore for a critical Rayleigh number R_0 given by

$$\frac{\sqrt{R_0 - R_c}}{P} = \text{constant}$$

for stress-free boundary conditions; thus $R_0 - R_c$ vanishes like P^2 in the limit $P \to 0$. This is connected with the galilean invariance of the Boussinesq equations, in the case of stress-free boundary conditions, for vanishing Prandtl number. For rigid boundary conditions, the galilean invariance is broken; however the oscillatory instability subsists, but occurs at a larger Rayleigh number. For moderate Prandtl number fluids, (figures 2b and 2c), the situation is much more complicated because the two non linear terms interacts to generate the knot (KN) and the skewed-varicose (SV) instabilities (Busse and Clever 1979, Siggia and Zippelius 1981, Cross 1982). One should also notice the stabilizating effect of the inertial term $\vec{v}.\vec{\nabla}\,\vec{v}$ on the zig-zag instability (ZZ). (The zig-zag instability bends the rolls in a sinusoidal fashion). When lateral boundaries are taken into account, the translational and orientational invariances connected with all these instabilities are broken; however, the instabilities occur, but at higher Rayleigh numbers, and the experimental observations have confirmed the numerical computations in all qualitative aspects.

The transition to time-dependent convection occurs, from a stationary three-dimensional pattern for high Prandtl number fluids, in the form of competing instabilities (skewed-varicose and oscillatory) for moderate Prandtl number fluids, and more simply from the basic two-dimensional rolls in the case of low Prandtl number fluids. One can therefore understand why low Prandtl number fluids are the best candidates for the study of the transition to chaos. In the experiments we shall describe, mercury is used. ($P = 0.025$ at room temperature).

1.5. Effect of a horizontal magnetic field

Mercury is also a fluid of interest because an applied magnetic field provides another experimental parameter. In laboratory experiments the characteristic diffusion time for magnetic field is much shorter than for heat and momentum. Consequently the magnetic field disturbances do not propagate as Alfven waves. Moreover, when the magnetic Reynolds number is small, one can neglect the magnetic field due to eddy currents. Therefore, the magnetic field effect is only to add some kind of anisotropic viscosity, that inhibits velocity variations along its direction. The importance of this effect is related to the Chandrasekhar number

$$Q = \frac{\sigma B^2 d^2}{\rho\ \nu}$$

where σ is the electrical conductivity, B is the magnetic field amplitude, and ρ is the fluid density. The Chandrasekhar number represents the ratio of the momentum diffusion term due to the magnetic field, to the momentum diffusion term due to viscosity. In the case of a horizontal magnetic field, we have shown that two-dimensional rolls parallel to the magnetic field direction are stabilized. (Fauve and Libchaber 1981). This result is in agreement with the experimental observations of Lehnert and Little (1957) and the numerical computations of Tabeling (1982). The oscillatory instability, corresponding to three-dimensional disturbances, is inhibited by the magnetic field; two different regimes exist. At low magnetic fields ($Q < 100$), the oscillatory instability onset, R_Q, is simply shifted towards larger Rayleigh numbers and follows a power law

$$R_Q - R_0 \propto Q^{1.2}$$

in good agreement with the numerical computations of Busse and Clever (1982). At higher magnetic fields, R_Q keeps increasing with Q, but with a smaller slope. This new regime is probably related to the presence of a stationary instability that competes with the oscillatory instability. The effect of the magnetic field is therefore twofold. Low magnetic fields just introduce an extra-damping for the oscillatory modes. High magnetic fields, by maintaining the flow two-dimensional at large Rayleigh numbers, lead to new instabilities for the transition to chaos. We shall show in the next sections that these two regimes correspond to different types of transition to chaos.

2. LOW MAGNETIC FIELDS : PERIOD DOUBLING CASCADE TO CHAOS

2.1. The period doubling scenario

Many experiments on a variety of non linear physical systems have revealed that a possible route to chaotic behavior is a cascade of successive half harmonics of a basic mode. This has been observed numerically in most current low dimensional dynamical systems : one dimensional non invertible maps, Hénon map, Lorenz equations, Duffing's equation, etc... (see for instance Ott 1981). Several experiments, on driven electrical anharmonic oscillators (Linsay 1981), and Rayleigh-Bénard experiments with, liquid helium (Maurer and Libchaber 1979), water (Giglio, Musazzi and Perini 1981), and mercury (Libchaber, Laroche and Fauve 1982), have exhibited the same behavior, in surprising agreement with computer iterations of such a simple non linear recursion relation as the logistic equation. This universality has been understood by Feigenbaum (1978, 1979, 1980) and, Coullet and Tresser (1978, 1980) in the framework of renormalization theory. In fact it is possible to construct an analogy with critical phenomena, and to derive critical exponents, that are in good agreement with experimental measurements. Therefore it has become customary to compare these experiments with the one dimensional map of the interval $[0,1]$ into itself, given by the logistic equation

$$x_{n+1} = a\, x_n(1-x_n) \qquad 0 < a < 4$$

describing a dynamical process, where the subscripts n and n+1 are assumed to represent two successive instants of time. For different values of the constraint a, the sequence generated by successive iterations can show steady state behavior, attracting periodic orbits of period 2^n, corresponding to pitchfork bifurcations at $a = a_n$, and chaotic behavior which is sensitive to the initial conditions, for many values of a larger than $a_\infty = 3.57$. Beyond this value, two remarkable phenomena occur. First, a reverse bifurcation sequence shows noisy period halving with increasing a. Second, within the chaotic region, a sequence of periodic states occurs, corresponding to tangent bifurcations (see Collet and Eckmann 1980). This pattern of

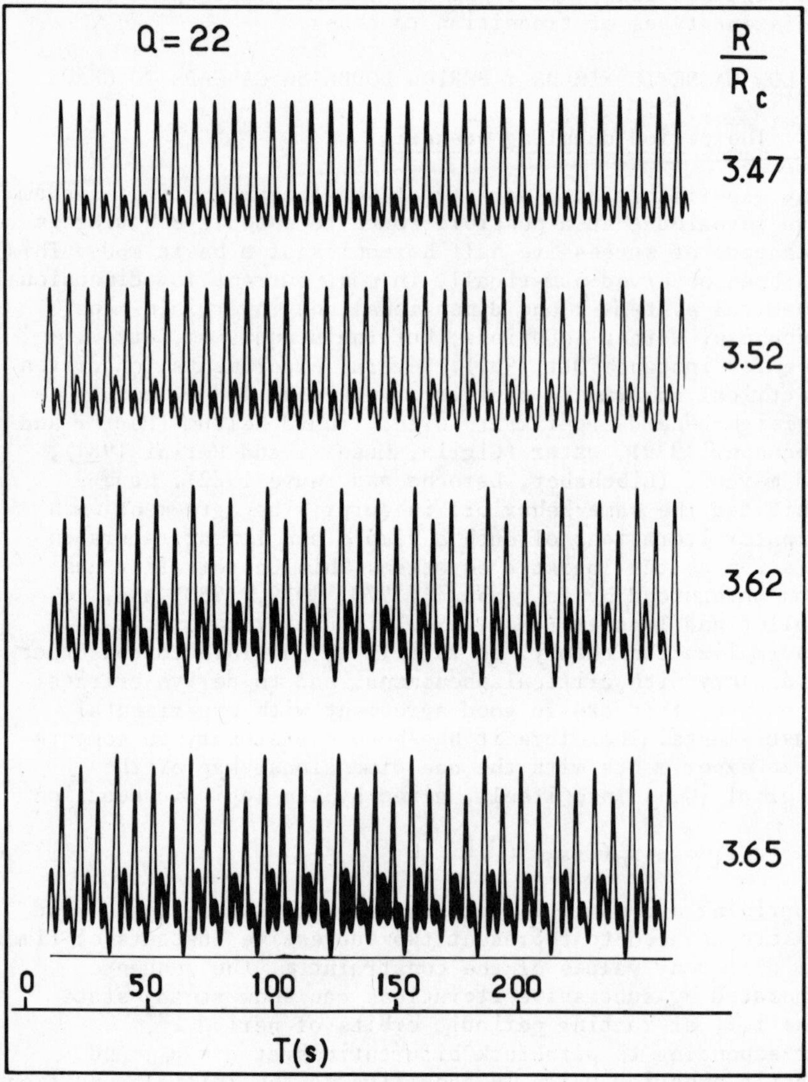

Fig. 4. Direct temperature recordings for the period doubling cascade in mercury.

bifurcations is shown in figure 3. The theory predicts that the values a_n of the constraint should asymptotically satisfy the recurrence relation

$$\frac{a_{n+1} - a_n}{a_{n+2} - a_{n+1}} \to \delta \quad \text{with } \delta = 4.669...$$

The other quantitative prediction is related to the ratio μ of the successive subharmonics Fourier amplitudes. When n is large

$$\mu \to 4.58$$

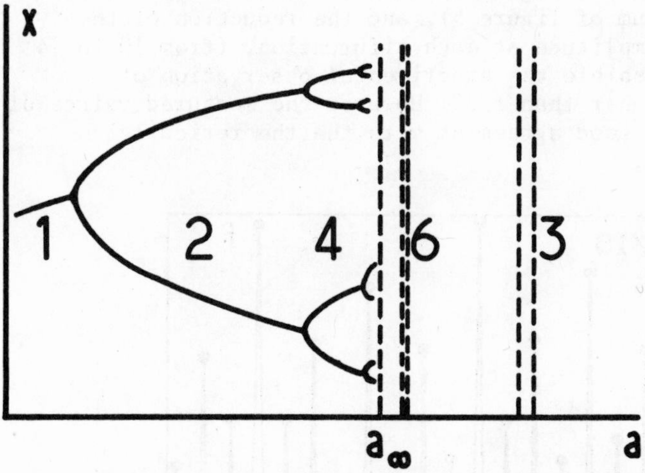

Fig. 3. Pattern of bifurcations for the logistic map.

2.2. The cascade of pitchfork bifurcations in mercury

For a Rayleigh-Bénard experiment with mercury at low magnetic field, the basic time dependent mode corresponds to the oscillatory instability. With a four convective roll experiment, the frequencies f and f/2 are present in the temperature spectrum at the onset of the time dependent regime. For a magnetic field amplitude of 270 gauss, the temperature recordings showing the bifurcations to the frequencies f/4, f/8 and f/16, are presented in figure 4. The Feigenbaum number δ, computed for the last three bifurcations is

$$\delta = \frac{R_8 - R_4}{R_{16} - R_8} = 4.4 \pm 0.1$$

The figure 5 shows the Fourier spectrum of the period doubling cascade developped up to f/16. One can see that the odd harmonics of f/16 have an amplitude depending on the order of the harmonic. We have therefore computed the ratio μ from the direct measurement of the signal amplitude and not from its Fourier spectrum. The last value of μ is

$$\mu \simeq 5$$

The finite ratio of signal to noise, (about 65 dB in the Fourier spectrum of figure 5), and the reduction of the subharmonics amplitude at each bifurcation, (from 10 to 14 dB), make impossible the experimental observation of bifurcation higher than f/32. However the measured values of δ and μ are in good agreement with the theoretically predicted ones.

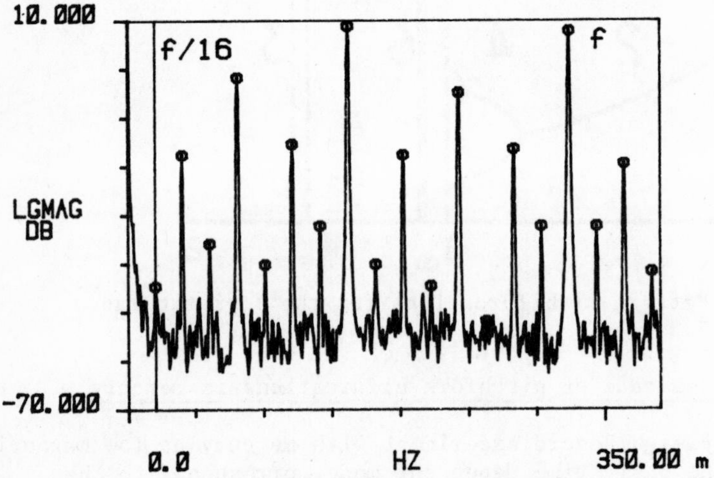

Fig. 5. Period doubling cascade in mercury.

2.3. The interrupted cascade

The period doubling cascade is often interrupted by another subharmonic bifurcation, especially for very low magnetic field amplitude. This scenario might be understood in the

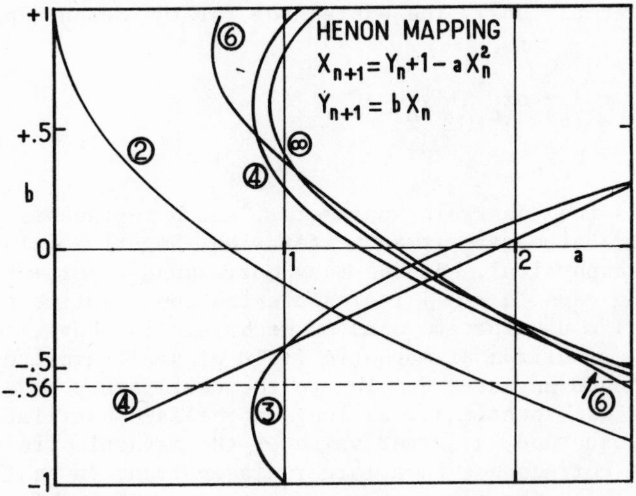

Fig. 6. Hénon mapping. Traces of some periodic attractors.

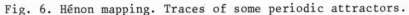

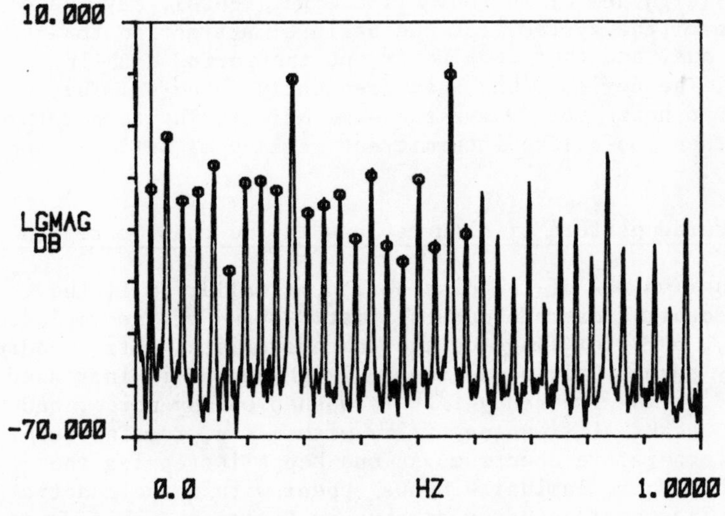

Fig. 7. Fourier spectrum for period 10.

framework of a two-dimensional mapping, the Hénon mapping (Arnéodo et al 1982). The expression for the Hénon mapping is

$$x_{n+1} = 1 - ax_n^{\ 2} + y_n$$

$$y_{n+1} = b\ y_n$$

where a is the constraint parameter, and b represents the area contraction rate at each iteration. Negative values of b are more physical, because they correspond to orientation preserving maps. $b = -1$ represents the conservative case, and $|b| < 1$ a dissipative one. As we have said above, the effect of a horizontal magnetic field of small amplitude is simply to add an extra damping to the oscillatory modes. Therefore we expect that, as long as the same oscillatory mode is concerned, a larger value of the magnetic field amplitude corresponds to a larger dissipation, (a smaller value of $|b|$). The figure 6 represents several stable dynamical states of the Hénon mapping for given values of a and b. When the dissipation is small, one can see competing basins of attraction, (the period 2 and the period 3 for instance). A non adiabatic change of the control parameter, (the Rayleigh number in convection experiments), may induce the jump of the system from one basin of attraction to another one, and therefore interrupt the period doubling cascade. The period 3 has been frequently observed. The background noise could have the same effect. The temperature signal then looks like intermittent regimes of period 2 and period 3.

2.4. Ordered pattern of bifurcations in the chaotic region

For higher values of the dissipation, (smaller $|b|$), the period-doubling cascade occurs first. Beyond its accumulation point R_∞, a mirror image of the cascade exists, corresponding to noisy period halving as the Rayleigh number is increased. At $R = R'_n$, the frequency $f/2^n$ is washed out by noise, and only the peaks at frequency $f/2^m$, with $m < n$, remain visible in the temperature spectrum. As one keeps increasing the Rayleigh number, laminar windows appear within the chaotic region. They correspond to domains in Rayleigh number where the temperature signal becomes periodic again. We have observed periods 10, 3 and 9, in the right order for the universal sequence of Metropolis, Stein and Stein (1973).

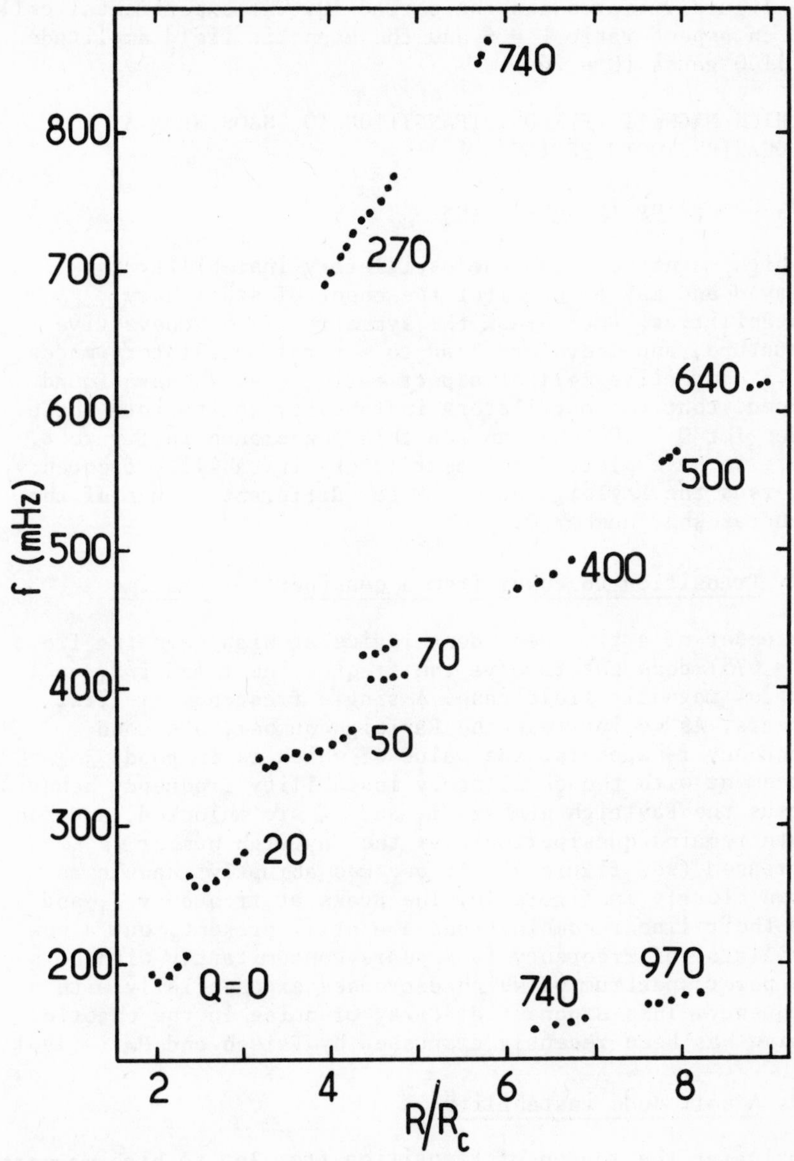

Fig. 8. Oscillatory instability frequency, f, versus the Rayleigh number, R, for different values of the Chandra--sekhar number, Q.

The figure 7 represents the period 10. The experimental cell has an aspect ratio $\Gamma = 6$ and the magnetic field amplitude is 1100 gauss ($Q = 260$).

3. HIGH MAGNETIC FIELD : TRANSITION TO CHAOS FROM A QUASIPERIODIC REGIME

3.1. Competing instabilities

At high magnetic field the oscillatory instability is delayed and may occur after the onset of stationary instabilities, that break the symmetry of the convective structure, and therefore lead to several oscillatory modes. For a convective cell of aspect ratio $\Gamma = 4$, we have found indeed, that the oscillatory instability splits into three modes for $Q > 100$. We can see this phenomenon in figure 8, where we have plotted the oscillatory instability frequency f versus the Rayleigh number R for different values of the Chandrasekhar number Q.

3.2. Transition to chaos from a quasiperiodic regime

The onset of a time dependent regime at high magnetic field ($Q = 675$) does not involve the frequencies f and f/2, as in the low magnetic field case. A single frequency f_1 first appears. As we increase the Rayleigh number, a second frequency f_2 appears, the value of which is in good agreement with the oscillatory instability frequency behavior versus the Rayleigh number. f_1 and f_2 are unlocked, and the state remains quasiperiodic as the Rayleigh number is increased (see figure 9). It becomes abruptly chaotic as shown clearly in figure 10. The peaks at frequency f_1 and f_2, and their linear combinations are still present, but a new oscillator of frequency f_3 appears concomitantly with noise, the power spectrum of which decreases exponentially with frequency. This exponential decay of noise in the chaotic regime has been recently discussed by Frisch and Morf (1981).

3.3. A soft mode instability

Looking at the region of transition from low to high magnetic field behavior, we have found another route to chaos. A secondary instability sets in as we increase the Rayleigh number from a state with the frequencies f and f/2. In the direct temperature time recordings of figure 11, we can see

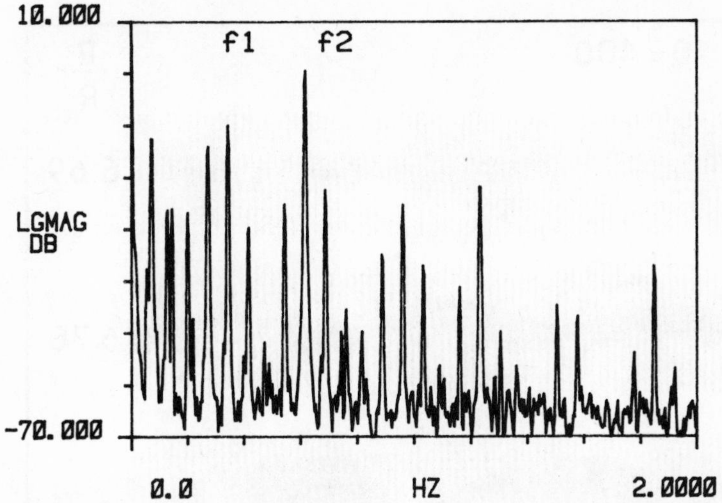

Fig. 9. Quasiperiodic state (R/R_c=8.91 ; Q=675).

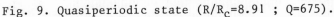

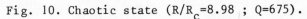

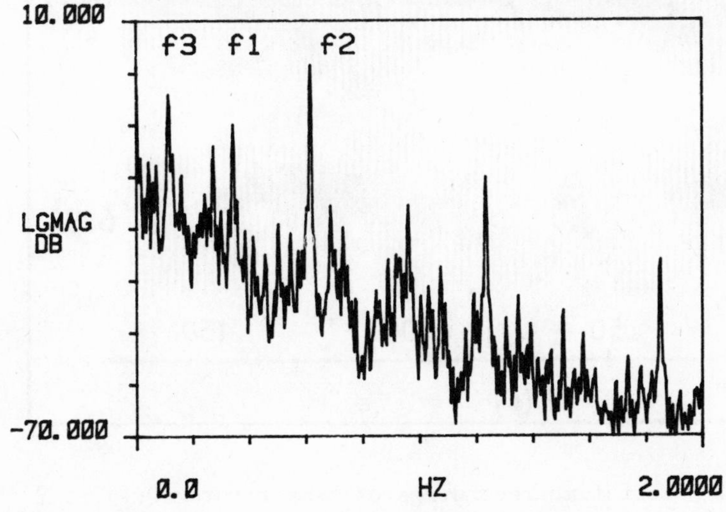

Fig. 10. Chaotic state (R/R_c=8.98 ; Q=675).

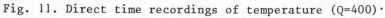

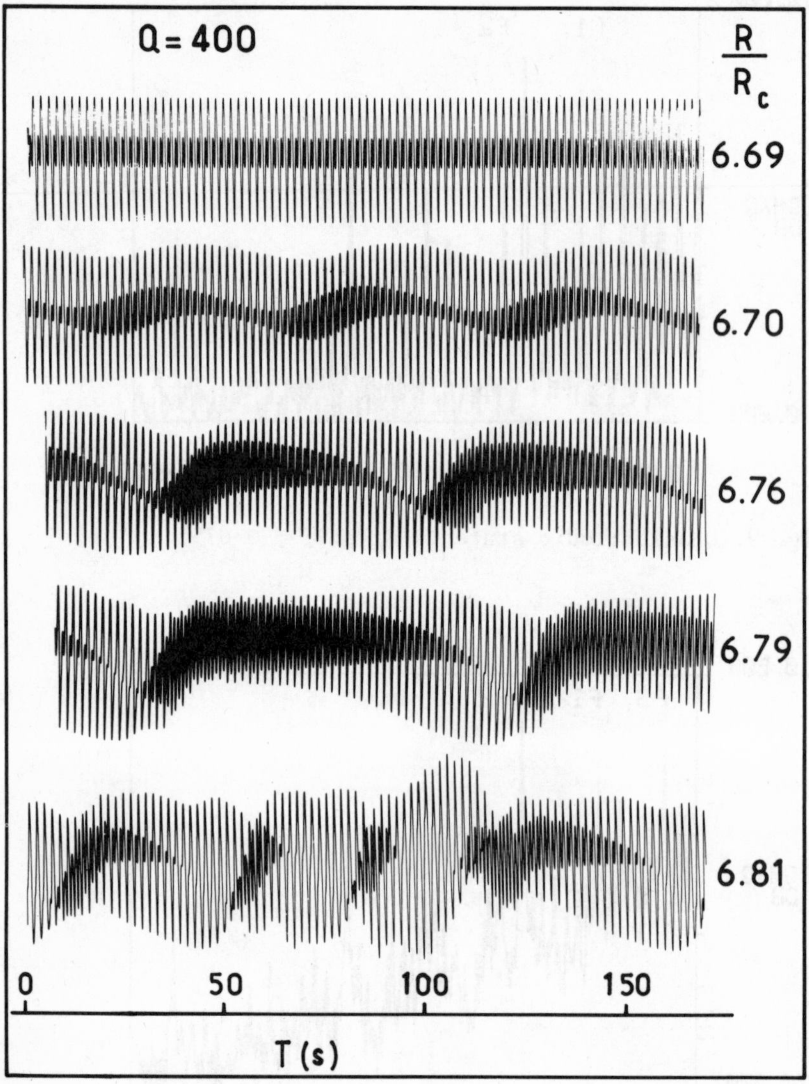

Fig. 11. Direct time recordings of temperature (Q=400).

that this secondary instability consists of a slow modulation that appears as the Rayleigh number is increased. It takes the form of a competition between the oscillators f and f/2. This behavior is in agreement with a mechanism proposed by Langford et al (1980), where a secondary instability associated with a low frequency mode, results from the competition between a symmetry-breaking linear instability and an oscillatory instability (see also Holmes 1980). When we increase the Rayleigh number the modulation period increases, which may be associated with the presence of an homoclinic (or heteroclinic) orbit. Finally, a chaotic state is reached with a very low frequency broad band noise.

CONCLUSION

Rayleigh-Bénard experiments in low aspect ratio containers have clearly demonstrated the relevance of dynamical system theory to temporal chaos in spatially ordered fluid systems. This experiment with mercury is the first attempt to study different routes to chaos in a two-dimensional space of physical parameters (R, Q). For low Q values, the essential effect of the magnetic field is to add an additional damping term to the oscillatory instability. For large Q values, it leads to the presence of competing instabilities. Unfortunately this occurs in a non linear regime, and no theory exists today. A better system, from this point of view, would present competing instabilities at the onset of convection. This is the case of rotating convection at low Prandtl number. Another extension of this work is the study of the transition to turbulence in large aspect ratio containers, when the magnetic field acts as an ordering field for the convective pattern. This might give insights into the interaction mechanism between temporal chaos and spatial patterns.

REFERENCES

Ahlers, G. : 1974, Physical Review Letters 33, 1185.
Arnéodo, A., Coullet, P.H., Spiegel, E.A. and Tresser C.: preprint.
Arnéodo, A., Coullet, P.H., Tresser, C., Libchaber, A., Maurer, J. and d'Humières, D., Physica D (in press).
Busse, F.H. : 1978, Rep. Prog. Phys. 41, 1929.
Busse, F.H. and Clever, R.M. : 1979, J. Fluid Mech. 91, 319.
Busse, F.H. and Clever, R.M., Preprint.

Busse, F.H. and Whitehead, J.A. : 1974, J. Fluid Mech. 66, 67.
Chirikov, B.V. : 1959, Atomnaya Energiya 6, 630.
Clever, R.M. and Busse, F.H. : 1974, J. Fluid Mech. 65, 625.
Collet, P. and Eckmann, J.P. : 1980, Iterated Maps on the Interval as Dynamical Systems, Birkhauser.
Coullet, P. and Tresser, C. : 1978, J. Physique 39, C5.
Coullet, P. and Tresser, C. : 1980, J. Phys. Lett. 41, L 255.
Cross, M.C., preprint.
Eckmann, J.P. : 1981, Rev. Mod. Phys. 53, 643.
Fauve, S. and Libchaber, A. : 1981, Chaos and Order in Nature, Springer Verlag.
Fauve, S., Laroche, C. and Libchaber, A. : 1981, J. Phys. Lett. 42, L 455.
Feigenbaum, M. : 1978, J. Stat. Phys. 19 (1), 25.
Feigenbaum, M. : 1979, J. Stat. Phys. 21 (6).
Feigenbaum, M. : 1980, Commun. Math. Phys. 77, 65.
Frisch, U. and Morf, R. : 1981, Phys. Rev. A23, 2673.
Giglio, M., Musazzi, S. and Perini, U. : 1981, Phys. Rev. Lett. 47, 243.
Gollub, J.P. and Swinney, H.L. : 1975, Phys. Rev. Lett. 35, 927.
Gollub, J.P. and Steinman, J.F. : 1981, Phys. Rev. Lett. 45, 505.
Hénon, M. : 1976, Commun. Math. Phys. 77, 50.
Holmes, P. : 1980, Ann. N.Y. Acad. Sci. 357, 473.
Langford, W.F., Arnéodo, A., Coullet, P., Tresser, C. and Coste, J. : 1980, Phys. Lett. A78, 11.
Lehnert, B. and Little, N.C. : 1957, Tellus 9, 97.
Libchaber, A., Fauve, S. and Laroche, C. : Physica D (to be published).
Libchaber, A., Laroche, C. and Fauve, S. : 1982, J. Phys. Lett. 43, L 211.
Linsay, P.S. : 1981, Phys. Rev. Lett. 47, 1349.
Lorenz, E.N. : 1963, J. Atm. Sci. 20, 130.
Maurer, J. and Libchaber, A. : 1979, J. Phys. Lett. 40, L 419.
Metropolis, N., Stein, M.L. and Stein, P.R. : 1973, J. Comb. Theory A15, 25.
Normand, C. : 1981, ZAMP 32, 81.
Ott, E. : 1981, Rev. Mod. Phys. 53 (4), 655.
Ruelle, D. and Takens, F. : 1971, Commun. Math. Phys. 20, 167.
Schluter, A., Lortz, D. and Busse, F.H. : 1965, J. Fluid Mech. 23, 129.
Siggia, E.D. and Zippelius, A. : 1981, Phys. Rev. Lett. 47, 835.
Tabeling, P. : 1982, J. Physique 43, 1295.

C. LOBRY and C. REDER.

MICRO COMPARTMENTAL SYSTEMS

We report here on a work in progress. We hope to show that the combination of combinatorial techniques with those of Non Standard Analysis may provide an alternative to the use of Partial Differential Equations in the study of spatial structurations of some distributed parameter systems.

Our study relies very much on an idea of J. HARTHONG in that it connects formal mathematical objects like infinitesimal and standard real numbers to physical concepts like microscopic and macroscopic . See [3] .

It relies also an old idea of WIENER an ROSENBLUETH who studied the cardiac muscle with the aid of a cellular automaton. See [9] .

The idea that Non Standard Analysis is not a game for logicians but merely the right language to describe physical objects in presence of two scales was imposed to us by G. REEB.

Thanks to recent results obtained by the second author in collaboration with J.P. ALLOUCHE we obtain here some explanation of observed target patterns in excitable media. To our knowledge this explanation is not accessible to the description by reaction diffusion equations.

This is the first relation of a long collaboration. We are not yet sure that the definitions are the best possible and we suspect that some technicalities may not be absolutely correct. Thus we ask the reader to take this just as a first draft.

Section one is devoted to one dimensional systems and section two to two dimensional ones. Non Standard Analysis is used in the terminology of NELSON . See [7] and [6] .

For a technical reason it was not possible to the second author to participate to the final version of this paper. Thus the first author is responsible for all mistakes and awkwardness in the form.

C. P. Bruter et al. (eds.), Bifurcation Theory, Mechanics and Physics, 277–297.

1. ONE DIMENSIONAL CONTINUOUS MICRO-COMPARTMENTAL SYSTEMS.

1.1. The formal mathematical object.

The motivations for the following definitions will become transparent later (we hope so !).

Let $h > 0$ be an infinitesimal (but different from 0) real number of the form :

$$h = \frac{1}{N}$$

where N is an infinitely large integer. Consider a partition of the interval $[0 , 1]$ in N sub intervals denoted by c_i :

$$c_i = [ih \, ; \, (i+1)h[\qquad i = 0,1,2...N-1$$

and call them "compartments". To each integer i associate the dynamical system:

$$\sum_i \left\{ \begin{array}{ll} \frac{dx_i}{dt} = f(x_i,u_i) & x_i \in R^n, \; u_i \in R^p \\ y_i = \phi(x_i) & y_i \in R^q, \quad (1 \leq i \leq N-1) \end{array} \right.$$

One recognizes the classical notations for an input output system. We emphasize that f, ϕ, n, p, q do not depend on i and thus we are concerned with N copies of the same system. We say that the compartment c_i supports the dynamic $\sum_i$. We assume on f and ϕ good regularity conditions to ensure existence, uniqueness and so on....

Each compartment c_i is coupled to his two neighbours (except for c_0 and c_{i})[i] through the input u_i . This is given by the mapping[i] :

$$K : R^n x R^n \longrightarrow R^p$$

and we have :

$$u_i(t) = K(x_{i-1}(t), x_{i+1}(t))$$

The compartments c_0 and c_{N-1} support specific dynamics and couplings specified below.

1.1.1. Definition

A micro-compartmental system is the system of ordinary differential equations:

$$\Sigma \begin{cases} \left\{ \begin{aligned} \frac{dx_i}{dt} &= f(x_i, u_i) \\ u_i &= K(x_{i-1}, x_{i+1}) \end{aligned} \right\} i = 1,2,..N-2 \\ \\ \begin{aligned} \frac{dx_\circ}{dt} &= f_\circ(x_\circ, u_\circ) \; ; & \frac{dx_{N-1}}{dt} &= f_{N-1}(x_{N-1}, u_{N-1}) \\ u_\circ &= K_\circ(x_1) & u_{N-1} &= K_{N-1}(x_{N-2}) \end{aligned} \\ \\ y_i = \phi(x_i) \qquad i = 0,\ 1,\ \ldots,\ N-1 \end{cases}$$

Provided that all the datas : f, K, $f_\circ, f_{N-1}, K_\circ, K_{N-1}, \phi$, are standard it is known that the system above is just an ordinary system of differential equations, even if N is an unlimited integer. Thus existence and uniqueness is given by the well known theorem.

Consider a micro-compartmental system Σ, and let y_i be the collection of observations at some instant t. Let m be a standard point and consider :

$$u(m,t,\omega_1,\omega_2) = \frac{1}{\omega_1+\omega_2+1} \sum_{i(m)+\omega_2}^{i(m)+\omega_1} y_i(t)$$

with :

$$i(m)h < m \leqslant (i(m)+1)h$$

1.1.2 Definition.

If there exists an integer ω such that for $\omega_1 > \omega$; $\omega_2 > \omega$ with $\omega_1 h$ and $\omega_2 h$ infinitesimal, the standard part:

$$°u(m,t,\omega_1,\omega_2)$$

of $u(m,t,\omega_1,\omega_2)$ does not depend on ω_1 and ω_2 we say that the macroscopic observation at point m an at time t exists. For $m = 0$ or $m = 1$ we prescribe that ω_1 or ω_2 is nul. The macroscopic observation, when it exists, is benoted by :

$$u(m,t) = °u(m,t,\omega_1,\omega_2)$$

Let us show on examples what the macroscopic observation is.

1.1.3 Examples.

a- Assume that if the integer i is odd then $y_i(t) = 0$ and otherwise $y_i(t) = 1$. Then one sees that $u(m,t)$ is well defined for every standard m and is equal to $\frac{1}{2}$. More generally this result ($u(m,t)$ is defined everywhere and independant of m) holds when $y_i(t)$ is periodic of period p with respect to i with p such that hp is infinitesimal.

b- Assume that if $i \leqslant \frac{N}{2}$ then $y_i(t) = 0$ and otherwise $y_i(t) = 1$. Then $u(m,t)$ is defined and equal to 0 if $m < \frac{1}{2}$, is defined and equal to 1 if $m > \frac{1}{2}$, but is not defined for $m = \frac{1}{2}$.

c- Define $y_i(t)$ in the following way:

$$y_i(t) = 0 \quad \text{if } 1+2+2^2+\ldots+2^{2p} < i \leqslant 1+2+2^2+\ldots+2^{2p+1}$$

$$y_i(t) = 1 \quad \text{if } 1+2+2^2+\ldots+2^{2p+1} < i \leqslant 1+2+2^2+\ldots+2^{2p+2}$$

Let ω_2 be an unlimited integer such that $\omega_2 h$ is infinitesimal and let $n(\omega_2)$ be defined by:

$$1+2+2^2+\ldots+2^{n(\omega_2)} < \omega_2 \leqslant 1+2+2^2+\ldots+2^{n(\omega_2)+1}$$

or equivalently:

$$2^{n(\omega_2)+1}-1 < \omega_2 \leqslant 2^{n(\omega_2)+2}-1$$

Because the number $(2^{n(\omega_2)+1}-1)h$ is infinitesimal, so is the number $(2^{n(\omega_2)+1})h$, and also the number $(2^{n(\omega_2)+r})h$ for each limited integer r. Then, a simple computation shows that, according to the parity of r the number:

$$u(o,t,2^{n(\omega_2)+r})$$

is equal to $\frac{1}{3}$ or $\frac{2}{3}$. Thus the macroscopic observation does not exist at 0 in this case.

The study of micro-compartmental systems is the study of the evolution in time of the macroscopic observation. From the knowledge of the microscopic structure:

the system $\sum$

we want to predict the macroscopic effects, i.e. the macroscopic observation. Before any comments on this we make more precise what we have in mind in the next section.

1.2. Micro-compartmental systems with linear diffusion.

We consider the case where:

$$\begin{cases} f(x_i, u_i) = X(x_i) + \frac{k^2}{h^2}(u_i^1 - x_i) + \frac{k^2}{h^2}(u_i^2 - x_i) \\ (u_1^2, u_i^2) = (x_{i-1}, x_{i+1}) \end{cases}$$

and for the two compartments of the end we assume that:

$$\begin{cases} f_o(x_o, u_o) = X(x_o) + \frac{k^2}{h^2}(u_o^2 - x_o) \\ u_o^2 = x_1 \end{cases}$$

$$\begin{cases} f_{N-1}(x_{N-1}, u_{N-1}) = X(x_{N-1}) + \frac{k^2}{h^2}(u_{N-1}^1, -x_{N-1}) \\ u_{N-1}^1 = x_{N-2} \end{cases}$$

we assume that x_i belongs to R, (i.e. n=1), that $u_i = (u_i^1, u_i^2)$ belongs to R^2, (i.e. p=2) and that the observation ϕ is just the identity: $y_j = x_i$. We assume that X is a standard vector field on $\mathbb{R}$ lipschitz :

$$|X(x_1) - X(x_2)| < \lambda |x_1 - x_2|$$

with a standard constant λ .

These assumptions define the system $\sum$, which we rewrite as $\sum_d$ in the more familiar form:

$$\Sigma_d \begin{cases} \frac{dx_i}{dt} = X(x_i) + \frac{k^2}{h^2}(x_{i-1} - 2x_i + x_{i+1}) \quad i=1,2,\ldots,N-2 \\ \frac{dx_o}{dt} = X(x_o) + \frac{k^2}{h^2}(x_o - x_1) \\ \frac{dx_{N-1}}{dt} = X(x_{N-1}) + \frac{k^2}{h^2}((x_{N-2}, x_{N-1}) \end{cases}$$

and in a more condensed form:

$$\Sigma_d \left\{ \frac{dx}{dt} = \mathcal{X}(x) + \frac{k^2}{h^2} \Lambda x \qquad x \in R^n \right.$$

where Λ is the ad-hoc tridiagonal matrix and $\mathcal{X} = \begin{pmatrix} X \\ \vdots \\ X \end{pmatrix}$.

One recognizes a discrete version of a partial differential equation and actually our objective is to prove that when the initial conditions of $\sum_d$ are regular (to be precised) then the macroscopic observation coincides for standard m and t with the solution of the partial differential equation:

$$\text{(P.D.E.)} \quad \begin{cases} \dfrac{\partial W(m,t)}{\partial t} = k^2 \dfrac{\partial^2 W(m,t)}{\partial m^2} + X(W(m,t)) . \\ \dfrac{\partial W(0,t)}{\partial m} = \dfrac{\partial W}{\partial m}(1,t) = 0 \end{cases}$$

We suspect that such a result is written somewhere by someone who has interest in Non Standard Analysis but we dont know where. We apologize for that.

We first need a technical lemma:

1.2.1. Lemma :

Let A be one of the two tridiagonal matrices:

$$\begin{pmatrix} -a & +1 & 0 & . & . & . & 0 \\ 1 & -2 & 1 & 0 & . & . & 0 \\ . & . & . & . & . & . & . \\ 0 & . & . & . & 1 & -2 & 1 \\ 0 & . & . & . & 0 & 1 & -a \end{pmatrix}$$

with $a = 1$ or $a = 2$. Consider the differential equation:

$$\frac{d\zeta}{dt} = A\zeta + F(\zeta,t) \qquad =(\zeta_i) \ i=1...N.$$

and assume that for each component F_i of F one has:

$$|F_i(\zeta,t)| < \lambda|\zeta_i| + g(t) \qquad \lambda>0 \quad g(t)>0$$

then the following majorization holds:

$$\max_{j=1,2,...N}\{ |\zeta_j(t)| \} < e^{\lambda t} \{\max_{j=1,2..N} |\zeta_j(0)| + \int_0^t e^{-s} g(s)ds\}$$

Proof: Consider $t \longrightarrow s(t)$ the solution of:

$$s'(t) = \lambda s + g(t)$$

$$s(0) = \max_{j=1,2...N}\{ |\zeta_j(0)| \}$$

and notice that :

$$\eta(t) = \zeta(t) - \begin{pmatrix} s(t) \\ s(t) \\ \vdots \\ s(t) \end{pmatrix}$$

is a solution of an ordinary differential equation that leaves invariant the negative cone which proves one side of the inequality, the other one being proved in the same way.

1.2.2 Existence of the macroscopic observation.

We prove that the macroscopic obsevation associated to the system $\sum_d$ is well defined provided that the initial condition satisfies the majorization:

$$|x_{i+1}(0) - x_i(0)| < \alpha h$$

for some limited positive real number α .

Recall that λ is the lipschitz constant for X. Consider:

$$\sum_d \quad \frac{d}{dt}\begin{pmatrix} x_0(t) \\ \vdots \\ x_i(t) \\ \vdots \\ x_{N-1}(t) \end{pmatrix} = \begin{pmatrix} X(x_0(t)) \\ \vdots \\ X(x_i(t)) \\ \vdots \\ X(x_{N-1}(t)) \end{pmatrix} + \frac{k^2}{h^2}\Lambda\begin{pmatrix} x_0(t) \\ \vdots \\ x_i(t) \\ \vdots \\ x_{N-1}(t) \end{pmatrix}$$

with its initial condition and observe that :

$$t \longrightarrow \zeta(t) = x_{i+1}(t) - x_i(t)$$

is just the solution of the differential system:

$$\frac{d}{dt}\begin{pmatrix} \zeta_0(t) \\ \vdots \\ \zeta_i(t) \\ \vdots \\ \zeta_{N-1}(t) \end{pmatrix} = \frac{k^2}{h^2}\Lambda\begin{pmatrix} \zeta_0(t) \\ \vdots \\ \zeta_i(t) \\ \vdots \\ \zeta_{N-1}(t) \end{pmatrix} + \begin{pmatrix} X(x_1(t))-X(x_0(t)) \\ \vdots \\ X(x_{i+1}(t))-X(x_i(t)) \\ \vdots \\ X(x_{N-1}(t))-X(x_{N-2}(t)) \end{pmatrix}$$

Consider the mapping :

$$(\zeta,t) \longrightarrow F(\zeta,t) = (F_i(\zeta,t) \quad ; i=1\ldots N-1$$

with $F_i(\zeta,t) = X(x_i(t)+\zeta) - X(x_i(t))$, one has :

$$F_i(\zeta,t) \;<\; \lambda|\zeta|$$

which, with the initial condition, are just the majorization we need to apply lemma 1.2.1 in order to obtain the next inequality.

$$\max_{i=0,1,\dots N-1} \{|x_{i+1}(t)-x_i(t)|\} < e^{\lambda t} \max_{i=0,1\dots N-1} \{|x_{i+1}(0)-x_i(0)|\}$$

Now we consider the standard number m $(0 < m < 1)$ and we compute the mean:

$$u(m,t,\omega_1,\omega_2) = \frac{1}{\omega_1+\omega_2+1} \sum_{j=i(m)-\omega_1}^{j=i(m)+\omega_2} x_i(t)$$

After the introduction of $x_{i(m)}$ in each term we get:

$$|u(m,t,\omega_1,\omega_2)-x_{i(m)}(t)| \;<\; \left|\frac{1}{\omega_1+\omega_2+1}\right| \sum_{j=i(m)-\omega_1}^{j=i(m)+\omega_2} |x_j(t)-x_{i(m)}(t)|$$

and from the trivial majorization:

$$|x_j(t) - x_{i(m)}(t)| \;<\; (\omega_1+\omega_2+1) \max_{j=0,1\dots N-1} \{|\, x_{j+1}(t)-x_j(t)|\}$$

we get:

$$|u(m,t,\omega_1,\omega_2)-x_{i(m)}(t)| \;<\; (\omega_1+\omega_2+1) \max_{j=0,1\dots N-1}\{|\, x_{j+1}(t)-x_j(t)|\}$$

which from the first majorization and hypothesis on the initial condition turns out to be:

$$|u(m,t,\omega_1,\omega_2)-x_{i(m)}(t)| \;<\; e^{\lambda t}(\omega_1+\omega_2+1)\alpha h$$

Provided $\omega_1 h$ and $\omega_2 h$ are infinitesimals, provided t is limited, the numbers λ and α being limited, the right member of the inequality is infinitesimal and thus the macroscopic observation is just the standard part of $x_{i(m)}(t)$.

1.2.3 Discretization of the solution of P.D.E.

We consider the cauchy problem:

$$(\text{P.D.E.})\begin{cases} \dfrac{\partial W(m,t)}{\partial t} = k^2 \dfrac{\partial^2 W(m,t)}{\partial t^2} + X(W(m,y)) \\ \dfrac{\partial W(0,t)}{\partial m} = \dfrac{\partial W(1,t)}{\partial m} = 0 \\ W(m,0) \text{ given} \end{cases}$$

Under the current assumptions made on X, if moreover we assume that W(m,0) belongs to $H^{2+\beta}(0,1)$ $\beta > 0, \beta$ standard, (which means that the second derivative of W with respect to m exists and is holder of order β) we know the:

Theorem: (P.D.E.) has a unique solution in $C^{2,1}[0,1] \times [0, +\infty]$

This seems to be a folklore result but unfortunately we dont know any classical explicit reference for it. See [8] for more details. Now, thanks to this result, a straightforward computation (Taylor formula) gives us the following:

Denote by:

$$t \longrightarrow W_i(t) = W(ih+\frac{h}{2}, t)$$

then there exist continuous mappings $t \longrightarrow \varepsilon_i(t)$ such that $(W_0(t), W_1(t), \ldots, W_{N-1}(t))$ is the solution of :

$$(\text{P.D.E.})_d \quad \frac{d}{dt}\begin{bmatrix} W_0(t) \\ W_1(t) \\ \cdot \\ \cdot \\ W_{N-1}(t) \end{bmatrix} = \frac{k^2}{h^2} \Lambda \begin{bmatrix} W_0(t) \\ W_1(t) \\ \cdot \\ \cdot \\ W_{N-1}(t) \end{bmatrix} + \begin{bmatrix} X(W_0(t)+k^2\varepsilon_0(t)) \\ X(W_1(t)+k^2\varepsilon_1(t)) \\ \cdot \\ \cdot \\ X(W_{N-1}(t)+k^2_{N-1}\varepsilon_{N-1}(t)) \end{bmatrix}$$

All the datas being standard in (P.D.E.) we also know that if h is infinitesimal (which is the case) , for limited t the real $\varepsilon_i(t)$ is also infinitesimal.

Now we can state and prove the result we anounced at the beginning of this section.

1.2.4. Theorem: Consider a standard mapping ϕ in $H^{2+\beta}|0,1|$ and the system $\sum_d$ with the initial condition:

$$x_i(0) = \phi(ih + \frac{h}{2}) \quad i = 0,1,\ldots N-1$$

Then under the current hypothesis the following holds:

i) The macroscopic observation $u(m,t)$ is defined for every standard m and limited t.

ii)The macroscopic obsevation coincides when it is defined with the solution of (P.D.E.) with initial condition $W(m.0) = \phi(m)$

Proof: Because ϕ is standard and C^2 there exist a standard constant λ such that :

$$|x_{i+1}(0)-x_i(0)| < \lambda$$

and thus the assumption made in 1.2.1. is satisfied which proves i).

Consider now the difference:

$$\eta_i(t) = W_i(t)-x_i(t)$$

It satisfies the differential equation:

$$\frac{d}{dt}\begin{bmatrix}\eta_o(t)\\ \eta_1(t)\\ \cdot\\ \\ \eta_{N-1}(t)\end{bmatrix} = \frac{k^2}{h^2}\Lambda\begin{bmatrix}\eta_o(t)\\ \eta_1(t)\\ \\ \\ \eta_{N-1}(t)\end{bmatrix} + \begin{bmatrix}X(W_0(t))-X(x_0(t)) + k^2\varepsilon_0(t)\\ X(W_1(t))-X(x_1(t)) + k^2\varepsilon_1(t)\\ \\ \\ X(W_{N-1}(t))-X(x_{N-1}(t)) + k^2\varepsilon_{N-1}(t)\end{bmatrix}$$

$$\eta_i(0) = 0$$

and from lemma 1.2.1. we have:

$$|\eta_i(t)| < e^{\lambda t}\int_0^t e^{-\lambda s}\varepsilon_i(s)ds$$

Thus $\eta_i(t)$ is an infinitesimal and hence $W_i(t)$ is equivalent to $x_i(t)$ for every limited t.

From the definition of i(m) it turns out that $W_{i(m)}(t)$ is equivalent to $W(m,t)$. Thus $W(m,t)$ is equivalent to $x_{i(m)}(t)$ which was proved to be equivalent to $u(m,t)$ in 1.2.2. This proves point ii) of the theorem.

1.3. Comments.

In section 1.1. we defined micro-compartmental systems and in 1.2. we proved that the macroscopic observation of the particular system $\sum_d$ coincides for standard values of the variables with the solution of some related partial differential equation.

No doubt that there are classes of micro-compartmental systems which are related to other kinds of differential operators, no doubt that homogeneization techniques have their counterparts in micro-compartmental systems. We shall not explore this direction because our goal is in some sense completely opposite. We want to provide alternative techniques to those of partial differential equations for the analysis of distributed parameter systems.

2. TWO DIMENSIONAL DISCRETE MICRO-COMPARTMENTAL SYSTEMS.

The formalism proposed in 1 is clearly extendable to higher dimensions. It is also possible to extend it to systems discrete with respect to time. We shall not give this general definition, we prefer to focus on a specific exemple.

2.1. The formal mathematical object.

The plane is divided into squares of size h. Each square is a compartment and denoted by c_{ij}. The state of each square is denoted by x_{ij} and belongs to some finite set :

$$\mathcal{S} = \{0,1,2,\ldots\ldots E,E+1,\ldots\ldots N-1\}$$

$$\mathcal{E} = \{1,2,\ldots,E\} \quad = \quad \text{Exitatory states}$$

$$\mathcal{R} = \{E+1,E+2,\ldots N-1\} \quad = \quad \text{Refractory states}$$

$$0 \quad = \quad \text{Neutral state}$$

The dynamics for each isolated compartment are given by:

$$\Sigma_{ij} \qquad x_{ij}(k\tau+\tau) = \begin{cases} x_{ij}(k\tau)+1 & \text{if } x_{ij}(k\tau) \neq 0 \\ 0 & \text{if } x_{ij}(k\tau) = 0 \end{cases}$$

We assume that each compartment c_{ij} is coupled to his four neighbours according to the rule: a state different from 0 is never affected, the state 0 is transformed into 1 if one of the neighbours is in an exitatory state. This gives the following rules.

$$\Sigma \qquad x_{ij}(k\tau+1) \begin{cases} = x_{ij}(k\tau)+1 \text{ if } x_{ij}(k\tau) \neq 0 \\ = 1 \quad \text{if } x_{ij}(k\tau) = 0 \text{ and one of } \begin{cases} x_{ij+1}(k\tau) \\ x_{ij-1}(k\tau) \\ x_{i+1j}(k\tau) \\ x_{i-1j}(k\tau) \end{cases} \\ \qquad \text{belongs to } \mathcal{E} . \\ = 0 \text{ if } x_{ij}(k\tau) = 0 \text{ and none of} \\ \qquad \text{the above neighbours belong to } \mathcal{E}. \end{cases}$$

These rules, with initial datas, describe completely the behaviour of the micro-compartmental system Σ .

Moreover we assume that the observation is just given by:

$$Y_{ij}(k\tau) = 1 \text{ if } x_{ij}(k\tau) \quad \varepsilon\ E$$

$$y_{ij}(k\tau) = 0 \text{ if } x_{ij}(k\tau) \quad \notin E$$

The macroscopic observation is defined for each standard point (m,n) in the plane as the mean over the halo of (m,n) of the observation. The halo of a point is not a set ; it is by definition the collection of those points which are infinitely close to this point, thus the meaning of "the mean over the halo" must be precised! The next definition is sufficient for the purposes of this paper:

Let Δ/infinitesimal rectangle containing the point/an (m,n) ; by definition such a rectangle is a rectangle of size rh x sh with r and h such that rh and sh are infinitesimals. Let us denote by u(m,n,kτ,Δ) the mean:

$$u(m,n,k\tau,\Delta) = \frac{1}{rs} \sum_{c_{ij} \in \Delta} y_{ij}(k\tau)$$

2.1.1. Definition: If there exists an infinitesimal rectangle Δ_0 such that for every infinitesimal rectangle which contains Δ_0 the standard part of the mean u(m,n,kτ,Δ) does not depend on Δ we say that the macroscopic observation is defined and we denote it by :

$$u(m,n,k\tau)$$

2.1.2. Definition: Microscopic initial condition. Let Δ_0 be an infinitesimal rectangle. An initial condition for the system $\sum$ which is equal to 0 for every compartment outside of Δ_0 is called a microscopic initial condition and denoted by:

$$x_{ij}(0) = \xi_{ij} \ ; \ c_{ij} \ \varepsilon\ \Delta_0$$

For simplicity we assume now that Δ_0 is centered at 0.

Our objective is to describe the macroscopic behaviour associated to microscopic initial conditions and to show that they are just of two types.

2.1 The evolution of $\sum$.

In what we called a micro-compartmental system one recognizes a cellular automaton. The study of the evolution of such an automaton has been done by the second author of this paper in collaboration with J.P. Allouche : [1] ; [2] .
In [2] the Théorème 3-9 gives a rather complete description of the solutions of $\sum$ in the case $E \geq 2$.

This description is a bit complicate and we apologize for not giving it here. We just give the picture of the next page which is proved to be typical in the following sense:

For each bounded observation window, after some transient mode there always exists a periodic regime of period N and two cases are possible:

i) Each compartment is in state 0

ii) We have a periodic pattern (target pattern) which looks like the one of next picture, (Th 3-9 is just the precise meaning of "looks like") see next page.

Consider now an initial condition which is 0 outside some small rectangle; Th 3-9 of [2] says that for a "living structure" (case ii)), except on the cross based on the initial rectangle of non zero initial conditions, the structure of the solution is always the one indicated on the picture; on the two orthogonal bands the solution depends on the initial conditions.

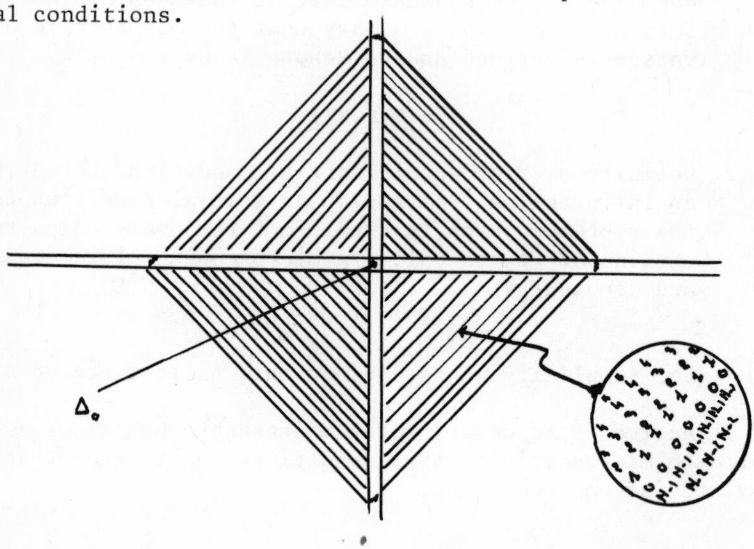

A solution of Σ obtained from a computer.

One sees that if the size of the initial condition is small (microscopic) it influences the solution just in a very small region. In non standard setting it leads us to the results of the next sections.

We emphasize that the proof from results of [2] is just a trivial reinterpretation. The non trivial part is contained in the proof (of combinatoric nature) of the results of [2].

2.3. Macroscopic behaviour (N limited).

We assume that τ is an infinitesimal such that the number:

$$v = \frac{h}{\tau}$$

is a standard real number . Then we claim the following:

For every microscopic initial condition:

$$x_{ij}(0) = \xi_{ij} \quad c_{ij} \in \Delta_\circ$$

For every time $t = k\tau$, such that $k \geq N^{\text{card } \Delta_\circ}$ there are exactly two possible macroscopic observations :

i) One observes nothing ! Which means that the macroscopic observation is defined and equal to 0 everywhere.

ii) The macroscopic observation is defined everywhere except on the lozenge L(t) defined on the picture and we have :

$$u(m,n,t) = \frac{E}{N} \quad \text{inside}$$
$$u(m,n,t) = 0 \quad \text{outside}$$

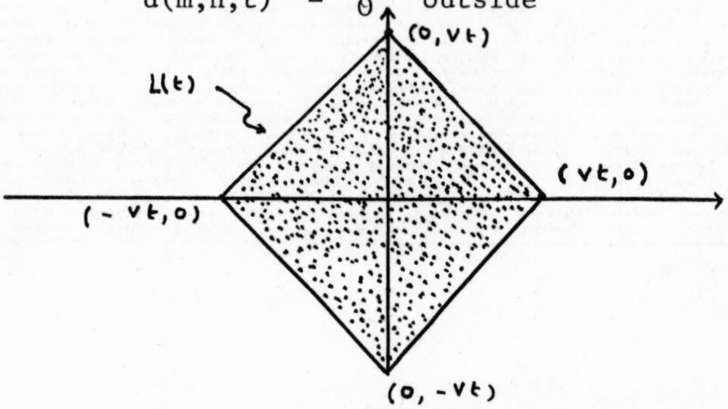

2.3. Macroscopic behaviour (N unlimited).

We assume that τ is an infinitesimal such that the number

$$v = \frac{h}{\tau}$$

is a standard real number.

We assume that N,E are unlimited integers such that:

$$Eh = \lambda_1 \quad (N-E)h = \lambda_2 \quad Nh = \lambda_1 + \lambda_2 = \lambda$$

are standard real numbers.

Then we claim that for any microscopic initial condition

$$x_{ij}(0) = \xi_{ij}\, c_{ij} \in \Delta_\circ$$

there are exactly two possible cases:

i) One observes nothing, which means that the macroscopic observation is defined and equal to 0 everywhere.

ii) One observes a target pattern picture progressing at velocity v described on the next picture.

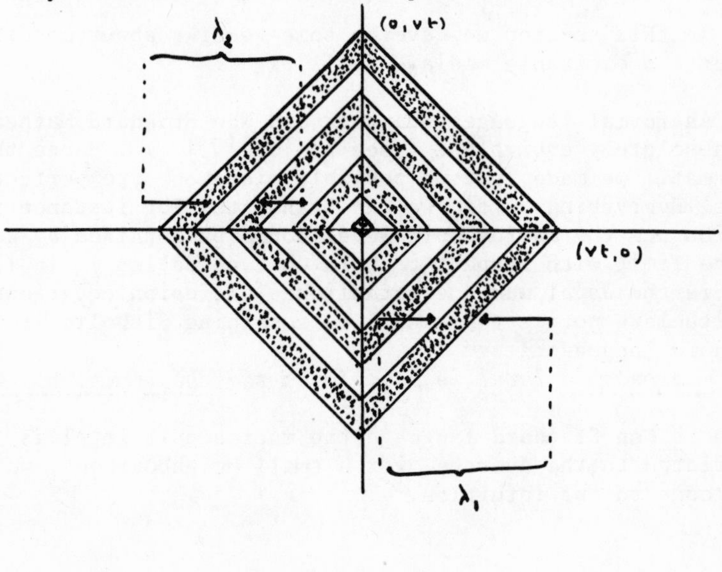

2.5. Comments.

In both cases one observes in the case of a "living structure" (second alternative) wave fronts progressing at velocity v.

Exemples in [2] show that the two alternatives are possible, but the caracterisation of initial conditions that lead to living structure is an open question.

In the case of limited N the majorization of the transient regime by $N^{\mathrm{card}\,\Delta_\circ}$ tels us that (at least if the size of $\Delta_\circ$ is rh x sh with r and s limited) the observation is valid for every time which is no equivalent to 0, but in the case of unlimited N we cannot say anything before some unlimited time $T_\circ$. The order of $T_\circ$ is very large:

$$T_\circ \simeq N^{\mathrm{card}\,\Delta_\circ}\,\tau \simeq N^{\mathrm{card}\,\Delta_\circ - 1}$$

(Because due to the order of N and τ which are necessary conditions to obtain finite velocity - v - and a visible wave length - λ - the order of τ is $\frac{1}{N}$) . Further investigations are necessary to understand to what extend this majorization is a good one.

3. HOW TO MODEL EXCITABLE MEDIA.

In this section we develop some remarks about modelling of certain excitable media.

As formal languages Standard and Non Standard Mathematics are absolutely equivalent (see [6] or [7]) . Hence the mathematic we made, except possible mistakes, are perfectly licit. Everything we did we can translate. For instance in section one the macroscopic level would be obtained by going to the limit with respect to h in what we called $\sum_d$ in 1.2. This is the usual way one establishes diffusion equations. Nevertheless notice that there is something diabolic in the Standard language:

"*The macroscopic level is associated to "the limit,* $h \to 0$".

While in Non Standard language the macroscopic level is associated to the average over a small neighbourhood, which is closer to our intuition.

Let us call an excitable medium any physical medium with the following properties:

-at a microscopic level the system has a stable equilibrium, and a small suitable deviation from this equilibrium makes the system have a large excursion before returning to the equilibrium.

The simple exemple is the grass. At the microscopic level one considers blades of grass and at the macroscopic one we consider the steppa seen from an aircraft. The problem is to study the propagation of fire in such a medium. This exemple was developped in [4].

More serious exemples are:

- Chemical reactions (Belousov Zhabotinskii)
- Cells and cardiac muscle
- Mass neurons in the brain
- Cells membranes
- Populations of amobea

These exemples are taken from Winfree [10] and are familliar to many scientists. One knows that experiments show at a macroscopic level spatial or temporal structurations. Some people try to explain this.

A classical approach is to model this via the:

$$\text{(P.D.E.)} \qquad \frac{\partial U}{\partial t} = X(U) + M\Delta U$$

where $U(x,t)$ is the state at a point x at time t, X is a vector field which models the microscopic dynamic, and $M\Delta$ (M diagonal positive matrix, Δ laplacian) models the interaction process .

This is equivalent, in our formalism, to what we called microcompartmental systems with linear diffusion. At this level it is quite transparent that linear diffusion is a model for the reality which is likely to be a crude approximation. It is also transparent that what we described in section 2. as discrete two dimensional micro-compartmental systems is also a possible model for propagation in excitable medium.

Which model is the best depends on two parameters:

- Adequation to well established physical laws
- Mathematical tractability

the second point is a necessary condition for a model to be usefull.

An important point in our formalism is that we can define the object of our search : the macroscopic observation, before any mathematical work and thus we can criticize our hypothesis. If we have good reasons to choose linear diffusion (one good reason could be some knowledge in P.D.E.) we can choose it and our problem turns out to be a problem in reaction diffusion equations. But if we have some good reason to choose another type of coupling which leads to an other chapter of mathematics - like combinatorics in section 2 - we can do it.

For the case of grass, cardiac muscle, cell membranes the model of section 2. is at least as much realistic than P.D.E. models and it turns out that:

- In the one dimensional case a complete description is possible and trivial (see [4])
- In the two dimensional case the description is not trivial, but possible via combinatorial techniques.

If we choose to model with a P.D.E. the corresponding result would consist in a general qualitative theory of FitzHug Nagumo type equations. This is still a challenge for mathematicians and we doubt that its achievement will provide great illumination in the knowledge of the mechanims of propagation in certain excitable media.

C. LOBRY

Departement de Math.

Université de Nice
Parc Valrose

06000 NICE (France)

C. REDER

U.E.R. Math. et Info.

Université de Bordeaux I
351 Cours de la Libération

33405 TALENCE (France)

REFERENCES.

[1] ALLOUCHE J.P. and C. REDER, "Spatio temporal oscillation of a cellular automaton in excitable media" to appear in Actes du colloque sur les rythmes en biologie, chimie, physique et autres champs d'applications. Springer series in Synergetics.

[2] ALLOUCHE J.P. and C. REDER, "Oscillations spatio - temporelles engendrées par un automate cellulaire" publications U.E.R. Mathematiques et Informatique, Université de Bordeaux,(351 Cours de la Liberation 33405 TALENCE) n° 8207.

[3] HARTHONG J. "Le Moiré", publications IRMA , Université de Strasbourg, 2° édition 1980.

[4] LOBRY C. "Burning Grass and Floating Corks" Lecture Notes in Control and Information Sciences n° 39 - 1982 .

[5] LOBRY C. "Faut il brûler les équations de Hodgkin Huxley" in les actes du colloque sur les rythmes en biologie, Université de Pau, Juin 1981.

[6] LUTZ R. and M. GOZE, "Nonstandard Analysis" Lecture Notes in Math. n°881.

[7] NELSON E. "Internal Set Theory" B.A.M.S. n° 83 (1977) pp. 1165-1198.

[8] REDER C. "Familles de convexes invariantes et equations de diffusion réaction" Annales de l'institut Fourier, XXXII n° 1 1982 pp.71-103.

[9] WIENER N. and A. ROSENBLUETH, The mathematical formulation of conduction of impulses in a network of connected excitable elements, specifically in cardiac muscle. Arch. Inst. Cardiologia de Mexico, 16, 205 - 265.

[10] WINFREE A."The geometry of biological time" Biomathematics Vol 8.

Paul Dousson

SOME MATHEMATICAL PROBLEMS ARISING FROM CHEMICAL KINETICS AND THERMODYNAMICS

1. PRELIMINARIES

1.1. Introduction

In this paper I would like to present some mathematical problems related to the following problem of macroscopic physics : *"Given a chemically reacting - or not reacting - system, to study its steady states and its evolution."*

Whenever a mathematician whishes to use his savoir-faire to study natural phenomena, he first encounters a translation problem. In these circumstances the first mathematical problem is modelling. In the present case, two mathematical models (not independent) have been conceived :
The first one for the determination and the study of equilibrium states,
The second one for the study of the chemical kinetics.

We then had to be absolutely certain of the reliability of these models. Since the experimental facts are pitiless judges, the natural or physical laws issuing from the experiment therefore constitued the tests.

Finally assured of the validity of the models, the use of miscellaneous mathematical techniques has allowed the first models exploitation which has chiefly led to the justification of admitted results and to some generalizations. A deeper exploitation will perhaps be able to satisfy (partly) the demand of Professor Leray : "The purpose of the mathematical Physics is to predict the phenomena".

1.2. Mathematical techniques and physical results.

Various mathematical techniques applied to the two aforementioned models have enabled us to get several results. The following is but a partial list :

C. P. Bruter et al. (eds.), Bifurcation Theory, Mechanics and Physics, 299–316.

A. From the "static model" :

A1. Elementary mathematical methods give :
A1.1. Gibbs-Duhem relations,
A1.2. Intrinsic chemical affinity.

A2. Optimization techniques yield :
A2.1. Conditions of equilibrium for the homogeneous isolated chemically reacting systems,
A2.2. Justification of the "activated complexes" theory,
A2.3. Conditions of equilibrium for composite systems,
A2.4. Justification and generalization of the lever rule,
A2.5. Justification - without entropy jump - of latent heat.

A3. Geometry and singularities theory give :
A3.1. Generalization of the phase rules.

A4. Work on the quadratic forms and the Legendre K-transform allow us :
A4.1. To study stability and metastability,
A4.2. To generalize the Le Chatelier principle.

B. From the "kinetic model"

B.1. The qualitative analysis of a differential system leads to several results which constitute the biggest part of this communication.

B2. The bifurcation theory allows us to study :
B2.1. The ramified chains,
B2.2. The oscillatory chemical reactions.

The biggest part of B2 is still under study therefore constituting an open problem.

2. THE KINETIC MODEL

One conjectures that an isolated reactive system $\mathcal{R}$, in homogeneous phase, is governed by the differential system :

$$\frac{dx_j}{dt} = \dot{x}_j = g(t;X)[x_j \partial_j S - \sum_{i=1}^{p} \xi_i(X)\beta_i^j x_j] , \quad 1 \leq j \leq n. \qquad (E)$$

where to $1 \leq j \leq n$ and $t \geq 0$, $x_j(t)$ is the mole number of chemical species (j) contained in $\mathbb{R}$ at time t. The mappings $t \to x_j(t)$ are assumed to belong to class C^k, $k \geq 1$, on $\mathbb{R}_+$. Furthermore, x_o (resp. x_{n+1}) is the internal energy (resp. volume) of $\mathbb{R}$. $\mathbb{R}$ being isolated, x_o and x_{n+1} are strictly positive constants; $x \in \mathbb{R}^n$ (resp. $X \in \mathbb{R}^{n+2}$) is the n-uple (resp. (n+2)-uple) $(x_1, \dots, x_n)$ (resp. $(x_o, x_1, \dots, x_n, x_{n+1})$).
S is the entropy of the phase [cf. appendix] to which $\mathbb{R}$ belongs, let $\mathcal{E}$ be the phase cone. β_i^j is the atoms number of the atomic species (i) contained in the chemical component (j), the matrix $\beta = (\beta_i^j)$ $1 \leq i \leq p$, $1 \leq j \leq n$, is the *inventory matrix* [3] and rank $(\beta) = p$. Let χ be the diagonal matrix

$$\chi = \begin{bmatrix} x_1 & & O \\ & x_2 & \\ O & & \ddots & \\ & & & x_n \end{bmatrix} \quad \text{and} \quad \mathbb{Q}(x) = \beta\, \chi\, {}^t\beta, \text{ put}$$

$\mathcal{E}^+ = \{X \in \mathcal{E},\ \det \mathbb{Q}(x) \neq o\}$ we have $\mathring{\mathcal{E}} \subset \mathcal{E}^+ \subset \mathcal{E}$.

$\xi = (\xi_1, \dots, \xi_p)$ is a mapping from $\mathcal{E}$ to $\mathbb{R}^p$, such as :

$\forall X \in \mathcal{E}^+$, $\xi(X) = [\mathbb{Q}(x)]^{-1} \beta\, \chi\, \nabla_n S(X)$ with $\nabla_n S = (\partial_1 S, \dots, \partial_n S)$

and ξ is extended by continuity to whole $\mathcal{E}$ [3].

$\mathbb{R}$ being an isolated reactive system, we know [3] that its invariant manifold K' is a convex and compact set in $\mathbb{R}_+^{n+2}$, as $\mathbb{R}$ is in homogeneous phase, we have : $K' \subset \mathcal{E}$. In this study, we assume that $K' \subset \mathring{\mathcal{E}}$.

Remarks

System (E) proceeds from the same philosophy as the differential system introduced by Marc and André Lichnerowicz [7] to study economical phenomena, viz : one supposes that Nature knows "to navigate in sight" and whatever the configuration of the invariant manifold may be, she keeps one's course toward a state of equilibrium.

3. WEI'S AXIOMATICS.

In [10] J. Wei stated the precepts which according to him, one should observe in order to build mathematical model validly representing the evolution of an isolated reactive system where every internal change is merely chemical and spontaneous (or autonomous).

(W1) The extensive parameters x_j, $j \in \{0,1,\dots,(n+1)\}$ are functions of the time t, and $t \to x_j(t)$ is at least C^1 on $[t_0, +\infty)$, t_0 being taken as time origin (in this paper, we take $t_0 = 0$).

(W2) $\forall j \in \{0,\dots,(n+1)\}$, $x_j(\mathbb{R}_+) \subset \mathbb{R}_+$.

(W3) The total mass of the system $\mathcal{R}$ is conserved :

$$\forall t \in \mathbb{R}_+ , \sum_{j=1}^{n} M_j x_j(t) = \text{constant},$$

where M_j is the molar mass of the component (j).

(W4) The atomic species are conserved :

$$\forall i \in \{1,\dots,p\} , \forall t \in \mathbb{R}_+ , \sum_{j=1}^{n} \beta_i^j x_j(t) = d_i = \text{constant},$$

d_i is atoms number of atomic species (i) contained in $\mathcal{R}$,

we put $d = (d_1,\dots,d_p) \in \mathbb{R}^p$ and

$$K = \{x \in \mathbb{R}_+^n ; \beta x = d, X \in K'\}$$

(W5.) The system $\mathcal{R}$ is "without memory" i.e. that it is governed by a quasi-autonomous system such as :

$$\forall j \in \{1,\dots,n\}, \quad \dot{x}_j = f_j(x)g(t,X) \qquad \text{(W5.1)}$$

with $f_j \in C^r(\Omega)$, $r \geq 1$, Ω : open neighborhood of K in $\mathbb{R}^n$ and g is a strictly positive continue function. The system (W5.1) reduces to an autonomous system :

$$\forall j \in \{1,\dots,n\}, \quad \frac{dx_j}{d\tau} = f_j(x) \qquad \text{(W5.2)}$$

where τ is a pseudo-time coordinate such as $d\tau = g(t;X)dt$, $\tau(o) = o$; the time scale is stretched and distorted but never halted or reversed.

(W6) There is a composition, corresponding to an equilibrium point, which does not change with time - at a certain point x^*, all $\dot{x}_j = o$.

(W7) The equilibrium point is stable - that is, when the system $\mathcal{R}$ at the equilibrium point is changed in composition by external causes, it changes its composition in such a way as to weaken the effect of this change (a form of the Le Chatelier-Braun principle).

(W8) The system $\mathcal{R}$ evolves in the course of time so as to reach an equilibrium point X^* defined by (W6) and (W7).

(W9) If $(x(t);\, x(o) = x^o)$ is a solution to the problem (W5.1) then :

$$\forall t \in \mathbb{R}_+ \, , \frac{dS}{dt}(X(t)) \geq o.$$

$H_S(X)$ denotes the hessian matrix of S in $X \in \mathcal{E}$ and M_k does the set $M_k = \{X \in \mathcal{E};\ \text{rank } H_S(X) = k\}$. According to (W6) and (W7), we adopt the

Définition D3.1. $\mathcal{R}$ being an isolated reactive system in homogeneous phase at $t_1 \geq o$:

i) $X(t_1)$ is a *chemically steady state* of $\mathcal{R}$ if $X(t_1) = o$.

ii) $X(t_1)$ is a *kinetic equilibrium state* if

a) $X(t_1) = o$ and b) $X(t_1) \in M_{n+1}$

(i.e. $X(t_1)$ is a intrinsically stable thermochemical equilibrium state).

4. CONFORMITY OF THE SYSTEM (E) WITH WEI AXIOMATICS.

From now one the reactive system R is assumed to satisfy the conditions worded in section 2 and $X(o) = X^{o} \in K' \subset \mathcal{E}$

♦ It is obvious that the system $\{(E);\ x^{o}\}$ verifies the five first conditions (W1) - (W5); the satisfaction of (W4) coming from the fact K is a positively invariant ensemble for (E).

♦ <u>(W6), (W9), De Donder relation</u>.

As K' is a compact set contained in $\mathring{\mathcal{E}}$, we know [3] that R has at least one thermochemical true equilibrium (t.t.e.), say Z, we have :

$$S(Z) = \max\{S(X),\ X \in K'\} \quad \text{and} \quad A(Z) = o,$$

where A is the chemical affinities field [cf. appendix]. Moreover R being isolated, we have :

$$\frac{d}{dt}[S(X(t))] = \dot{S}(X(t)) = \langle \nabla_n S(X(t)),\ \dot{x}(t)\rangle$$

by (W4) $\dot{x}(t) \in \operatorname{Ker} \beta$, therefore :

$$\langle \nabla_n S(X(t)),\ \dot{x}(t)\rangle = \partial_o S(X(t))\langle A(X(t)),\ \dot{x}(t)\rangle .$$

As (E) can again write :

$$\nabla_n S(X(t)) = {}^{t}\beta\xi(X(t)) + [g(t;\ X(t))\chi(t)]^{-1}\ \dot{x}(t)$$

we have : $\forall t \in \mathbb{R}_+,\ S(X(t)) = \langle \dot{x}(t),\ D(t)\dot{x}(t)\rangle$

where $D(t) = [g(t;X(t))\chi(t)]^{-1}$ is a positive definite diagonal matrix, then :

<u>Proposition P.4.1.</u> : R *being the reactive system defined in section 2, let* x^{o} *be an arbitrary point of* K, *then the solution* $x(t)$ *to* $\{(E);\ x(o) = x^{o}\}$ *verifies* :

i) $\forall t \in \mathbb{R}_+,\ \dot{S}(X(t)) \geq o,$

ii) $\forall t \in \mathbb{R}_+$, $\langle A(X(t)), x(t)\rangle \geq o$,

iii) The equality $\dot{S}(X(t_1)) = o$ *is realized if and only if* $X(t_1)$ *is a chemically steady state of* $\mathcal{R}$.

i) is (W9); ii) generalizes the De Donder relation; if we take $X^o = Z$ then iii) yields (W6).

♦ Asymptotic behavior of the solutions of (E) - (W8).

x^o denotes always an arbitrary point of K, let Γ be the orbit of the solution of the Cauchy problem : $\{(E); x(o) = x^o\}$, we put : $\Gamma' = \{X(t) = (x_o, x(t), x_{n+1}); x(t) \in \Gamma\}$. In considering on the one hand $\int_{\Gamma'} dS$ and on the other hand the Laplace transform :

$$\int_0^{+\infty} e^{-zt} S(X(t))dt$$ it is established the

Proposition P4.2. $\mathcal{R}$ *reaches a chemically steady state when t increases indefinitely, let* X^∞ *be this state :*

$$X^\infty = \lim_{t\to+\infty} \{X(t); X(t) \in \Gamma'\}.$$

By (W9), we have $S(X^\infty) = \max\{S(X); X \in \Gamma'\}$ but this is not sufficient to conclude that X is a thermochemical equilibrium state of $\mathcal{R}$[cf. appendix] . In order to show that $S(X^\infty) = \max\{S(X); X \in K'\}$ we use the Lyapunov function introduced by M. and A. Lichnerowicz in [7] and [8], we get, moreover, information on the stability of this equilibrium. $z = (z_1, \dots, z_n)$ being the element of K such that $Z \in K'$

$Z = (x_o, z, x_{n+1})$ where Z denotes a (t.t.e.) of $\mathcal{R}$, we consider :

$$V(x(t);z) = \sum_{k=1}^{n} (x_k(t) - z_k - z_k \operatorname{Log} \frac{x_k(t)}{z_k})$$

V is a Lyapunov function for the system (E). The concavity of S, the properties of the affinities field and these of the function V allow to establish the

Proposition P4.3. *Let* x° *be an arbitrary element of* K *and* $x(t)$ *a solution to problem* $\{(E);x(o) = x^{\circ}\}$ *then* :

i) $\forall t \in \mathbb{R}_+$, $V(x(t);z) \geq o$,

ii) $V(x(t);z) = o \Longleftrightarrow x(t) = z$,

iii) $t \in \mathbb{R}_+$, $\frac{d}{dt}[V(x(t);z)] = \dot{V}(x(t);z) \leq o$,

iv) $\dot{V}(x(t);z) = o \Longleftrightarrow S(X(t)) = S(Z)$.

which yields :

Corollary C4.1. *i) Every chemically steady state of* ℞ *is a (t.t.e.) of* ℞.

ii) $\forall x^{\circ} \in K$, *the process* $(x(t);x^{\circ})$ *solution to the problem;* $\{(E);x(o) = x^{\circ}\}$ *tends to a (t.t.e.) when* t *increases to the infinite.* [*i.e.* (E) *satisfies* (W8)].

Moreover, in using the concavity of S, the Lyapunov theorem about the asymptotic stability [6] and the results about the intrinsic [3, ch.5], we get :

Proposition P4.4. *i)* $Y \in K'$ *is an asymptotically stable equilibrium of* ℞ *if and only if* Y *is the single equilibrium of* ℞

ii) If Y *is an intrinsically stable equilibrium of* ℞ *then* Y *is asymptotically stable.*

iii) If Y *is the unique equilibrium of* ℞ *then for all* $x^{\circ} \in K$ *the orbit of* $\{(E);x(o) = x^{\circ}\}$ *converges to* Y [3].

Remarks about the condition (W7).

(W7) is a *Moderation law* as the Lenz law in electromagnetism or the Leontieff law in econometrics, besides it is implicitly expressed in the "intrinsic stability criterion" ([3], I.T5.2.1), by following (W7) is verified if and only if we impose : "*Every chemically steady state of* ℞ *belongs to* M_{n+1}."

This is not very satisfactory for that implies, at least locally, that the entropy is a quasi strictly concave function (it is the frame of the Gibbs theory) and this too strong condition forbids the existence of the bifurcating phenomena such as the oscillatory chemical reactions. But Wei wrote his paper in 1961, before Zhabotinski's experiment (1964) [5]; now, in 1961, people thought a chemically reacting system should reach one equilibrium state. Wei specified, moreover, that this distinguished a chemical system from a biological system. In the following, we do not impose the condition (W7) the experiment will show if the entropy must or must not be degenerate.

5. MONOREACTIVE SYSTEMS : SOME RESULTS.

Definition D5.1. $\mathcal{R}$ is said to be a *monoreactive system* if $\dim(\operatorname{Ker}\beta) = 1$; likewise $\mathcal{R}$ is said *multireactive* if $\dim(\operatorname{Ker}\beta)>1$.

For all closed reactive systems, it is easy to verify that the following vectors belong to Ker β :

$(x(t) - x^{o})$, $\dot{x}$, A and the stoichiometric vectors.

Let us suppose that $\mathcal{R}$ is an isolated monoreactive system governed by the differential system $\{(E);\ x(o) = x^{o}\}$, let $\nu = (\nu_1,\dots,\nu_n)$ be the stoichiometric vector of the reaction taking place in $\mathcal{R}$, we have :

$$x(t) - x^{o} = \nu\zeta(t) \qquad \text{and} \qquad \dot{x} = \nu\dot{\zeta}$$

ζ is the advancement (or extent) of the reaction, in considering the ν-directional derivative of S, we introduce the affinity $\mathcal{A}$ of the reaction :

$$\mathcal{A} = T\,\frac{\partial S}{\partial \zeta} = \langle \nu, A\rangle$$

where T is the absolute temperature of $\mathcal{R}$: $\partial_o S = 1/T$.
In the case of an isolated monoreactive system of which invariant manifold K' is contained in $\overset{o}{\mathcal{E}}$, the differential problem $\{(E)\ ;x(o) = x^{o}\}$ becomes :

$$\sum_{j=1}^{n} \nu_j(\dot{x}_j/x_j) = g(t;X)[\,\mathcal{A}(X)/T(X)] \tag{5.1}$$

As S belongs to the class $C^2(\overset{o}{\mathcal{E}})$, $\dfrac{\mathcal{A}}{T} : X \to \dfrac{\mathcal{A}(X)}{T(X)}$ is integrable on the compact set K', furthermore K' is positively by (E), hence (5.1), gives : $\forall t \in \mathbb{R}_+$,

$$\prod_{j=1}^{n} [x_j(t)/x_j^o]^{\nu_j} = \exp\left(\int_o^t g(s;X(s))(\mathcal{A}(X(s))/T(X(s)))\,ds\right) \qquad (5.2)$$

As $\mathcal{R}$ is isolated we shall bring in the concentrations in the following of this section. For $j \in \{o,1,\dots,n\}$, we put

$$c_j = x_j/x_{n+1} \quad , \; c(t) = (c_1(t),\dots,c_n(t)) \text{ and } C(t) = (c_o,c(t),1).$$

5.1. Relations between concentrations and affinity - Equilibrium constant.

S being a homogeneous first-order function, A/T, $\mathcal{A}/T$ and g are homogeneous zero-order functions of X hence, with notation abuse, the relation (5.2) writes : $\forall t \in \mathbb{R}_+$

$$\prod_{j=1}^{n} [c_j(t)/c_j^o]^{\nu_j} = \exp\left(\int_o^t g(s;C(s))[\mathcal{A}(C(s))/T(C(s))]\,ds\right)$$

We know that there is $t_1 \geq o$ such as $X(t_1)$ is a chemically steady state of $\mathcal{R}$ therefore a (t.t.e) of $\mathcal{R}$ let

$$K_c = \prod_{j=1}^{n} [c_j(t_1)]^{\nu_j}$$

K_c is the *equilibrium constant* relative to the concentrations; we have :

$$\forall t \in \mathbb{R}_+, \quad \prod_{j=1}^{n} [c_j(t)]^{\nu_j} = K_c \exp\left(-\int_t^{t_1} g(s;C(s))\,\frac{\mathcal{A}(C(s))}{T(C(s))}\,ds\right)$$

As X^∞ is a (t.t.e.) of $\mathcal{R}$, it comes : $\forall t \in \mathbb{R}_+$,

$$\prod_{j=1}^{n} [c_j(t)]^{\nu_j} = K_\infty \exp\left(-\int_t^{+\infty} g(s)\,[\mathcal{A}(C(s))/T(C(s))]\,ds\right)$$

K_∞ is the *equilibrium constant* relative to $\lim_{t\to+\infty} C(t)$

5.2. The case of the perfect gas.

We assume the chemical components of the monoreactive system

$\mathcal{R}$ are perfect gas. We know [2] that in this case chemical potentials are given by :

$$\mu_j = RT \operatorname{Log} c_j + \eta_j(T) \quad , \quad j \in \{1,\dots,n\} \ ,$$

where R is the perfect gas constant and η_j a function depending only on the temperature T. Let be $\mu = (\mu_1,\dots,\mu_n)$, $\eta = (\eta_1,\dots,\eta_n)$ and

$$L(t) = \sum_{j=1}^{n} \nu_j \operatorname{Log}(c_j(t));$$

As $\mathcal{A} = -\langle \nu,\mu \rangle$ we get :

$$\mathcal{A} = - RT\, L(t) - \langle \nu,\eta(T)\rangle . \tag{5.3}$$

If at $t = t_1$ $\mathcal{R}$ has a steady state then $C(t_1)$ is a (t.t.e.) of $\mathcal{R}$ hence $A(C(t_1)) = o$ and $\mathcal{A}(C(t_1)) = o$ but the (5.3) yields:

$$K_c = \exp[- \langle \nu , \eta(T) \rangle / RT] \tag{5.4}$$

The relation (5.4) is the *mass-action law* : "In steady states, the equilibrium constant of the perfect gas depends only on the system temperature."

Proposition P5.2.1. *If the chemical components of the isolated monoreactive system* $\mathcal{R}$ *are perfect gas, then :*

i) $\eta(T)/T$ *depends actually on time,*

ii) $\eta(T)/T$ *and* $\frac{d}{dt}(\eta(T)/T)$ *are not orthogonal to* ν *, in particular these two vectors do not belong to the range of* ${}^t\beta$.

Proof. We have : $\forall t \in \mathbb{R}_+$,

$$\frac{d}{dt}\left\{ e^{R\tau}\left(L + \frac{\langle \nu,\eta(T)\rangle}{RT}\right)\right\} = e^{R\tau}\,\frac{d}{dt}\left\{\frac{\langle \nu,\eta(T)\rangle}{RT}\right\} \tag{5.5}$$

where τ is the pseudo-time coordinate described in section 3, (W5) : $\dot{\tau} = g(t;C(t))$. If $\mathcal{R}$ has a equilibrium state at time $t = t_1$, (5.3) and (5.5) give : $t \in \mathbb{R}_+$,

$$e^{R\tau(t)}\left\{L(t) + \frac{\langle \nu,\eta(T(t))\rangle}{RT(t)}\right\} = \int_{t_1}^{t} e^{R\tau(s)}\,\frac{d}{ds}\left\{\frac{\langle \nu,\eta(T(s))\rangle}{RT(s)}\right\} ds \tag{5.6}$$

so that if $\eta(T)/T$ does not depend on time, or if $\eta(T)/T$ (resp. $\frac{d}{dt}(\eta(T)/T)$ is orthogonal to ν, we have by (5.3) and (5.6)

$$\forall t \in \mathbb{R}_+, \quad \mathcal{A}(C(t)) = 0$$

but as $\dim(\mathrm{Ker}\,\beta) = 1$ then $A = \|\nu\|^{-2} \mathcal{A}\nu$ therefore the previous conditions imply : $\forall t \in \mathbb{R}_+$, $A(C(t)) = 0$ which is absurd since this means that system $\mathcal{R}$ is perpetually in a state of equilibrium whereas it is reactive by hypothesis ■

Remarks : As $\mathcal{R}$ is monoreactive, there is one function and only one, say α: $T \to \alpha(T)$, such as :

$$\mu_j = RT[\mathrm{Log}\, c_j + \sum_{i=1}^{n} \lambda_i(T)\beta_i^j + \alpha(T)\nu_j].$$

The Proposition P5.2.1 says $\alpha(T)$ must depend on the time and the relation (5.6) shows this function α determines the behavior of the system $\mathcal{R}$.

Corollary C5.2.1. *Let* t_1 *be,* $0 < t_1 < +\infty$, *if* $\mathcal{R}$ *has a steady state in* $t = t_1$ *and if* $\alpha(T) : t \to \alpha(T(t))$ *is monotonous and belongs to* $C^1([t_1, +\infty[)$ *then after the finite time* t_1, $\mathcal{R}$ *can't evolve.*

Proof. As X is a (t.t.e.) the relation (5.6) can be written : $\forall t \in \mathbb{R}_+$,

$$\frac{e^{R\tau(t)}}{R\|\nu\|^2 T(t)} \mathcal{A}(C(t)) = \int_t^{+\infty} e^{R\tau(s)} \frac{d}{ds}[\alpha(T(s))]\, ds \qquad (5.7)$$

(5.7) and the corollary hypothesis imply : $t \geq t_1$, $\mathcal{A}(C(t)) = 0$ ■

6. SOME RESULTS ABOUT MULTIREACTIVE SYSTEMS.

$\mathcal{R}$ is always an isolated reactive system in homogeneous phase of which the invariant manifold K' is contained in $\overset{\circ}{\mathcal{S}}$. Now we assume that $\dim(\mathrm{Ker}\,\beta) = n - p > 1$.

Let $\{b^1,\ldots,b^{n-p}\}$ be a base of Ker β, B denotes the $n\times(n-p)$-dimensional matrix of which the columns are the $b^i \in \mathbb{R}^n$, $1 \leq i \leq n-p$. There are two mappings :

$$y : t \in \mathbb{R}_+ \to y(t) \in \mathbb{R}^{n-p} \quad \text{and} \quad a : t \in \mathbb{R}_+ \to a(t) \in \mathbb{R}^{n-p}$$

Such as $x(t) - x^o = By(t)$ and $A(X(t)) = Ba(t)$.

As $K' \subset \mathring{\mathcal{E}}$, the problem $\{(E);\ x(o) = x^o \in K\}$ reduces to :

$$y = \frac{g}{T}\,[\,{}^tB\,\chi^{-1}B]^{-1}\,{}^tBBa \quad \text{and} \quad y(o) = o \tag{6.1}$$

Proposition P.6.1. (Reciprocity Theorem). *The reaction rates field* $\dot{x}$ *and the chemical affinities field* A *being referred to the same base of* Ker β, *if the component* $\dot{y}_i$ *of* $\dot{x}$ *is influenced by the component* a_j *of* A *then the component* y_j *of* $\dot{x}$ *is influenced by the component* a_i *of* A *and this by means of the same phenomenological coefficient* ℓ_{ij}.

(i.e. $\dot{y} = La/T$ *where* $L = (\ell_{ij})$ *is a symmetrical matrix*).

Proof. The proposition P6.1. is the physical interpretation of the relation (6.1), it is sufficient to put

$$L = g[\,{}^tB\,\chi^{-1}B\,]^{-1}\,{}^tBB\,,$$

it is obvious that L is symmetrical positive definite ∎

P6.1. is a form of the Onsager theorem, but there the coefficients ℓ_{ij} are not constant, they depend on X and t. However, P6.1. is not only valid in a neighborhood of an equilibrium state but over the whole manifold K.

As matrix B we can take
$$B = \begin{bmatrix} -[\beta^{I}]^{-1}\,\beta^{II} \\ I_{n-p} \end{bmatrix}$$

where β^I is a square submatrix of β such as $\text{rank}(\beta^I) = p$, after eventual permutation of the columns of β, we have :

$\beta = [\beta^{I} : \beta^{II}]$, the components of y are then the (n-p) components $(x_j(t) - x_j^0)$ correspondent to the *guide variables* [3] of the system $\mathcal{R}$.

If we could show that (n-p) independent intermediate reactions took place in $\mathcal{R}$, then their stoichiometric vectors : $\nu^1, \nu^2, \ldots, \nu^{n-p}$ constitute a base of Ker β so that :

$$x(t) - x^0 = \sum_{i=1}^{n-p} \zeta_i \nu^i \quad \text{and} \quad \dot{x} = \sum_{i=1}^{n-p} \dot{\zeta}_i \nu^i$$

ζ_i is the extent of the i° intermediate reaction, $\dot{\zeta}_i$ is the rate of the i° reaction; $\dot{\zeta} = (\dot{\zeta}_1, \ldots, \dot{\zeta}_{n-p})$

$\mathcal{A}_i = \langle \nu^i, A \rangle$ is the affinity of the i° reaction,

$\mathcal{A} = (\mathcal{A}_1, \ldots, \mathcal{A}_{n-p})$.

In putting $B = [\nu^1 \ldots \nu^{n-p}]$ we have $\mathcal{A} = {}^tB\,A$ and the problem $\{(E);\ x(o) = x^0 \in K\}$ becomes :

$$\dot{\zeta} = g\,[\,{}^tB\,\chi^{-1}\,B]^{-1}\,\frac{\mathcal{A}}{T} \tag{6.2}$$

Proposition P6.2. $\mathcal{R}$ *being a multireactive system such as* dim(Ker β) = n-p, *if we know a reaction mechanism which includes* (n-p) *independent intermediate reactions, then the rates and the affinities of every intermediate reaction are related by :*

$$\dot{\zeta}_i = \frac{1}{T} \sum_{j=1}^{n-p} m_{ij}(\zeta(t))\,\mathcal{A}_j(\zeta(t)) \qquad 1 \le i \le n-p,$$

where $m_{ij}(\zeta(t))$ *are the coefficients of the matrix*

$M = g\,[{}^tB\,\chi^{-1}\,B]^{-1}$ *which is symmetrical positive definite.*

Remarks. If a reactional scheme offers q independent intermediate reactions and if $q < n-p$ the previous result is not valid.

The proposition P6.2. is to compare Van Rysselberghe's results [9].

7. CONCLUDING REMARKS.

The subject is far from being completely treated, but one must know when to stop. Among the points not evoked there is the function g; in the case of the monoreactive system one can give approximate form of g making act on the relaxation time. In conclusion, I shall go back to the last remark of section 6, for though it is obvious for the mathematician, it does not seem to be so for the chemist. Indeed, one often reads that :

Assertions (A) : *"If the affinities (resp. rates) of all independent intermediate reactions are null, then the system $\mathfrak{R}$ is in a state of equilibrium (resp. steady state)."*

These assertions are manifestly erroneous from the frame chemist's perspective, for this contrives to find reaction mechanisms without ever paying attention to the inventory matrix, whose existence and use he knows nothing about Therefore he can't know if $q = (n-p)^{-}$. This is indeed a strange way for an experimenter to proceed, for the reaction mechanisms are only unverifiable conjecture, as for the matrix β can be better determined as long as the methods and the means of chemical analysis are surer.

However, certain authors perceived that $\dot{\zeta} = o$ does not imply $\dot{x} = o$ and $\mathcal{A} = o$ does not involve $A = o$, so Wei [10] added one condition in his axiomatics :

Condition (W10) : *"A kinetics system must satisfy microscopic reversibility postulate".*

This condition can be translated by :

" $\dot{x}$ belongs to linear subspace begotten by the stoichiometric vectors $\nu^1, \dots, \nu^q$ of the independent intermediate reactions".

By means of (W10) the assertions (A), become correct; indeed, if ν denotes the matrix $\nu = [\nu^1 \dots \nu^q]$

then $\dot{x} = \nu\dot{\zeta}$ and $[{}^t\nu \overset{-1}{\chi} \nu]\dot{\zeta} = g\mathcal{A}/T$

where $\mathcal{A} = {}^t\nu A = (\mathcal{A}_1, \dots, \mathcal{A}_q)$; hence if $\mathcal{A}(\zeta(t_1)) = o$

we have $\dot{\zeta}(t_1) = o$ (for $[{}^t\nu \overset{-1}{\chi} \nu]$ is a regular matrix), therefore $\dot{x}(t_1) = o$ and so $X(t_1)$ is a steady state of $\mathbf{R}$ hence $X(t_1)$ is a (t.t.e.) of $\mathbf{R}$ and $\mathbf{A}(X(t_1)) = o$.

I were not to trespass on the rights of others I would recommend that my chemist co-workers use the inventory matrix because :

1. It was created from the experiment,
2. It avoids contriving reactional schemes,
3. It allows to replace the extent of the intermediate reactions by the guide-variables (the first are not controllable by experiment whereas the second are),
4. It allows us to do away with the microscopic reversibility postulate,
5. It allows us to define the chemical affinities field.

APPENDIX

a1. Phase-entropy - Thermochemical equilibrium state.

The chemically reacting systems studied in this paper are isolated, partitionless (i.e. without an internal restrictive wall), in the homogeneous phase so that the entropy which is adapted for them, is the *phase-entropy*, is the simplest. We do not give the general definition of the *system-entropy* [3] for it is not useful to us in this study and is complicated.

Definitions D.1. To every homogeneous thermodynamical phase are associated a mapping S and a set $\mathcal{E}$ satisfying :

$$\mathcal{E} = \{X \in \mathbb{R}_+^{n+2} \; ; \; X \neq o \; , \; S(X) \geq o\} \; ,$$

(H). S is a homogeneous first-order function and S is C^k, $k \geq 2$,

(C). S is concave upon $\mathbb{R}^{n+2}$,

(N). $\forall X \in \mathcal{E}$, $\partial_o S(X) > o$ and $\lim\limits_{\partial_o S(X) \to +\infty} S(X) = o$.

The absolute temperature T is given by $T = (\partial_o S)^{-1}$.

Definitions D.2. Let $\mathcal{R}$ be a thermodynamical system of which K' is the invariant manifold,
(i) $Y \in K'$ is an equilibrium state of $\mathcal{R}$ if and only if $S(Y) = \max\{S(X);\ X \in K'\}$
(ii) $Y \in K'$ is a thermochemical true equilibrium (t.t.e.) of $\mathcal{R}$ iff Y is an equilibrium state of $\mathcal{R}$ and every component y_j, $o \leq j \leq n+1$, of Y differs from zero.

a2. Chemical affinities field.

$\mathcal{R}$ denotes an isolated reactive system in the homogeneous phase and β is the inventory matrix of $\mathcal{R}$.

Definitions D.3. The chemical affinities field of $\mathcal{R}$ is the vectors field **A** defined on the invariant manifold K' by : $\forall X \in K'$, $[\partial_o S(X) A(X)]$ is the orthogonal projection of $\nabla_n S(X)$ upon $\operatorname{Ker} \beta$.

From Kuhn-Tucker theorem, we get :

Proposition. (i) $X \in K'$ *is an equilibrium state of* $\mathcal{R}$ *if and only if*

$$\forall Y \in K', \ \langle A(X), (y-x)\rangle \leq o$$

with $(x,y) \in K^2$ *and* $X = (x_o, x, x_{n+1})$, $Y = (x_o, y, x_{n+1})$

(ii) *If* X *is a (t.t.e.) of* $\mathcal{R}$ *then* $A(X) = o$.

Remarks : $A(X) = o$ assure X is an equilibrium state of $\mathcal{R}$ but this does not imply generally that X is an (t.t.e.) of $\mathcal{R}$, however, in this paper, as $K' \subset \mathring{\mathcal{E}}$ then $A(X) = o$ implies X is a (t.t.e.) of $\mathcal{R}$.

REFERENCES.

1. Callen, H.B. : 1960,. Thermodynamics, John-Wiley, New-York.
2. De Groot and Mazur : 1969, Non Equilibrium Thermodynamics, North-Holland, Amsterdam.
3. Dousson P. : 1980, Etude mathématique de la cinétique chimique et de la thermodynamique chimique, Thèse d'Etat, Saint-Etienne.
4. Gavalas, G. : 1968, Non linear differential equations of chemically reacting systems, Springer-Verlag, Berlin

5. Glansdorff et Prigogine : 1971, Structure, stabilité et fluctuations, Masson Paris.
6. La Salle and Lefschetz : 1961, Stability by Lyapunov's direct method, Academic Press, New-York.
7. Lichnerowicz, M. and Lichnerowicz, A. : 1971, "Economie et Thermodynamique : un modèle d'échange économique", Cahiers de l'I.S.E.A., V, n°10, pp. 1640-86, Paris.
8. Lichnerowicz, A. : 1970, Point-selle et systèmes differentiels, C.R. Ac. Sc., Paris, 271, Serie A, pp.1123-8.
9. Van Rysselberghe, P : 1958, "Reaction rates and affinities", Journal of Chemical Physics 29, pp. 640-2.
10 Wei, J. : 1962, "Axiomatic treatment of chemical reaction systems", Journal of Chemical Physics 36, 1578-84.

F. Alberto Grünbaum

THE LIMITED ANGLE PROBLEM IN TOMOGRAPHY AND SOME RELATED MATHEMATICAL PROBLEMS

1. INTRODUCTION

Consider the problem of recovering a compactly supported function $f(\underline{x})$, $\underline{x} \in R^n$, from a collection of its "one dimensional projections" given by

$$Pf(w,t) = \int_{\langle x,w \rangle = t} f$$

$$w \in S^{n-1} .$$

Here S^{n-1} denotes the unit sphere in R^n , and $\langle x,w \rangle$ denotes the usual inner product in R^n .

The most common applications call for $n = 2$ (X-ray tomography) or $n = 3$ (Nuclear Magnetic Resonance Imaging) and therefore we will occasionally specialize our considerations to these cases.

One of the main concerns in this lecture is the effect that a limitation of the range over which the directions w can be chosen has on the quality of the reconstruction.

If one had no restriction on the choice of directions w , more precisely if one knew $Pf(w,t)$ for all $w \in S^{n-1}$ one could use the formulas of Radon [1]. In the two dimensional case we have

$$f(x,y) = \frac{1}{2\pi^2} \int_0^{\pi} d\theta \ \text{p.v.} \int_{-\infty}^{\infty} \frac{1}{t-s} \frac{\partial}{\partial s} Pf(\theta,s) .$$

with $t = x \cos\theta + y \sin\theta$.

C. P. Bruter et al. (eds.), Bifurcation Theory, Mechanics and Physics, 317–329.

The formula depends on the dimension n but some features of it are independent of n : the projections are <u>first</u> "filtered" by the application of some operator -- differential if n is odd, integral if n is even -- <u>then</u> for each $w \in S^{n-1}$ the result is "backprojected" to get a function of $\underline{x}$ -- i.e. the function is made constant along the hyperplanes $\langle \underline{x},\underline{w} \rangle = t$, <u>finally</u> these backprojections are added up with uniform weight over S^{n-1} .

Thus Radon's formula has the form of a sum of filtered backprojection. The history behind the formula is interesting. Although Radon, in 1917, certainly proved it first for general n , the Dutch physicist H. Lorentz knew of this formula for $n = 3$ around the end of the century [2]. A paper in 1906 gives credit to Lorentz [3]. Later on Uhlenbeck [4], Cramer and Wold [5], Bracewell [6], Cormack [7] -- and probably many others had to rediscover this result on their own. An independent attack, not related to Radon's formula was conceived and carried out by Hounsfield [7].

2. THE LIMITED ANGLE PROBLEM

A problem of practical interest is that of recovering f from its projections $Pf(w,t)$ in a given limited range. For convenience the following discussion deals with R^2 and we use the notation

$$Pf(\theta,t)$$

for the projections.

Although one can prove that knowing $Pf(\theta,t)$ for $-\alpha < \theta < \alpha$, with $\alpha > 0$ arbitrary, determines f uniquely it is clear that the "quality" of the reconstruction should depend on α . We have the case of an ill-posed inversion problem and the degree of ill-conditioning will depend seriously on the range of views $-\alpha < \theta < \alpha$.

Assume that all the projections $Pf(\theta,t)$ $-\frac{\pi}{2} < -\alpha \leq \theta \leq \alpha < \frac{\pi}{2}$ are known. The problem is to recover f. It is convenient to visualize the data as the result of applying a linear operator A to the unknown function f. For mathematical expediency we take $f \in L^2(D)$ which gives $Pf(\theta,t) \in L^2(I,w)$, for each θ. Here D denotes the unit disk in R^2, $L^2(D)$ the space of square integrable functions on D, and $L^2(I,w)$ the space of functions on the unit interval I, square integrable with respect to some appropriate weight w.

In this fashion the operator A maps $L^2(D)$ into $\oplus L^2(I,w)$, with the direct sum extending over all $-\alpha \leq \theta \leq \alpha$. If b -- the data -- denotes the set of all available projections we have

$$Af = b \quad .$$

It is now convenient to bring in the singular value -- singular vector decomposition of A.

Take ψ_i to be the orthonormal eigenvectors of AA^*, with positive real eigenvalues μ_i^2, $\mu_i > 0$. Then set $\phi_i = A^*\psi_i/\mu_i$ and get for the smallest f which minimizes the error $Af - b$ the expression

$$\tilde{f} = \Sigma \frac{(b,\psi_i)}{\mu_i} \phi_i$$

A crucial role is played by the spectral decomposition of AA^*. This is an operator from $\oplus L^2(I,w)$ to itself. This space can be expressed as a discrete direct sum

$$\oplus L^2(I,w) = \sum_{n=0}^{\infty} V_n$$

where the subspace V_n is made up of functions of the factorized form

$$k(\theta)\, C_n'(\cos t)\sqrt{1 - t^2},$$

with $k(\theta) \in L^2(-\alpha,\alpha)$ and $C_n'(\cos t)$ a Gegenbauer polynomial.

One can see that AA^* leaves each one of the spaces V_n invariant, and that its action on V_n is given by

$$((AA^*)_n\, k)(\theta) = \int_{-\alpha}^{\alpha} \frac{\mathbb{C}_n^1(\cos(\theta - \zeta))}{\mathbb{C}_n^1(1)}\, k(\zeta)\, d\zeta$$

If $k_{i,n}(\theta)$ denote the eigenfunctions of $(AA^*)_n$ then the eigenfunctions of AA^* are given by

$$k_{i,n}(\theta)\, \mathbb{C}_n'(\cos t)\sqrt{1 - t^2}$$

For all the missing proofs the reader can see [9, 10, 11].

Now we consider the family of integral operators given by $(AA^*)_n$. Since

$$\frac{C_n^1(\cos\theta)}{C_n^1(1)} = \frac{\sin(n+1)\theta}{(n+1)\sin\theta}$$

we are -- by a stroke of luck -- in the case considered by Slepian in [12].

Slepian found that a remarkable phenomenon holds true for this integral operator, to wit: a second order differential operator exists which has the same eigenfunctions as $(AA^*)_n$. This is in the same spirit of the pioneering papers of Whittaker [13] and Ince [14].

In the same paper Slepian considers the matrix

$$\frac{\sin 2\pi\alpha(i - j)}{\pi(i - j)}$$

and shows that a tridiagonal matrix exists which commutes with it. This holds for any value of the parameters n,α .

In summary, the computation of the singular values and singular vectors of PP^* can be reduced to a managable numerical problem since we are in the presence of the remarkable miracle described above. For the actual implementation of this program in the tomography context see [15, 16].

3. COMMUTING INTEGRAL AND DIFFERENTIAL OPERATORS

The phenomenon mentioned above has appeared before. Already in Ince [14] we see examples of this situation.

In the sixties Slepian, Landau and Pollak came up with a remarkable series of papers dealing with the problem of time and band limited functions [17-20]. At the core of the paper is the fact that if A denotes the finite Fourier transform

$$(Af)(\lambda) = \int_{-T}^{T} e^{i\lambda t} f(t)\, dt \ , \ \lambda \in [-\Omega,\Omega]$$

then the integral operator given by A^*A , namely

$$(kf)(t) = \int_{-T}^{T} \frac{\sin \Omega(t - s)}{t - s} f(s)\, ds \quad t \in [-T,T]$$

admits a second order differential operator commuting with it, namely

$$(Df)(t) = ((T^2 - t^2)f')' - \Omega^2 t^2 f \ .$$

Slepian proved that this same result holds if R^1 is replaced by R^n [20], and then in [12] -- mentioned earlier in connection with tomography -- extended this fact for functions defined on the circle or on the integers. We become interested in this problem and tried to extend its range. In [21] one finds an extension to the case of the nth roots of unity and then in [22] one finds a classification of all Toeplitz matrices which have a tridiagonal matrix, with single spectrum, in its commutator.

In [23] this property is seen to hold for the Hilbert matrix and in [24] one finds that the property (essentially) holds for the Discrete Fourier Transform matrix.

Another example not reported before is given by the matrix

$$G_{m,n} = \frac{\sin(n-m)\theta}{n-m} - \frac{\sin(n+m)\theta}{n+m} \qquad 1 \le n,m \le N$$

which comes about as the Gram matrix for the functions $\sin n\xi$, $n \le N$.

It is not hard to see that in this case a commuting tridiagonal matrix with single spectrum is given by

$$T_{ii} = 2(i^2 - 1)\cos\theta$$

$$T_{i,i+1} = T_{i+1,i} = (2+N)(N-1) - (2+i)(i-1)$$

It is a bit harder to see that in order for the commutativity property mentioned above to hold for a matrix of the form

$$\frac{\sin(n-m)\theta}{n-m} + \alpha\,\frac{\sin(n+m)\theta}{n+m}$$

one needs $\alpha = \pm 1$ or the previously known case $\alpha = 0$ [12]. These three instances correspond to the Gram matrices for the functions $\sin n\xi$, $\cos n\xi$, $e^{in\xi}$.

Other examples of Gram matrices with this property have been given in [25].

In trying to deal with the "limited angle" problem in R^3, we were led to consider the extension of Slepian's [12] results to the case of S^2, or more generally S^n. This was found to be true and is reported, along with other results in [26].

Since the property holds for R^n and S^n it is only natural to expect that it should hold for hyperbolic space H^n. Indeed in [26] the result is shown to hold for H^3, see also [27] for a much more transparent proof.

And yet the result is not true for any H^n, $n \geq 2$, except when $n = 3$ [28].

The eigenfunctions for the radial part of the Laplacian in R^n are given by

$$\phi(\lambda,r) = \frac{J_{\frac{n-2}{2}}(\lambda r)}{(\lambda r)^{\frac{n-2}{2}}}$$

We make here the remark that as functions of λ these functions satisfy a second order differential equation of the form

$$D_\lambda \phi(\lambda,r) = \Theta(r)\,\phi(\lambda,r) \qquad (1)$$

The eigenfunctions fo the Laplacian for H^3 are

$$\phi(\lambda,r) = \frac{\sin \lambda r}{\lambda \sin hr}$$

and we have

$$\frac{1}{\lambda^2}\frac{d}{d\lambda}\left(\lambda^2 \frac{d}{d\lambda}\phi(\lambda,r)\right) = -r^2\phi(\lambda,r)$$

It is now a fact that for any $n \neq 3$, $n \geq 2$, the radial eigenfunctions of the Laplacian for H^n do not satisfy an equation of the form (1).

This fact and other evidence from [25] made us consider equation (1) an important ingredient in the existence of a differential operator of "band and time limiting." More on this in the next section.

4. DIFFERENTIAL EQUATIONS IN THE SPECTRAL PARAMETER

Given a second order differential operator

$$L \equiv -\frac{d^2}{dx^2} + V(x)$$

denote by $\phi(\lambda,x)$ a family of eigenfunctions of L, $L\phi = \lambda^2\phi$. The previous considerations lead us to the following

<u>Question</u>. For what $V(x)$ does there exist a differential operator D_λ such that

$$D_\lambda\,\phi(\lambda,x) = \Theta(x)\,\phi(\lambda,x)$$

We have attacked this question in collaboration with H. Duistermaat [29].

We give below some few examples of this situation leaving a more complete discussion for [29].

(a) Take $V(x,t) = \dfrac{6x(x^3 - 2t)}{(x^3 + t)^2}$,

then if $\phi(\lambda,x) = \dfrac{(\lambda^2x^3 + 3i\lambda x^2 - 3x + \lambda^2 t)}{\lambda^2(x^3 + t)}\, e^{i\lambda x}$

we've

$$\left(-\frac{d^2}{dx^2} + V(x,t)\right)\phi(\lambda,x) = \lambda^2\phi(\lambda,x)$$

and

$$\left(\left(\frac{d}{d\lambda} - \frac{2}{\lambda}\right)\left(\frac{d}{d\lambda} + \frac{2}{\lambda}\right)\right)^2 \phi - 4it\,\frac{d\phi}{d\lambda} = (x^4 + 4tx)\phi$$

(b) Take $V(x,t,s) = \dfrac{12x^{10} + 324sx^5 + 450t^2x^4 + 300t^3x + 162s^2}{(x^6 + 5tx^3 + 9sx - 5t^2)^2}$

then for an appropriate family of eigenfunctions $\phi(\lambda,x)$ we've

$$\left[i\left(\frac{d}{d\lambda} - \frac{3}{\lambda}\right)\left(\frac{d}{d\lambda} - \frac{2}{\lambda}\right)\left(\frac{d}{d\lambda} - \frac{1}{\lambda}\right)\frac{d}{d\lambda}\left(\frac{d}{d\lambda} + \frac{1}{\lambda}\right)\left(\frac{d}{d} + \frac{2}{\lambda}\right)\left(\frac{d}{d\lambda} + \frac{3}{\lambda}\right)\right]\phi \; +$$

$$+ \frac{63}{2}s\left(\frac{d}{d\lambda} - \frac{1}{\lambda}\right)\left(\frac{d}{d\lambda} + \frac{1}{\lambda}\right)\phi \; - \; 35it^2\frac{d}{d\lambda}\phi$$

$$+ \frac{35}{4}t\left[\left(\left(\frac{d}{d\lambda} - \frac{2}{\lambda}\right)\left(\frac{d}{d\lambda} + \frac{2}{\lambda}\right)\right)^2 - \frac{36}{\lambda^4}\right]\phi$$

$$= (x^7 + \frac{35}{4}tx^4 - \frac{63}{2}sx^2 - 35t^2x)\,\phi$$

We make the observation that the function $V(x,t)$ in example (a) is a solution of the Korteweg-deVries equation. Also $V(x,t,s)$ in example (b) can be written in the form

$$e^{tX_1+sX_2}\left(\frac{12}{x^2}\right)$$

where X_1, X_2 denote the Kdv flow and the fifth order differential operator given by $[(-L^{3/2})_+, L]$ and $[(-L^{5/2})_+, L]$ respectively, (up to scaling). See [30 - 35].

As one may suspect this works for the whole "hierarchy." More interesting still not all examples are obtained this way.

(c) The simplest example which is not obtained in this fashion is the following.

Take

$$V(x,t) = \frac{15x^4 - 18tx^2 - t^2}{4x^6 + 8tx^4 + 4t^2x^2}$$

with eigenfunctions

$$\phi(x,k) = -\sqrt{x}\ J_2(xk) + \frac{2t}{(t + x^2)kx}\sqrt{x}\ J_1(xk)$$

One then has

$$\left[\left(-\partial_K^2 + \frac{15}{4k^2}\right)^2 + 2t\left(-\partial_K^2 - \frac{1}{4k^2}\right)\right]\phi = (x^4 + 2tx^2)\phi$$

It is interesting to notice that for the first family of examples one has

$$\Theta'(x) = \tau(x) \quad \text{where}$$

$$V(x) = -2(\log \tau(x))''$$

Department of Mathematics
UC Berkeley 94720 USA

REFERENCES

1. Radon, J.: 1917, 'Über die Bestimmung von Funktionen durch ihre Integral werte längs gewisser Mannifaltigkeiten', Berichte Saechsische Akademic der Wissenschaften 69, 262-277.

2. Uhlenbeck, G., personal communication.

3. Bockwinkel: 1906, 'On the propagation of light in a two-axial crystal around a center of vibration'. Versl. Kon Akad. v. Wet XIV 2, 636-651.

4. Uhlenbeck, G.: 1925, 'Over Eeen Stelling Van Lorentz En Haar Uitbreding Voor Meerdimensionale Ruimten'.

5. Cramer, H. and Wold, H.: 1936, London Math. Soc. 11, 290.

6. Bracewell, R.: 1956, 'Strip integration in radio astronomy', Austr. J. Physics 9, 198.

7. Cormack, A.: 1963, 'Representation of a function by its line integrals with some radiological applications', J. Appl. Physics 34, 2722.

8. Housenfield, G.: 1972, British Patent No. 1283915.

9. Davison, M.E. and F. A. Grünbaum: 1979, 'Convolution algorithms for arbitrary projection angles'. IEEE Transactions on Nuclear Science NS-26, 2670-2673.

10. Davison, M.E. and F. A. Grünbaum: 1981, 'Tomographic reconstruction with arbitrary directions', Communications on Pure and Applied Mathematics 34, 77-120.

11. Grünbaum, F.A.: 1980, ' A study of Fourier space methods for "limited angle" image reconstruction', Numerical Functional Analysis and Optimization 2 (1), 31-42.

12. Slepian, D.: 1978, 'Prolate spheroidal wave functions, Fourier analysis and uncertainty', Bell System Tech. Journal 57, no. 5, 1371-1430.

13. Whittaker, E.T.: 1915, Proc. London Math. Soc. (2) 14, 260-268.

14. Ince, E.L.: 1922, 'On the connection between linear diff. systems and integral equations', Proc. of the Royal Society of Edinburgh (42), pp. 43-53.

15. Davison, M.E.: to appear, 'The ill conditioned nature of the limited angle tomography problem', SIAM J. Applied Math.

16. Davison, M.E.: 1981, 'A singular value decomposition for the Radon transform in n-dimensional euclidean space', Numer. Funct. Anal. and Optimiz. 3(3), 321-340.

17. Slepian, D. and H.P. Pollak: 1961, 'Prolate spheroidal wave functions, Fourier analysis and uncertainty I.', Bell System Tech. Journal 40, no. 1, 321-340.

18. Landau, H.J. and H.P. Pollak: 1961, 'Prolate spheroidal wave functions, Fourier analysis and uncertainty II', Bell System Tech. Journal 40, No. 1, 65-84.

19. Landau, H.J. and H.P. Pollak: 1962, 'Prolate spheroidal wave functiions, Fourier analysis and uncertainty III', Bell System Tech. Journal 41, No. 4, 1295-1336.

20. Slepian, D.: 1964, 'Prolate spheroidal wave functions, Fourier analysis and uncertainty IV', Bell System Tech. Journal 43, No. 6, 3009-3058.

21. Grünbaum, F.A.: 1981, 'Eigenvectors of a Toeplitz matrix: discrete version of the prolate spheroidal wave functions', SIAM J. Alg. Disc. Math. 2(2), 136-141.

22. Grünbaum, F.A.: 1981, 'Toeplitz matrices commuting with a tridiagonal matrix', Linear Algebra and its Applications 40, 25-36.

23. Grünbaum, F.A.: 1982, 'A remark on Hilbert's matrix', Linear Algebra and its Applications 43, 119-124.

24. Grünbaum, F.A.: 1982, 'The eigenvectors of the discrete Fourier transform: a verion of the Hermite functions', J. Math. Anal. Applic. 88, 355-363.

25. Grünbaum, F.A.: to appear, 'A new property of reproducing kernels for classical orthogonal polynomials', J. Math. Anal. Applic.

26. Grünbaum, F.A., Longhi, L. and M. Perlstadt: 1982, 'Differential operators commuting with finite convolution integral operators: some nonabelian examples', SIAM J. Applied Math.

27. Grünbaum, F.A.: 1982, 'Doubly concentrated functions for three hyperbolic space', IHES preprint.

28. Grünbaum, F.A.: 'The Slepian-Landau-Pollak phenomenon for the radial part of the Laplacian'.

29. Duistermaat, J. and F.A. Grünbaum: in preparation, 'Differential equations in the eigenvalue parameter'.

30. Lax, P.: 1968, 'Integrals of nonlinear equations of evolutions and solitary waves', Communications on Pure and Applied Mathematics 21, 467-490.

31. Gelfand, I. and Dikii: 1976, 'Fractional powers of operators and Hamiltonian systems', Funkts. Anal. Prilozhen. 10, 4, 13-39.

32. H. Airault, H. McKean, and J. Moser: 1977, 'Rational and elliptic solutions of the Korteweg-deVries equation and a related many body problem', Communications in Pure and Applied Math.(30), 95-148.

33. D. Chudnovski and G. Chudnovski: 1977, 'Pole expansions for nonlinear partial differential equations', Nuovo Cimento 40 B , 339-353.

34. M. Adler and J. Moser: 1978, 'On a class of polynomials connected with the Korteweg-deVries equation', Communications in Mathematical Physics (61), 1-30.

35. M. Ablowitz and H. Airault: 1981, 'Perturbations finies et forme particuliere de certaines solutions de l'equation de Korteweg-deVries. C.R. Acad. Sci. Paris t. 292, 279-281.

Maurice KLEMAN

GEOMETRICAL ASPECTS IN THE PHYSICS OF DEFECTS

1. INTRODUCTION

Geometry has been since long a subject of reflection and admiration. Philosophers all agree on this opinion that, apart their intrinsic beauty, geometrical concepts are, together with concepts in other parts of mathematics, the touchstone of a good theory, in any domain of science. Although many often forget that this prominent role might prove dangerous when geometry is the master, not the servant (think for example of the vast and hopeless effort devoted to explain the solar system in terms of Platonic solids) there is little doubt that geometry plays a major role in most sciences, and in particular in the physical sciences. Concerning recent achievements, we can quote the development of gauge field theories in particle physics[1], which is based on the geometry of fibre bundles, of astrophysics (Misner et al.), which uses extensively the properties of geodesics in various manifolds, and of the theory of defects which, notwithstanding its plebeian origins (the metallurgical properties of solids), has found answers to some of the questions it raises only in the most elaborate theorems of algebraic topology, and in some other non-trivial branches of geometry. My examples are certainly too scarce: others can come to the mind; also there is no big risk in foreseeing other major advances in physics in the years to come, with the help of very up-to-date advances in geometry. For it is a very remarkable fact that many recent results in geometry have found applications in physics[2].

But other sciences or disciplines, even far from physics, show the same trend to use advanced mathematics (and in particular geometry) to a large extent. This is true for the problem of morphogenesis in biology or the modeling of the syntactic structure of sentences in semantics (Thom, 1974), or even the topology of the unconscious in psychoanalysis[3], for example; no doubt these problems have

C. P. Bruter et al. (eds.), Bifurcation Theory, Mechanics and Physics, 331–355.

nothing to do with those encountered in physics, and we cannot expect all these attempts to prove successful. But it is a fact that such attempts are done, and this might well indicate, as R. Thom hopes, a revival of natural philosophy (R. Thom, 1979). R. Thom's theory of catastrophes is precisely the geometrical frame which is used in many of these attempts; catastrophe theory is the theory, to quote Zeeman, which deals with "singular behaviours produced by continuous causes". We quote this definition here because such a character has something in common with the types of problems we encounter in the physics of defects. While the mathematics are quite different, it is not without interest anyway to develop this point in this introduction. (Note incidentally that catastrophe theory has already proven important in its applications to physical problems, in particular in the study of light caustics (Berry)).

Both theories (of catastrophes and of defects in ordered media) are theories of forms and/or of complexity and "imperfections", in the sense that they put the stress on the existence of <u>singularities</u> whose individual features bring the essentials in the description of a global phenomenon. In other words these <u>local</u> singularities provide <u>global</u> answers, at least at some spatio-temporal scale of description. For example the complexity of the plastic deformation of a piece of crystal can be entirely analyzed in terms of dislocations whose geometrical characteristics belong to a limited set; similarly the structure of light caustics, however complex might be the geometries of the diffracting region and of the light source, can be analyzed in terms of elementary catastrophes; the magnetic field produced by a magnet is the effect of the arrangement of very simple Weiss domains, etc...

In both cases, the singularities have <u>spatio-temporal persistance</u>. This means, in the case of defects in ordered media, the existence of <u>topological invariants</u> attached to each type of defect, in the case of catastrophes, the existence of canonical <u>bifurcation sets</u>. Slight variations to the external conditions, or due to a dynamic process, do not modify the singularities in an essential way. Large variations might lead to events in which the singularities interplay, multiply, disappear,... but still obeying the constraints imposed by the actual invariants.

In both cases, finally, the discontinuous effects represented by the singularities are produced by continuous causes (in the case of ordered media, this notion of continuity implies that we consider the medium at a scale larger than the lattice parameter).

One might notice the similarity between such concepts and those which appeared last century in Riemann's theory of functions of a complex variable. These functions are entirely defined by their singularities and the conditions at the boundary of their domain of definition (boundaries are in fact special types of singularities, too). It is worth mentionning the resistance of Weierstrass and of most of the prominent physicists of the time to the uprising of Riemann's ideas; this might help us to understand why it took so long for the physics of defects to obtain the status of a reasonable scientific discipline (metallurgists were for many years the only physicists to take interest in dislocations, sometimes to the scorn of their colleagues, who found the study of imperfections impure).

But, apart these (important) similarities in spirit, the physics of defects adresses itself to a completely different problem that catastrophe theory. Its object is the study of local breakings of order in ordered media like solid crystals (regular packings of atoms, correlated in position over large distances), liquid crystals (regular packing of anisotropic molecules correlated in angle over large distances), magnetic systems (spins angularly correlated over large distances), etc... In all these cases the medium is invariant under a symmetry group H, which is a subgroup of the euclidean group E (we just mention for the sake of completeness the case of superfluids, whose "gauge" group of invariance acts on hidden variables, not molecules -so that the group has no reason to be related to E- but whose defects can be discussed by methods similar to ordinary ordered media); these symmetry groups of crystals have been extensively studied and classified in the second half of the last century; but the problem of symmetry breaking is a problem which is very active to-day [4], either under its global version (the problem of second order phase transitions), or its local version (defects). We shall try in this paper to give a rapid overview of the various mathematical problems which occur in the last case. The reader interested in a larger account of these problems can consult Kléman (1982) and the other (specialized to various topics) quoted references.

The paper is organized as follows. Section 2 is devoted to the study of the isolated defects, in particular their topological classification by homotopy methods. Stress is put on the difference between local and global properties. Section 3 treats of densities of defects, and the related mathematical models, i.e. either Cartan spaces with torsion and/or curvature, or caustics, or gauge theories. Section 4 discusses of various examples where the classical theory of two dimensional surfaces has proven useful. Section 5, finally, discusses of recent attempts to describe glassy (non-ordered) materials in terms of curved spaces, and how the notion of defects is naturally involved in the mapping between a curved space and a flat (euclidean) space.

2. ISOLATED DEFECTS

The classical view on defects concerns line defects only, i.e. dislocations, which break symmetries of translation, and disclinations, which break symmetries of rotation.

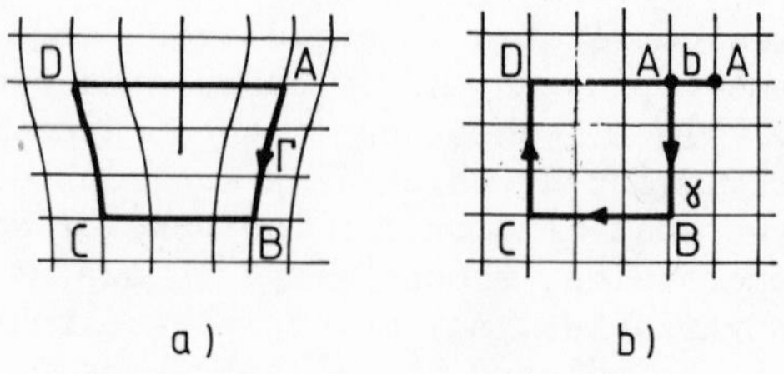

Fig. 1

Burgers'circuit Γ, its image γ, and the Burgers'vector $\underset{\sim}{b}$

The Burgers circuit, which permits to define an invariant in the case of dislocations, is illustrated fig.1 : it consists in a mapping $p: \Gamma \to \gamma$ into the perfect medium of a circuit Γ drawn in the deformed medium around the presumed defect. The closure failure $\underset{\sim}{b}$ of the image circuit is an element of the group of translations $\underset{\sim}{b} = n_1 \underset{\sim}{a}_1 + n_2 \underset{\sim}{a}_2 + n_3 \underset{\sim}{a}_3$ (n_1, n_2, n_3 integers) of the crystal and an invariant of the line: it does not depend on the precise form or location of the circuit Γ; consequently, when two lines $\underset{\sim}{b}_1$ and $\underset{\sim}{b}_2$ meet, they form a line of Burgers'vector $\underset{\sim}{b}_1 + \underset{\sim}{b}_2$ (we do not discuss here about some problem concerning the orientation of the line, in defining the sign of $\underset{\sim}{b}$). The topological nature of the process just defined is clear.

But it is not so easy to define an equivalent invariant for the case of disclinations; fig.2 illustrates the case of a disclination in a nematic when the molecules are constrained to sit in planes perpendicular to the line: in that case the rotations which bring the molecules located on a circuit Γ from their actual direction to a fixed one are all commutative, and there is no difficulty in drawing an hodograph of these rotations

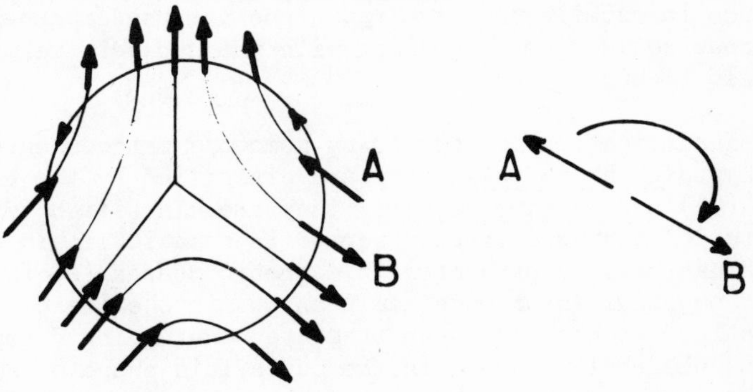

Fig. 2

Disclination of angular strength π

a) Scheme of the molecular distribution (lines of forces of the molecules) and Burgers'circuit. b) hodograph of the Burgers'circuit on which an angle of π is measured.

which enable us to define an invariant. But the general case of a disclination does not prove so easy, and the question of the relevant invariant is solved only by using the methods of algebraic topology. The invariants in question are therefore all defined as the classes of equivalence of the elements of the homotopy group $\Pi_1(V)$, where $V = E/H$ is the quotient manifold of the euclidean group by the group of symmetry of the ordered medium. Disclinations and dislocations, in fact all lines defects including mixed lines (dispirations), are classified together by the same group.

The origin of the homotopy classification is clear; $V = E/H$ is the orbit of local states of the deformed medium. The induced mapping of the circuit Γ on V

$$v : \Gamma \to \gamma_V \qquad (1)$$

is a circuit on V. If this circuit can be continuously deformed on V to a point, the defect surrounded by Γ is said to be topologically unstable. Otherwise γ_V belongs to one of the homotopy classes of $\Pi_1(V)$, which classes clearly do not depend on the precise form and location of Γ around the defect, which is then topologically stable. We do no want to discuss in details this theory, since the reader can find now numerous review papers dealing with the subject (Mermin, Michel, Trebin).

The classification of defects by homotopy methods extends easily to point defects (they are classified by the elements of $\Pi_2(V)$) and surface defects (they are classified by $\Pi_0(V)$). Points defects were first observed in nematic liquid crystals, after having been predicted in magnetic phases (Döring), where their physical importance has been put into evidence only recently. It has also been predicted that point defects are not topologically stable in the suprafluid phase of He^3, a result which could not have been obtained without the powerful tool of homotopy (Toulouse and Kléman). Let us note also that homotopy has predicted the existence of "configurations", classified by $\Pi_3(V)$, which lead to smooth differentiable distortions of the ordered medium which are not homotopic to a constant; it is fair to say that these configurations were understood in terms of homotopy much before homotopy was applied to defects (Finkelstein).

How does the global distortion of a given sample of an ordered phase relates to the isolated defects present inside ? The solution of this problem requires the knowledge of the dynamical equations, elasticity constants, etc... of the ordered phase. In the case of dislocations (of translation) in solids, it is usual to distinguish between applied distortions (or stresses) and internal distortions (or stresses); the latter refer to the distortions (stresses) attending the dislocations when the boundaries are free from stresses; the former refer to the distortions (stresses) due to applied efforts on the boundaries. Applied distortions can be described independently of the presence of defects in a linear theory (quadratic free energy) but this is not so in general. However a definition of applied efforts, distortions, appears necessary, since we have to describe changes in the ordered medium which conserve the topological invariants of the various defects when they move or change shape, or to deal with topological interactions between defects when they interact.

A general answer is with geometrical gauge field theory. Consider for example a dislocation of translation; it creates a distortion field β_{ij} such that the variation in the displacement function $u_i(r)$ between two neighboring points $\underset{\sim}{r}$ and $\underset{\sim}{r} + d\underset{\sim}{r}$ reads

$$du_i = \sum_j \beta_{ji}\, dx_j \qquad (2)$$

This 1-form is not closed, since its summation on a closed circuit surrounding the line yields by definition the Burgers vector $\underset{\sim}{b}$

$$\oint du_i = b_i \qquad (3)$$

This invariant is clearly conserved if β_{ij} is changed to

$$\beta_{ij} \rightarrow \beta_{ij} + \frac{\partial \lambda_j}{\partial x_i} \qquad (4)$$

where $\lambda_j(\underset{\sim}{r})$ is a variable displacement which is an element of the full group of translation, subgroup of E. Equ.4 expresses a gauge transformation; we qualify it of "geometrical" gauge transformation because, contrary to its use in particle physics, we do not require the total energy to be invariant in such a gauge transformation, but only the invariants of defects to be conserved (or modified in a suitable way).

The case of a dislocation of translation in an ordered medium on which acts the full gauge group of translation is particularly simple (and not extremely fascinating). The gauge group is commutative. Any applied distortion is then a permitted gauge transformation and reciprocally. The extension to the dynamical case (when the dislocation moves) does not present any kind of difficulty (Dzyaloshinskii). More interesting considerations happen when the group is not commutative (like the group of rotations) or the defects more sophisticated than dislocation lines. In particular, if the homotopy group which classifies the defects is not isomorphic to Z, the group of integers, then there are no analytical topological invariants attached to each defect, which fact makes the methods of exterior calculus useless. Some steps in the direction of a general analysis have been made (Dzyaloshinskii), but much remains to be done; the problem can be stated as follows : find and classify the gauge transformation types which conserve the topological invariants of a given type of defect. A similar problem has already been solved in electrodynamics of abelian and non-abelian gauge groups, concerning the gauge types of the Dirac monopoles (Wu and Yang).

Let us give an example of the considerations which can be of interest. The singular points in Heisenberg ferromagnets are classified by

$$\Pi_2 (S^2) = Z \qquad (5)$$

The "order parameter" can be represented by a vector $\mathbf{S}(\mathbf{r})$ of constant magnitude, the spin; the manifold of internal states V is therefore a sphere S^2. Suppose a sample contains one topologically stable singular point P (whose class of homotopy is not the trivial class of $\Pi_2(V)$), and look for the rotations $\boldsymbol{\omega}(\mathbf{r})$ acting on $\mathbf{S}$ which do not change the topological invariant of the defect. A general gauge transformation $\boldsymbol{\omega}(\mathbf{r})$ can be represented by attaching to $\mathbf{r}$ an orthonormal frame with three rectangular unit vectors $\mathbf{S}/S$, $\mathbf{a}(\mathbf{r})$, $\mathbf{b}(\mathbf{r})$: any rotation which acts non-trivially on $\mathbf{S}$ reads $\omega_a \mathbf{a} + \omega_b \mathbf{b}$. The orthonormal frames themselves are representative of an order parameter whose manifold is V = SO(3). We have $\Pi_2(SO(3)) = 1$. Therefore the point defect P is not topologically stable with respect to any set of frames $\mathbf{S}$, $\mathbf{a}$, $\mathbf{b}$, whatever $\mathbf{a}$ and $\mathbf{b}$ might be. Since the defect is preexistent in the field of $\mathbf{S}$ (and topologically stable), the paradox can only be solved by imagining that there is a singular string which starts from P and goes to infinity, along which the set ($\underset{\sim}{S}$, $\underset{\sim}{a}$, $\underset{\sim}{b}$) is singular. (This string is similar to a Dirac string, but with a different physical interpretation). Therefore the only gauge transformations which conserve P must have a vanishing $|\underset{\sim}{\omega}|$ along a string. Otherwise, the point defect "extends" along the part of the string which carries a nonvanishing rotation. This is certainly what happens in a dynamic regime when P moves.

3. DENSITIES OF DEFECTS

It is easy to define defect densities when the homotopy group $\Pi_i(V)$ is isomorphic to Z; the densities so defined have a physical meaning when considering the ordered medium at a scale greater than the mean distance between defects, and one can go to the continuous limit. For example, starting from equ.(2), one can write

$$d^2u_i = \sum_{j,k} \partial_k \beta_{ji} \, dx_k \wedge dx_j \tag{6}$$

and the quantity

$$\alpha_{ij} = \sum_{k,\ell} \varepsilon_{ik\ell} \, \partial_k \beta_{\ell j} \tag{7}$$

is clearly a density of dislocations of translations of direction i, with a Burgers'vector component along j. For an isolated dislocation one gets

$$\alpha_{ij} = b_j \, t_i \, \delta(L) \tag{8}$$

where $\underset{\sim}{t}$ is a unit vector along the line L. Extensions of these notions can be done to point defects. But we shall keep the discussion at the level of dislocation lines. There are three types of geometrical interpretations of the quantities we have introduced in equ.7.

3.1. Gauge field theories and fibre bundles (Dzyaloshinskii, Edelen).

These are the more recent interpretations, we have already discussed shortly of their interests and limitations. Note also that there is an increasing use of gauge theories to describe phase transitions driven by defects (Kleinert), but in that case the gauge invariance which is considered applies to the free energy density.

3.2. Geometrical relations in dislocated crystals (Nye)

One usually distinguishes in the distortion effect due to a density of dislocations, a part of pure strain, the other one of pure rotation. Let us analyze the case of pure rotation. Fig.3 represents a case where in the limit of a vanishing lattice parameter, there is no strain, but a pure rotation; the lattice planes have "glided" in a constant direction and form a family of parallel cylinders; in order to keep constant distances between atoms, densities of parallel

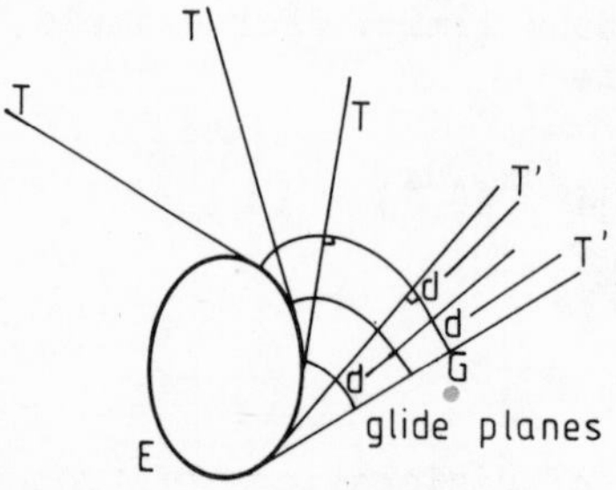

Fig. 3

Single glide, two-dimensional case. E is the focal surface enveloped by the orthogonal trajectories T of the deformed glide planes G. If T and G represent atomic planes, extra planes T' have to be introduced in order to conserve the density of atoms. Hence a density of dislocations d.

dislocations, all of the same sign, have to be introduced. These densities α_{31} are proportional to the curvature of the crystal

$$\alpha_{31} \propto \frac{1}{R} \tag{9}$$

This simple case of single glide in a two dimensional crystal can be extended to multiple glide, as long as the rotations are small. This results in the definition of a so-called <u>tensor of contortion</u> K_{ij} which describes the rotation of the lattice, through the 1-form

$$d\,\omega_i = \sum_j K_{ji}\, dx_j \tag{10}$$

and which is related to α_{ij} by

$$K_{ij} = \frac{1}{2}\,\delta_{ij}\,(\alpha_{11} + \alpha_{22} + \alpha_{33}) - \alpha_{ji} \tag{11}$$

The 1-form (11) is closed, except near the singular surfaces along which the curvature of the glide surfaces becomes infinite, i.e. on the focal surfaces (caustics) of the orthogonal trajectories of the deformed glide planes, one will notice that the focal surface, in the case of single glide in a two-dimensional crystal, has clearly the character of a defect which breaks the symmetries of rotation; if the focal surface E is closed, the crystal rotates by an angle of 2π about it. The focal surface plays therefore the role of a disclination line, at long distance from it; the lattice planes are involutes of E.

The phenomenon of polygonization in crystals is related to the analysis of two-dimensional glide. Polygonization is a process by which dislocations, all of the same sign, gather in a plane which separates the crystal in two parts rotated one with respect to the other (fig.4).

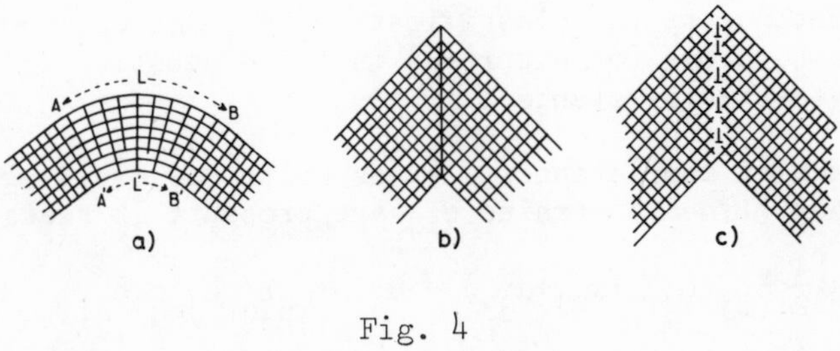

Fig. 4

Polygonization: the dislocations, equally dispersed in a), gather on a surface on which the crystal orientation suffers a jump $\omega \simeq b/\ell$ (b, Burgers'vector modulus, ℓ distance between dislocations).

The angle of rotation is directly related to the(surface) density of dislocations in the wall. Polygonization walls are tilt walls (the axis of rotation from one sub-crystal to the other being in the plane of polygonization), or twist walls (axis of rotation perpendicular to the wall), or of mixed character. Polygonization walls are limited by disclinations, according to our last remark on caustics.

This analysis of the geometrical meaning of dislocations densities can be applied with success to liquid crystals like nematics (Kléman, 1973). In such media, the only non-trivial defects are disclinations and point defects; the mass density is constant in any deformation, the strain tensor is strictly vanishing, and it is of big interest to describe the distortions of the medium, outside the defects, as due to true densities of dislocations in the above sense, since the distortions are pure molecular rotations. This allows for a definition of disclinations and point defects in terms of "caustics" (focal lines, points, or surfaces) of the integral lines of the molecular directions.

In a way, the application of the geometrical relations involved with densities of dislocations to liquid crystals and similar media (where distortions relax viscously to a state of vanishing strain) is perhaps more important than their application to ordinary solid crystals, where strains are always present. There is certainly much to be done in this direction, in particular to develop the relation with the theory of catastrophes (its interpretation in terms of caustics) and to compare the classification of defects by homotopy methods to the classification by caustics. However it remains to prove (or disprove) that the caustics in nematics are generically stable.

For the sake of completeness, let us write how equ.11 generalizes when (linear) strains e_{ij} are present; it reads:

$$K_{ij} = \frac{1}{2}\,\delta_{ij}\,(\alpha_{11}+\alpha_{22}+\alpha_{33}) - \alpha_{ji} + \sum_{p,q} \varepsilon_{jpq}\, e_{qi,p} \qquad (12)$$

The 1-form (10) is still closed, but the two contributions to the rotation gradient K_{ij}, the geometrical one and the elastic one, are not gradient like separately (Kröner, 1981).

3.3. Differential geometry of defects in crystal lattices.

This subject has been developed in the fifties, after Nye's paper, by Kondo, Kröner, and Bilby and coworkers. The basic mathematics used there are Elie Cartan's works, on Riemannian and non-Riemannian geometries. It consists in a description of a dislocated crystal, in the limit when the lattice parameter goes to zero without disappearance of the lattice itself (in order that local frames can still be defined), as a non-Riemannian space endowed with torsion (representing dislocation densities) and vanishing curvature. This theory applies to large deformations. Its similarity (in spirit) with generalized relativity (vanishing torsion, non vanishing curvature) is conspicuous. Note that torsion has been introduced in recent relativistic theories (Hehl and al.). Note also that it is possible to extent such types of theories to ordered media with complex microstructures (spin systems, micropolar media) by using more sophisticated geometries, like Finsler geometry. The present theory applies only to point lattice crystals.

The essential idea can be stated, in the very pictorial manner of Kröner, as follows. The properties of a deformed crystal containing dislocations can be tested separately : the topological properties by an <u>internal observer</u> who counts lattice steps and changes of directions when walking along the lattice rows, but does not recognize the changes of lengths and angles with respect to the undeformed crystal, the elastic properties by an <u>external observer</u> who can measure lengths and angles, hence differentiate the various elastic fields which can attend a given set of defects densities. Hence the internal observer can follow by <u>parallel displacement</u> any lattice vector, with components $v_h(\underset{\sim}{r})$ measured by the external observer, and not realize that these components have changed by a quantity dv_ℓ which, by hypothesis in this theory, is expressed as an <u>affine connection</u>

$$dv_\ell = + \Gamma_{m\ell k} v^k dx^m \qquad (13)$$

but he can recognize the existence of a dislocation by parallely displacing two infinitesimally small lattice vectors $\underset{\sim}{u}(r)$ and $\underset{\sim}{v}(r)$, the first one along $\underset{\sim}{v}(r)$, the second one along $\underset{\sim}{u}(\underset{\sim}{r})$. If there is a dislocation in the vicinity, the Cartan's circuit so constructed $\vec{u}$, $\vec{v}$, $-\vec{u}-d\vec{u}$, $-\vec{v}-d\vec{v}$, does not close. The infinitesimal closure failure, which can be counted by the internal observer in units of lattice vectors, reads for the external observer :

$$db^h = - \Gamma_{m\ell}{}^h \, dS^{m\ell} \tag{14}$$

where $dS^{m\ell}$ is the oriented area span by the circuit (the fact that the circuit does not exactly enclose an area element is discussed in Kröner's paper [21]).

$dS^{m\ell}$ is antisymmetric; the antisymmetric part of the affine connection (the so-called Cartan's torsion) can therefore be interpreted as a dislocation density, the relation with α_{ij} being

$$T_{m\ell}{}^h = \Gamma_{[m\ell]}{}^h = - \frac{1}{2} \varepsilon_{nm\ell} \, \alpha^{nh} \tag{15}$$

Now the total rotation of a lattice vector is expressed by equ.13, in which one can differentiate a contribution of the antisymmetric part of $\Gamma_{m\ell h}$, which according to equ.15 is purely geometrical (and akin to the contribution of the contortion tensor (equ.11) of the linear theory), and a contribution of the symmetric part, which describes the effect of the infinitesimal change over dx_m of the strain tensor $e_{\ell k}$; writing $d\beta_{\ell k} = de_{\ell k} + d\omega_{\ell k}$, one gets

$$de_{\ell k} = \Gamma_{m(\ell k)} \, dx^m \quad ; \quad d\omega_{\ell k} = \Gamma_{m[\ell k]} \, dx^m \tag{16}$$

Still in the spirit of Cartan's theory, the external observer can measure (and the internal observer recognize) the change in orientation of a vector v^h when parallely transported along a closed circuit. If the medium does not contain disclinations, and since the lattice must be univocally defined everywhere (law of distant parallelism), we have

$$\oint dv_h = 0$$

which also reads

$$R_{nm\ell}{}^k = 2 \left(\partial_n \Gamma_{m\ell}{}^h - \Gamma_{mp}{}^h \Gamma_{n\ell}{}^h\right)_{[nm]} = 0 \tag{17}$$

The Riemann-Christoffel tensor, which measures the curvature, vanishes in a dislocated crystal.

Now, what would mean, from the point of view of the theory

of defects, a space with non-vanishing curvature? It is tempting to answer that curvature represents disclination densities. According to Kröner, it cannot be so, because any disclination can be constructed as a sum of dislocations; this is true at least in a "well-behaved" crystal (we meet in section 5 crystals which are not well-behaved). Kröner proposes to attach curvature to the presence of extra matter (interstitials) which changes the metric locally. However the question does not seem to be settled yet. An interesting problem in this respect would be to describe in the present frame the liquid crystals and various anisotropic fluids for which the distortion field e_{ij} is strictly zero. In that case it is quite probable that localized curvature would have to be introduced to represent isolated disclinations, in the spirit of the remarks we did in the last subsection (curvature appearing as the singularity set of the torsion field).

4. SURFACES

This will be a very rapid overview of the subject. A large literature exists on the subject, due to the manifold works on the smectic phases of liquid crystals, whose layered character was first recognized by G. Friedel from his observations of defects in the A-smectic phase [6]. A subsection is also devoted to columnar liquid crystals.

4.1. Focal conics

This very particular arrangement of layers is due to the easiness with which layers can curve, compared to their difficulty to change thickness. Therefore most deformations of smectic phases consist in stackings of parallel layers. In this geometry, the normals to the different layers are straight; comparison with a set of light rays, the layers being wave fronts, comes immediately to the mind.

This situation would suggest that the singularities of these deformations are similar to caustics, and that Thom's theory should apply, as it applies to light. But this is not so, and energetic considerations come still into play to restrain the possible solutions to those where the focal surfaces are degenerated to focal lines. So that the solutions are not generic in the sense of Thom's theory. It has been proven long ago [7] that the focal lines are then necessarily cofocal conics, and that the layers are folded into Dupin cyclides. This is exactly what is brought by multiple observations of smectic phases in optical microscopy.

Note however that some rare situations, closer to genericity, have been observed, probably because of the very peculiar experimental conditions (C.E. Williams), in which the defects of the A-smectic phase appeared when cooling the sample from its nematic phase, in which there were some straight disclinations which transformed to irregular zigzagging lines in the smectic phase (fig.5).

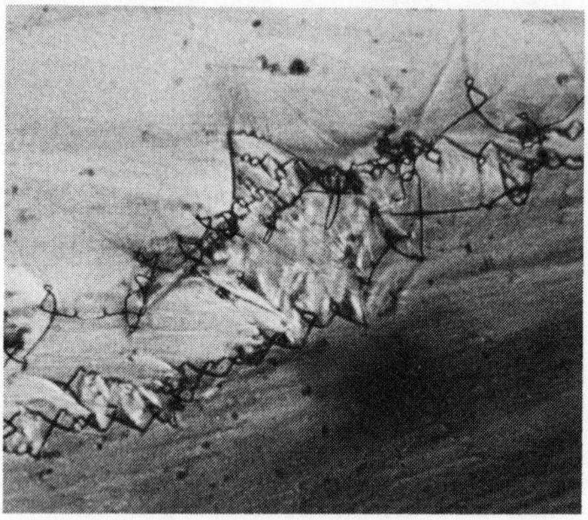

Fig. 5

Zigzagging and helical focal lines in a sample of A-Smectic cooled from the nematic phase (with courtesy of C.E. Williams).

Letting the sample anneal (i.e. transforming irreversibly to a situation of lower energy), the zigzagging lines turned continuously to pairs of helical lines whose interpretation proved easier [8]: these helical lines are the cusps of the focal surfaces pertaining to a set of layers (wave fronts) parallel to a right helicoïd. Since the (non-generic) helical lines are built by a continuous process from the zigzagging lines, one can infer that these zigzagging lines are also the cusps of some focal surface, and are generic.

But this cannot be the whole story; why is it that only the cusp of the focal surface is visible? To understand this point, we refer the reader to Wright's paper, which discusses the geometry of cusp catastrophes in waves and rays. He notices that the region inside the cusp is triply covered by the system of waves (fig.6), which therefore interfere and produce a system of wavefront dislocations. In our case,

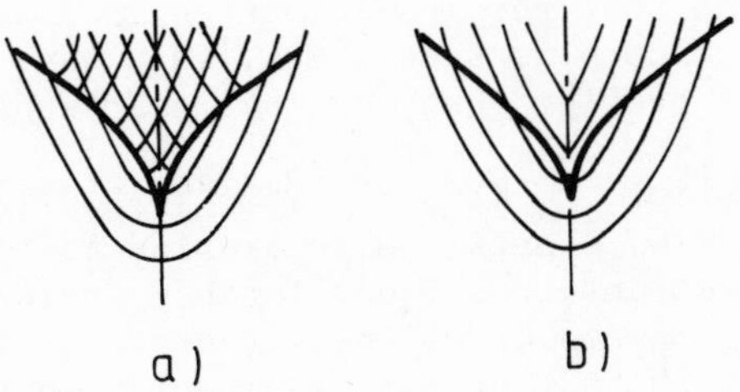

Fig. 6

Cusp catastrophe. a)triple covering of the region inside the cusp. b)physical case: two of the coverings remain and meet on a dislocation wall.

it is physically impossible that three systems of molecular layers be present simultaneously in the same region of space. The chosen solution is that the two of those systems which are in continuity with the systems outside the cusp penetrate inside the cusp (making the focal surfaces disappearing) and meet on a wall of dislocations which constitute the analog of the wavefront dislocations, but gathered on a surface (fig.6). One can persuade oneself that this surface is a singularity which is not much visible optically.

In fact, this example of the presence of dislocations topologically necessary near a focal line is not isolated in the physics of layered systems. A discussion of this point is made in C.E. Williams'thesis and in Bourdon et al. paper on C-smectic phases.

4.2. Developable domains

These singular objects should appear in columnar liquid crystals, in which the molecules like to stack in infinite parallel columns which arrange themselves in an ordered two-dimensional pattern. In the distorted state, configurations are energetically favoured in which the curved columns keep parallel and such that the local lattice in a section perpendicular to the columns is undeformed. In a way, this is for columns the same problem than for layers in subsection 4.1.

The general solution (M. Kléman, 1980; Y. Bouligand, 1980) consists in a set of planes, perpendicular to the curved columns. This set envelops a developable surface which is the locus of the points where the curvature of the columns become infinite; it is therefore a singular surface for the material.
All the developable domains can be classified by classifying their cuspidal edges. This is discussed in M. Kléman (1980).

Developable domains whose developable surfaces are cylinders have been observed (P. Oswald and M. Kléman); they are not topologically different from disclinations. Since the concept of developable domain is new, there is still a limited amount of results, but the concept seems general enough to permit numerous interpretations of future experimental results.

5. DEFECTS IN AMORPHOUS MEDIA; CURVED SPACES

Amorphous media, like silica glass, glycerine, or metallic systems rapidly quenched from the melt, offer a challenge to present-day solid state physicists. They are characterized by a disordered state, which can be put into evidence by a number of experimental methods (like X-Ray diffraction) whose interpretation relies on the hypotheses that the repartition of atoms is "at random". But what is randomness ? There are many reasons to believe that randomness is not absence of structure; physical properties of all amorphous media, whatever the nature of bonding between atoms (metallic, covalent, etc...) display describable universal features (Kléman, 1981, for a short review) which do not exist in ordered media.

Should amorphous bodies display universality properties if disorder were structureless? Moreover, there are all reasons to believe that there is not only one type of structure for amorphous media, since atoms bound together by very different types of forces can display an amorphous state. Can these various structures of disorder be classified?

There have been a number of attempts to describe the structure of amorphous bodies like a structure resulting from the presence of a huge quantity of defects in a primitively ordered material. Numerous discussions have taken place and continue to take place on the nature of these defects (P.W. Anderson, 1978; F. Spaepen, 1980). Experiments cannot brings an answer, since there is no possibility to have a direct look at atoms. This is among others, a good reason why numerical simulations of amorphous bodies are so numerous. These simulations seem to corroborate (to explain?) X-ray diffraction patterns; hence, they should be a reliable basis to describe the structural properties of disorder. But, interestingly enough, there is no consensus on the manner these structural properties should be described, but rather several ways of approaching them (Zallen, 1979). We propose here, very shortly, some geometrical principles which could help in discovering the laws of disorder (Kléman and Sadoc, 1979; Kléman, 1981, 1982). The notion of defect will appear, but in a context quite different from that one used up to now, but not less interesting.

There is at least one well-known fact, about disorder, it is the predominance of <u>five fold symmetry</u>. This was first proposed by Bernal for liquids; later Frank showed on the basis of energetical arguments that a cluster of 12 atoms about a central atom, bound by a Lennard-Jones pair potential, should pack with icosahedral symmetry. Many numerical simulations display the same feature (Srolovitz et al.), which has also been recognized long ago in froths (Coxeter). While this five fold symmetry seems to hold for systems which are not far (like froth) from the dense random packing of hard spheres, the covalent random network of tetracoordinated atoms (like silicium) display numerous sevenfold rings of bonds (Steinhardt and Chaudhari). Five-and sevenfold symmetries are not euclidean symmetries, in the sense that it is not possible to tile the euclidean space with a pattern having one of these symmetries. The only symmetries permitted in an euclidean space pertain to one of the 230 crystallographic groups. Therefore disorder in amorphous materials is easier to understand: it is not possible to repeat regularly the above symmetries. But how such a process does break regularity ?

Let us consider a n-dimensional space of constant curvature, spherical S_n or hyperbolic H_n. To such a space belongs a property which it shares with euclidean space, viz. the property of free mobility; there are continuous groups under wich S_n and H_n are transitive. Similarly also to the euclidean space, these spaces can be tessellated by the repetition of a pattern whose symmetry group is a subgroup of the continuous group. Therefore one can classify the "crystallographic" groups of S_n and H_n. This is well know for S_n ; the patterns are the polytopes of E_{n+1}, among which one finds the regular polytopes. The classification for H_3 has been done by Coxeter and Whitrow, but not studied in detail. An important fact is that there is no duplication of any of these crystallographic groups of H_n, S_n and E_n. Therefore any of the local symmetries which lead to the tessellations of S_3 and H_3 is not permitted, as the source of tessellation, in the euclidean space. In particular 5 and 7-fold symmetries are permitted in S_3 and H_3.

We have therefore here the basis of a classification of the local structure of amorphous bodies, if we know how to "map" a spherical or hyperbolic "crystal" on the usual flat space. This mapping must be very peculiar because the distances and angles between atoms have to be conserved as much as possible, for energetical reasons; for it would have no sense to construct a spherical or hyperbolic crystal whose local properties (symmetry , but also orientations and lengths) would not reproduce at best a state of low energy. Such constraints impose a mapping in which the curvature of S_3 and H_3 is suppressed by adding matter (S_3) or removing matter (H_3), which can be done by introducing disclinations. In a way, the process we are imagining is similar to the process which, for the mapping of a cone on a plane, consists in cutting it along a generatrix, flattening it without further distortion on the plane, and filling the void with matter. Hence the (positive) curvature concentrated at the apex is completely removed. We do such a process in S_3 or H_3 by introducing a number of disclinations whose angle of rotation in an angle of symmetry of the curved perfect crystal. Of course, since we cannot go to the continuous limit of densities of disclinations, we will have to introduce some small distortions, and the resulting amorphous material will be stressed. But the local symmetries will be conserved everywhere, except along the cores of the disclinations. Note that, since there are no symmetries of translations in a crystal on S_3 or H_3, we do not have dislocations in the usual sense.

A few specific models have been achevied using this method, most of them for S_3 (Mosseri and Sadoc, 1982). There is however an important difficulty in the use of S_3 : it is a finite crystal, and in order to fill a large volume of euclidean space it is necessary to add together, by some operation of glueing, the mappings of many S_3's. Therefore there are not only disclinations, but also walls, in the resulting amorphous body. However this situation might corresponds to some physical cases. The hyperbolic crystal has been less studied, but has the advantage to be drawn in H_3, which is diffeomorphic to the euclidean space. We can therefore expect interesting results. A general model of the resulting amorphous body has been established by Kléman (1982), which already permits to make some predictions on the physical properties at low temperature of elastic vibrations, which appear to be in good agreement with the universality properties already known at low temperature.

Laboratoire de Physique des Solides, Université Paris-Sud, 91405-Orsay (France)

NOTES

1. Wu and Yang (1975, p.3856) write: "it is a widely held view among mathematicians that the fiber bundle is a natural geometrical concept. Since gauge fields, including in particular the electromagnetic field, are fiber bundles, all gauge fields are thus based on geometry. To us it is remarkable that a geometrical concept formulated without reference to physics should turn out to be exactly the basis of one, and indeed may be all, of the fundamental interactions of the physical world.

2. d'Alembert states the same remark in 1759: "La Géométrie, en reculant ses limites, a porté son flambeau dans les parties de la Physique qui se trouvaient le plus près d'elle; le vrai système du monde a été connu, développé et perfec-

tionné" (Essai sur les Elémens de philosophie, in Oeuvres philosophiques, vol.2, J.F. Bastien, Paris, an XIII (1805) p.9).

3. Scilicet 2/3 (Collection dirigée par Jacques Lacan), Editions du Seuil, Paris 1970, p.169 et ssq.

4. See for example the proceedings of the colloquium on "Symmetries and broken symmetries (in condensed matter physics)", edited by N. Boccara, IDSET, Paris, 1981.

5. Media where the "order parameter" can be represented locally by an oriented frame enter this categorie. There is therefore much similarity between the media we are discussing here and those studied by the Cosserat brothers at the beginning of the century. See E. and F. Cosserat, Théorie des Corps Déformables, Hermann, Paris, 1909.

6. A detailed study of the theory of defects in layered media can be found in Kléman, 1982 (Chapt 5 and 7). For Columnar Systems, see Kléman (1980).

7. The proof is due to Dupin; the theorem was known to G. Friedel, which enabled him to infer the layered structure of smectics from observations in the optical microscope, where systems of focal conics (of gigantic size, compared to the layer thickness) are conspicuous.

8. A complete study of this phenomenon can be found in Kléman, 1982, Chapt.5, in fine.

RÉFERENCES

Amari, S.: 1962, 'A theory of deformation and stresses of ferromagnetic substances by Finsler geometry', RAAG Memoirs, 3.3.

Anderson, P.W.: 1979, 'Lectures on amorphous systems', in R. Balian, R. Maynard and G. Toulouse (ed.), ill-condensed matter, Les Houches, session XXXI, North-Holland, Amsterdam, p.162.

Bernal, J.D.: 1959, 'A geometrical approach to the structure of liquids', Nature, 183, 141.

Berry, M.: 1981, 'Singularities in waves and rays', in R. Balian, M. Kléman and J.P. Poirier (ed.), Physics of defects, Les Houches, session XXXV, North-Holland, Amsterdam, p.456.

Bilby, B.A.,:1960, 'Continuous distributions of dislocations', Prog. Sol. Mech., 1, 329.
Bouligand, Y.: 1980, 'Defects in hexagonal phases of discotic molecules', J. de Phys., 41, 1297.
Bourdon, L., Sommeria, J. and Kléman, M.: 1982,'sur l'existence de lignes singulières dans les domaines focaux en phases SmC et SmC*',Jour. de Phys., 43, 77;
Cartan, E.: 1923,1924,1925, 'Sur les variétés à connexion affiné et la théorie de la relativité généralisée', Ann. Ecol. Norm. Sup., 40, 325; 41,1; 42,17.
Coxeter, H.S.M.: 1958, 'Close packing and froth', Ill J. Math., 2, 746.
Coxeter, H.S.M. and Whitrow, G.J.: 1950, 'World-structure and non-euclidean honeycombs', Proc. Roy. Soc. London, A201, 417.
Döring, W.: 1968, 'Point singularities in micromagnetism', J. Appl. Phys., 39, 1006.
Dzyaloshinskii, I.: 1981, 'Gauge theories and densities of topological singularities' in R. Balian, M. Kléman and J.P. Poirier 'ed.), Physics of defects, Les Houches, session XXXV, North-Holland, Amsterdam, 317.
Edelen, D.G.B.: 1980, 'A gauge theory of dislocations and disclinations', Int. J. Eng. Sci., 18, 1095.
Finkelstein, D.: 1966, J. Math. Phys., 7, 1218.
Frank, F.C.: 1952, 'Supercooling of liquids', Proc. Roy. Soc. London, 215A, 43.
Friedel, G.: 1922, 'Les états mésomorphes de la matière', Ann. Phys. Paris, 9, 273.
Hehl, F.W., Von der Heyde, P., Kerlick, G.D. and Nester, J.M.: 1976, 'General relativity with spin and torsion: foundations and prospects', Rev. Mod. Phys. 48, 393.
Kleinert, H.: 1982, 'Gauge theory of dislocation melting', Phys. Lett. 89A, 294.
Kléman, M.: 1973, 'Local chiral axis in a directional medium and wedge component of a disclination', J. de Phys., 34, 931.
Kléman, M.: 1980, 'Developable domains in the columnar phases of discotic molecules', J. Phys., 41, 737.
Kléman, M.: 1981, 'Crystallography of amorphous bodies' in O. Brulin and R.K.T. Hsieh (ed.) Continuum models of discrete systems 4, North-Holland, Amsterdam, p.287.
Kléman, M.: 1982, 'Geometry of disorder and its elementary excitations', J. Phys., September, in press.
Kléman, M.: 1982, Points, lines and walls (in liquid crystals, magnetic systems and various ordered media), Wiley, Chichester (in press).

Kléman, M. and Sadoc, J.F.: 1979 'A tentative description of the crystallography of amorphous solids', J. de Phys. Lettres, 40, L-569.

Kondo, K.: 1955, 1958 'Non-Riemannian geometry of imperfect crystals from a macroscopic viewpoint, RAAG Memoirs, 1, 459; 2,227.

Kröner, E.: 1958 'Kontinuumstheorie der Versetzungen und Eigenspannungen', Springer, Heidelberg.

Kröner, E.: 1981, 'Continuum theory of defects', in Physics of defects, Les Houches, session XXXV, North-Holland, Amsterdam, p.215.

Mermin, N.D.: 1979, 'The topological theory of defects in ordered media', Rev. Mod. Phys. 51, 591.

Michel, L.: 1980, 'Symmetry defects and broken symmetry configurations. Hidden symmetry', Rev. Mod. Phys. 52, 617.

Misner, C.W., Thorne, K.S. and Wheeler, J.A.: 1973, 'Gravitation', Freeman, San-Francisco.

Mosseri, R. and Sadoc, J.F.: 1982, 'Order and disorder in amorphous tetracoordinated semiconductor: a curved space description', Phil. Mag., B45, 467.

Nye, J.F.: 1953, 'Some geometrical relations in dislocated crystals', Acta Met., 1, 153.

Oswald, P. and Kléman, M.: 1981 'Défauts dans une mésophase hexagonale discotique: disinclinaisons et parois', J. Phys., 42, 1461.

Spaepen, F.: 1981, 'Defects in amorphous metals' in Physics of Defects, Les Houches, session XXXV, North-Holland, Amsterdam, p.133.

Srolovitz, D., Maeda, K., Takeuchi, S., Egami, T. and Vitek, V.: 1981, 'Local structure and topology of a model amorphous metal', J. Phys. F: Metal Phys., 11, 2209.

Steinhardt, P.J. and Chaudhari, P.: 1981, 'Point and line defects in glasses', Phil. Mag., A44, 1375.

Thom, R.: 1974, 'Modèles mathématiques de la morphogenèse', 10/18, Paris.

Thom, R.: 1979, 'A revival of natural philosophy', in W. Güttinger and H. Eikemeier (ed.), Structural stability in Physics, Springer, Berlin, p.1.

Toulouse, G. and Kléman, M.: 1977, 'Classification of topologically stable defects in ordered media', J. Phys. Lettres, 38, L-195.

Trebin, H.R.: 1982: 'The topology of non-uniform media in condensed matter physics', Adv. in Phys. (in press).

Williams, C.E.: 1975, 'Helical disclination lines in smectics A',Phil. Mag., 32, 313.

Wright, F.J.: 1979, 'Wavefront dislocations and their analysis using catastrophe theory' in Structural stability in Physics, Springer, Berlin.

Wu, T.T. and Yang , C.N.: 1975, 'Concepts in non-integrable phase factors and global formulation of gauge fields', Phys. Rev., D12, 3845.

Zallen, R.: 1979, 'Stochastic geometry', in W. Montroll and J.L. Lebowitz (eds), Fluctuation phenomena, North Holland, Amsterdam.

Zeeman, C.: 1979, 'Catastrophe theory' in Structural stability in Physics, Springer, Berlin.

Yves BOULIGAND

SOME GEOMETRICAL AND TOPOLOGICAL PROBLEMS IN LIQUID CRYSTALS

1. INTRODUCTION

Liquid crystals or mesomorphic states are anisotropic liquids (see de Gennes, 1974). The theory of defects presented by Kléman (1983) applies to these ordered media. The observation of defects such as dislocations, focal curves, disclinations and other singularities leads to study an extreme variety of problems and we shall briefly consider a narrow selection. These questions concern smectic, nematic and cholesteric liquids. A brief recall of their structure is presented in figs 1 to 6.

In smectic liquid crystals, the elongated molecules form layers. Molecules lies normally to layers in smectics A (fig. 1) or are tilted in smectics C (fig. 2). In chiral smectics C, obtained with optically active components, the tilt orientation progressively rotates, from layer to layer, as indicated on the right by the conic indicatrix (fig. 3). In nematic liquids, the order is purely orientational, without layers (fig. 4). In cholesteric liquids (or twisted nematics), the mesophase contains optically active components and molecules, represented lying parallel in horizontal planes, rotate progressively from plane to plane. The orientations varies continuously (fig. 5). Discoidal molecules also form liquid crystals (Chandrasekhar et al., 1977). Piles of discs arrange into an hexogonal array in cross section and columns can glide one past the other (fig. 6).

The kind of order adopted by molecules often corresponds to a need of least encumbrance, the first step corresponding to a local parallel orientation. Weak interactions between dipoles for instance can favour the differentiation of layers. Asymmetry in the configuration of molecules often results in

C. P. Bruter et al. (eds.), Bifurcation Theory, Mechanics and Physics, 357–381.

1
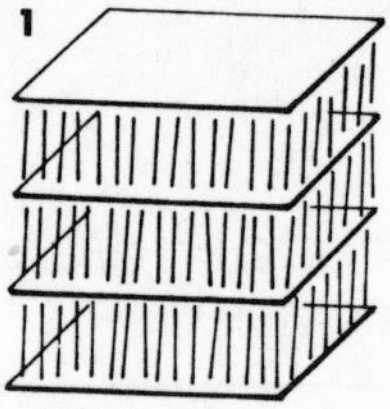

2
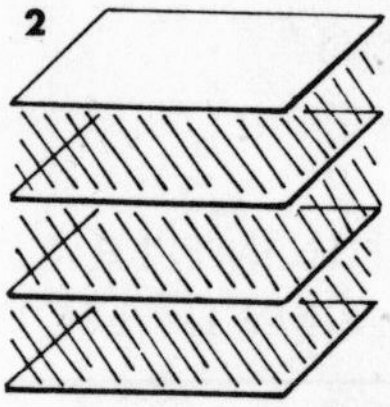

3
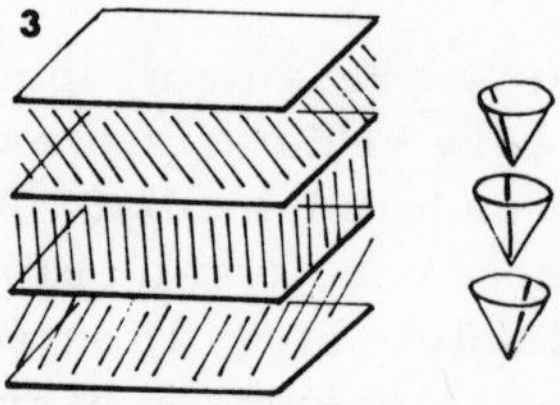

4
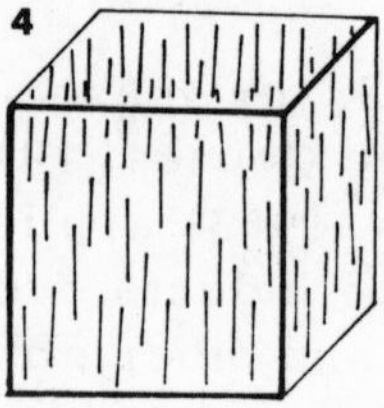

5
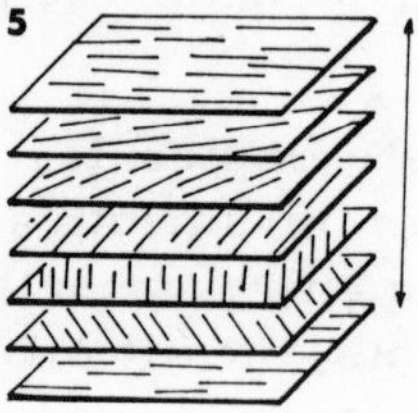

6

Figures 1 to 6 -

1 : smectic liquid crystal with elongated molecules lying normally to layers (smectic A). 2 : smectic liquid crystal with molecules tilted with respect to the layers (smectic C). 3 : twisted smectic C, with molecules rotating from layer to layer as indicated on the right, in successive cones. 4 : nematic liquid crystal ; elongated molecules are aligned but do not form layers. 5 : twisted nematic liquid. Molecules rotate from one level to the following one by a small angle. The successive planes do not exist but facilitate the drawing ; p/2 is the half helicoidal pitch. 6 : discoidal molecules forming a columnar phase, with uncorrelated position of molecules in neighbouring piles.

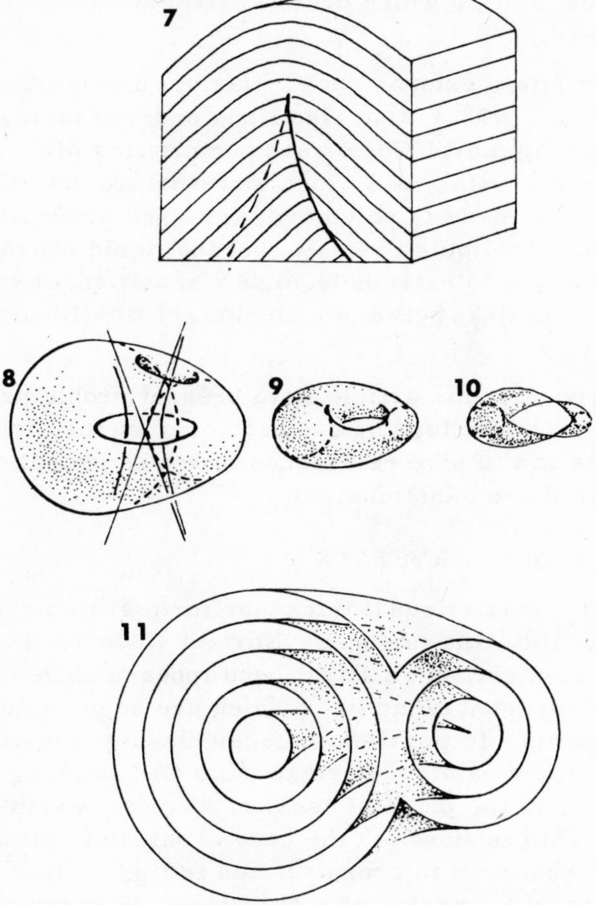

Figures 7 to 11 -

7 : cylindrical parallel layers of equal thickness forming a caustic surface. One of the two parts of the caustic is virtual and there is an asymmetry of the arrangement of layers on each side of the materialized caustic.
8, 9, 10 : three different shapes of Dupin's cyclides.
11 : series of nested Dupin's cyclides, with virtual parts which are not represented.

the formation of twisted liquid crystals such as cholesterics and chiral smectics C, which exhibit extremely large form rotatory powers.

Liquid crystals exist as such in various biological materials (Bouligand, 1978a). One also finds ordered biological materials showing morphologies and symmetries of certain classes of liquid crystals and which however are not liquid. Such systems are made of polymers, arranged according to the patterns found in liquid crystals, but the liquid character is abolished, owing to extremely high viscosities or to the presence of cross-links between molecular chains (Bouligand, 1978b, 1981a).

The purpose of this article is to present problems or sets of geometrical or topological problems in the field of liquid crystals and to give references of works providing a good access to these questions.

2. SMECTIC LIQUID CRYSTALS

Superimposed layers of equal thickness form systems of parallel and equidistant surfaces. Normal lines to these surfaces are straight, but a slight bend appears when some layers are interrupted along lines which are edge-dislocations. Normal lines to parallel surfaces envelop caustics in general, but one easily understands that the stacking of layers creates, in the place of caustics, certain asymmetrical walls. This is shown in the case of parallel cylindrical surfaces, observed in cross section in fig. 7. Such walls give rise to a 'vernier effect' between the two domains they separate and one observes layers forming a right angle with the wall in one domain, whereas the angle is variable in the opposite domain (Bouligand, 1972a). Such asymmetries are not compatible with the liquid character of the system and the energy of such walls would be very high. Accordingly, in such liquids, the layers form parallel surfaces whose normals pass through two different curves, representing degenerated caustics. The well known solutions of this problem are the cyclides of Dupin, forming a set of parallel surfaces with an associated system of two focal conics. The three main examples are represented in figs. 8, 9, 10. (See also Rosenblatt et al., 1977, for Dupin's cyclides associated with focal parabolae. In Dupin's cyclides which cut one of the two conics, one can distinguish two regions separated by two conic singular points (fig. 8 and 10). One of the two regions remains virtual in liquid crystals, since materializing both regions would be incompatible with the stacking of layers as shown in fig. 11. Note that planes, cylinders, tores and spheres are particular cases of Dupin's cyclides.

When a smectic phase grows in an isotropic liquid or in another phase, several domains appear and generate systems of parallel layers in the form of Dupin's cyclides. Each domain has its own system depending on eight parameters : the six coordinates of the two foci of one of the two conics, the length of the corresponding axis of this conic and, say, the angle of one of the symmetry planes. This system will not develop at infinity. Several germs of smectic phase form and meet along boundaries, which change, since the system is a liquid, and finally stabilizes, when a minimum of energy is reached. Observations due to Friedel and Grandjean (1910) show the existence of domains limited by revolution cones as indicated in fig. 12. In such domains, molecules lie along straight segments joining any points M_1 and M_2 belonging to the two focal conics. A narrow subdomain can be considered, built on small segments of conics, centered in M_1 and M_2. A line running normally to molecules at the surface of this subdomain closes and forms no helices. We shall see below that one also finds in smectics large domains, where this property is not verified.

Domains limited by revolution cones are tangent along common generators. The arcs of focal conics, mainly ellipses, which limit the revolution cones, are tangent at one extremity of these generators (fig. 13). Narrow domains, enclosed by three such conical domains, can be filled by spherical layers centered on the common apex of domains (Sethna and Kléman, 1982) or may contain another conic domain tangent to the three first ones. This principle can

12

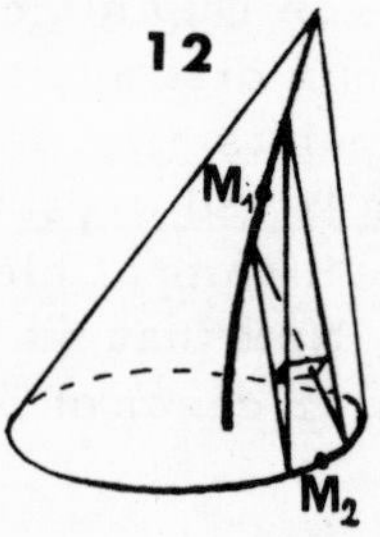

14

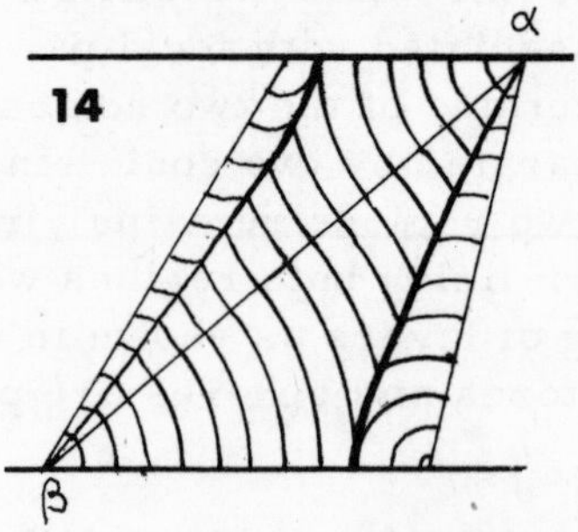

13

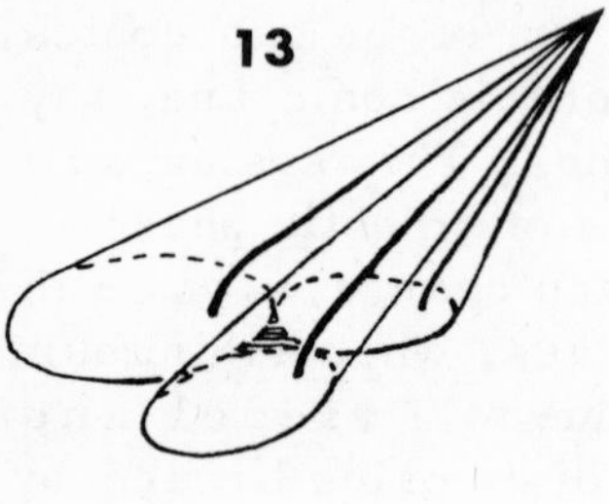

15

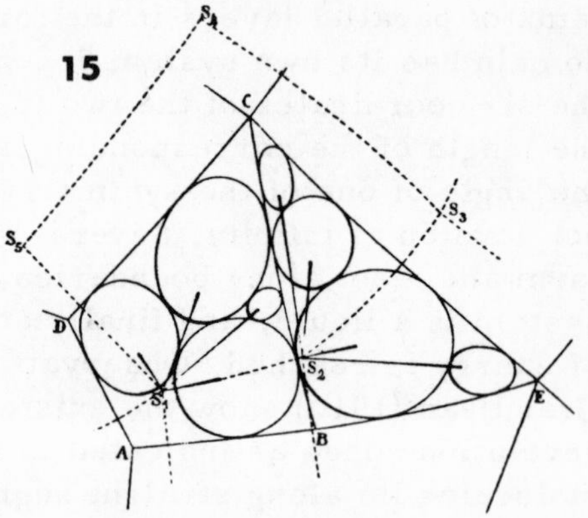

16

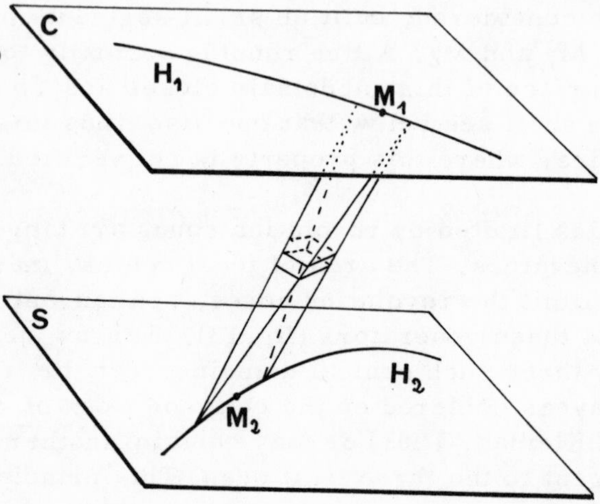

Figures 12 to 16 -

12 : conic domain limited by an ellipse and a point at the extremity of the focal hyperbolic arc ; a small subdomain is considered in the vicinity of the segment M_1 M_2 ; a line running parallel to layers along the surface limiting such a subdomain closes itself. 13 : section of two tangent domains along a common generator $\alpha\beta$; layers are conical along the two hyperbolic arcs ; they present a tangent plane along $\alpha\beta$, but show however an abrupt change of curvature along this line. 14 : set of three tangent conical domains. 15 : polygonal structure. Polygons extend on each side of the preparation. All the ellipses of a given polygon ABCD, at the coverslip level, have their hyperbola which join at the corner S_1 of several different polygons of the mosaic observed at the slide level. The ellipses of the lower side are not represented (after Grandjean and Friedel, 1910). 16 : two arcs of hyperbola limiting polygones at the coverslip level (H_1) and at the slide level (H_2). A small subdomain is represented. Since H_1 and H_2 are not focal conics, a line running parallel to layers, at the surface limiting this domain form an helix. A density of screw-dislocations is necessary in such domains.

Figures 17 and 18 -

Observations, at the coverslip level (17) and at the slide level (18), of a polygonal field with ellipses in a smectic phase. Cyanooctylbiphenyl + a small amount of Canada balsam. (Polarizing microscopy).

be iterated to the resting smectic phase, but such a fractal process is rarely repeated a great number of times. A planar section passing through a series of common generators, associating two tangent domains is shown in fig. 14.

The boundaries of the mesophase and, in particular, the limiting slide and coverslip exert strong distortions on smectic liquid crystals. Interfaces, such as those separating an isotropic phase from the smectic liquid, create similar disturbances. The elastic energy of the deformations due to the presence of defects is considerably lowered when these defects stick at the limiting surface of the phase. The domains of integration of the elastic energy, extremely high in the vicinity of defects, is divided by a factor not far from 2. One observes frequently polygonal fields or mosaics, between slide and coverslip. Each polygon is limited by arcs of hyperbolae, which lie in the parallel planes of the two limiting glasses of the preparation. The smectic phase is enclosed between two polygonal lattices lying at the lower and upper levels of the smectic slab (Friedel and Grandjean, 1910). These two curvilinear mosaics form a dual system. Polygon centres of the upper system superimpose vertically with the corners, where join several polygons of the lower system and conversely. This is sketched in fig. 15. Dotted lines represent the lower mosaic at the slide level. Certain domains limited by revolution cones are observed in such polygons. Their apices join at the common corners of polygons, whereas their elliptical bases lie in the plane of the opposite glass slide. These ellipses are tangent one another and to the arcs of hyperbola forming polygons. They form frequently lattices of tangent ellipses, an extremely remarkable texture as shown in figs. 17 and 18 ; the two views correspond to two different foci, one at the level of the lower lattice, the second one to the upper level, that of the coverslip. The polygonal edges of both lattices cross each other in horizontal projection. Pyramidal domains are built on each pair of these hyperbolic arcs, which differ from focal conics. In these domains, the parallel layers form necessarily screw-dislocations (Bouligand, 1972a). One can isolate narrow subdomains in such regions and verify that

$\underset{\sim}{n} . \underset{\sim\sim\sim}{\text{curl}}\ \underset{\sim}{n} \neq 0$, $\underset{\sim}{n}$ being the unit vector normal to layers. Along the limiting surface of this subdomain, a line running normally to the molecular orientation forms an helix (fig.16).

When the anchoring conditions correspond to molecules lying normally to the glass boundaries, layers are horizontal or subhorizontal and the only possible defects are translation dislocations as shown in fig. 19. Dupin's cyclides and the corresponding domains form in preparations which are thick enough (fig. 20). As one deduces easily from fig. 14, molecules lie horizontally in the limiting interface of the mesophase. If the smectic phase is thin, with horizontal anchoring conditions for molecules, layers lie vertically in the whole thickness of the preparation and the defects observed are mainly disclinations (fig. 21).

There are three types of textures : fan-textures with disclinations, focal conics and dislocations ; focal textures with focal conics and dislocations and, finally, planar textures with only dislocations. There is a graduation in the energy of defects. Dislocations have a low energy, focal curves a mean one and disclinations a high one. It appears accordingly that fan-textures will be textures of the highest energy and will be found in the vicinity of phase transitions which increase the symmetry. Planar textures on the contrary correspond to textures favoured by low temperatures.

A complete mathematical treatment of these textures does not exist. I just tackled the job in my paper of 1972 and from a purely geometrical point of view. There are interesting topological questions as indicated below. Energy calculations show that Dupin's cyclides do not correspond exactly to a minimum (Kléman and Parodi, 1975) and several observations clearly indicate the presence of these differences. For instance, layers do not form conical points in the vicinity of conics but show a somewhat rounded apex at this level (fig. 22). Screw dislocations superimpose with focal lines (fig. 23) etc... Finally, one finds two levels of observations .

19
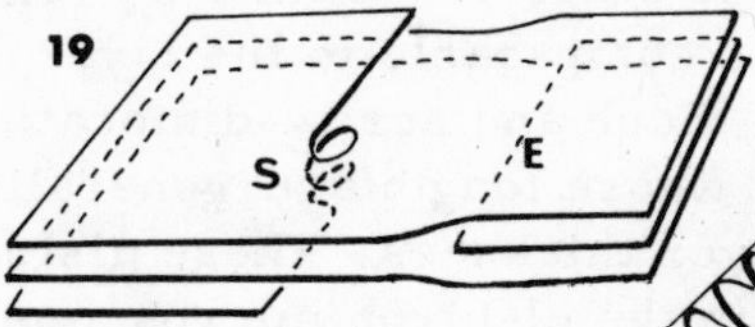

20
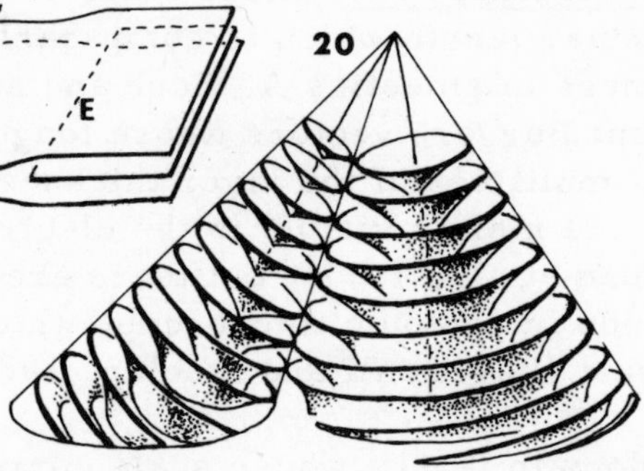

21

22
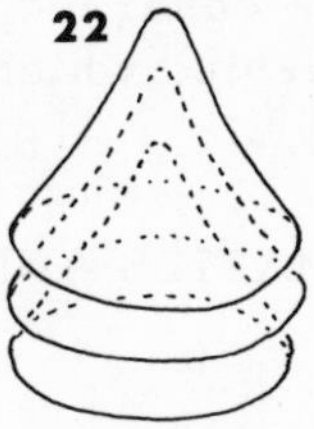

23
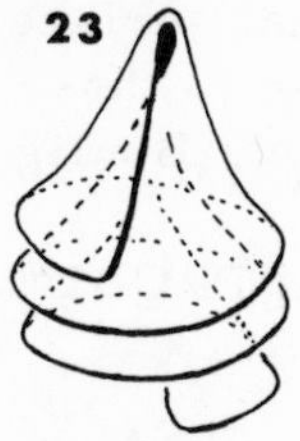

Figures 19 to 23 -

19 : sketch of a screw-dislocation (S) and of an edge dislocation (E) in a layered liquid. 20 : arrangement of layers in conical domains of a smectic liquid.
21 : arrangement of layers with a $+\pi$ disclination, a $-\pi$ one and an edge-dislocation. 22 : rounded shape of layers along a focal curve (ellipse or hyperbola).
23 : screw-dislocations often superimpose with focal curves.

2.1. <u>The molecular level</u> whose scale is defined by the molecular length which is comparable to the layer thickness in smectics A. Edge and screw-dislocations present Burgers vectors whose length are generally small multiples of the layer thickness. These dislocations are mainly visible in the electron microscope (Kléman et al., 1977), but there are however certain methods in photonic microscopy which allow a good contrast (Lagerwall and Stebler, 1981).

2.2. <u>The domain level,</u> whose scale corresponds to the length of certain arcs of conics or disclinations. One observes that defects such as screw dislocations are concentrated in certain domains as indicated above. Certain parts of these domains correspond mainly to right handed helices, whereas complementary subdomains are filled with left handed ones. These segregations and, more generally, the differentiation of textures deserve the collaboration of mathematicians. Layers are often deformed with respect to the Dupin cyclides model and their shape is often reminiscent of that of minimal surfaces, since the observed textures are very close to those of cholesterics, which clearly show this trend (Bouligand, 1972b).

3. GEOMETRICAL PROBLEMS IN NEMATICS AND CHOLESTERICS

The difference between smectic and nematic liquid crystals is that the parallel molecules form or do not form layers. A remarkable phenomenon obtained with nematic liquid crystals is the progressive formation of a layered aspect by the simple addition of certain soluble and optically active components to a nematic phase. Actually the obtained phase is not smectic, but cholesteric. This means that the phase has been twisted by the addition of the chiral molecules. These twisted nematic liquids were first observed in cholesterol derivatives ; this is the origin of the term cholesteric. A great number of biological molecules and polymers form such phases, due to their helical configurations (nucleic acids, α-helices in proteins, etc...)

The cholesteric structure has been presented in fig. 1, but one can adopt a representation in a plane lying normally to layers. Molecules parallel to the drawing plane are represented by segments of length 1. Those normal to this plane are viewed as points. Those lying obliquely are represented by nails, whose length varies as the projection onto this plane of a segment of length 1 ; the sharp extremity of nails corresponds to the molecular extremity pointing towards the observer. These conventions are used to represent the passage from a nematic structure to a cholesteric one by the addition of an active compound, indicated by black spheres (figs. 24, 25, 26). The helicoidal pitch p corresponds to the distance perpendicular to layers, allowing a rotation by 2π of molecules. The periodicity is actually $+\frac{p}{2}$, since neighboring molecules are parallel or antiparallel. A first question is to find the mechanism which leads to the presence of such a twisted layering. It was shown by Franck (1958) and de Gennes (1974) that the elastic energy dF of a uniaxial liquid crystal in a volume dv writes :

$$dF = \frac{1}{2}\left[K_1 \,(\mathrm{div}\, \underset{\sim}{n})^2 + K_2 \,(\underset{\sim}{n} . \underset{\sim}{\mathrm{curl}}\, \underset{\sim}{n} - \frac{2\pi}{p})^2 + K_3 \,(\underset{\sim}{n} \wedge \underset{\sim}{\mathrm{curl}}\, \underset{\sim}{n})^2\right] dv \quad (1)$$

$\underset{\sim}{n}$ is a unit vector representing the local orientation of molecules ; K_1, K_2 and K_3 are three elastic constants and concern three different kinds of deformation, which are respectively ; splay, twist and bend. The elastic energy is zero, when each of the three terms is zero. This means that one has respectively

$$\mathrm{div}\, \underset{\sim}{n} = o \quad (2), \quad \underset{\sim}{n} . \underset{\sim}{\mathrm{curl}}\, \underset{\sim}{n} = \frac{2\pi}{p} \quad (3) \text{ and } \underset{\sim}{n} \wedge \underset{\sim}{\mathrm{curl}}\, \underset{\sim}{n} = o \quad (4)$$

Equation (4) shows that integral lines are straight and from both equations (2) and (4), it can be proved that these straight lines are parallel to a constant plane direction (Bouligand, 1974a). If a referential 0 x y z is chosen so that 0 x y lies parallel to this plane, one sees that n, normal to 0 z, depends only on z. When $\underset{\sim}{n} . \underset{\sim}{\mathrm{curl}}\, \underset{\sim}{n} = \mathrm{cte} \neq 0$ (3), one finds the cholesteric layering with a constant twist. I had presented previously these results, but A. Chenciner prepared a cleaner demonstration, which is summarized in the first annex. Some other related problems are also briefly proposed in this annex.

In cholesteric liquid crystals, the normal to layers is called cholesteric axis. This axis can be represented by a unit vector $\underset{\sim}{P}$ whose sense is chosen arbitrarily, since there is no polarity, as for the sense of molecules.

In general, div $\underset{\sim}{n}$ and $\underset{\sim}{n} \wedge \operatorname{curl} \underset{\sim}{n}$ are different from zero but remain much smaller than the twist $\underset{\sim}{n} . \operatorname{curl} \underset{\sim}{n}$. It follows that, geometrically, the layers are not always defined. Indeed, consider a field of unit vectors $\underset{\sim}{n}$, which represents the continuous and derivable distribution of molecular orientations. At a point M_o, the director is $\underset{\sim}{n}_0$. Let us adopt a local frame $M_o x\,y\,z$, with $M_o x$ parallel to $\underset{\sim}{n}_0$ and $M_o x\,z$ chosen arbitrarily (see Annex II). Let be $M_o Y$ an axis in the plane $M_o y\,z$ forming an angle θ with $M_o y$. One can define the twist along $M_o Y$, at point M_o, as $t_Y = - \frac{\partial n_Z}{\partial Y}$ which is a function of θ. It is shown in the second annex how t_Y is a circular function of 2θ and therefore presents two pairs of diametrically extrema, oriented at right angle. This means that one can choose for $0z$ a direction corresponding to a maximum of twist (in absolute value), $0y$ representing then the direction of minimum twist (in absolute value). The twist axis is thus, defined in a nematic liquid, when there are distortions, even in the absence of any optically active components. Both domains showing either a right-handed twist or a left-handed one, can be found. This axis of maximum twist Mz probably coincides with the vector $\underset{\sim}{P}$ defined when the cholesteric structure is perfect, without any deformation. In general, in the distorted helical nematic liquid, the cholesteric unit vector $\underset{\sim}{c}$, parallel to Mz, can form a more or less solenoidal field with

$$\underset{\sim}{c} \cdot \operatorname{curl} \underset{\sim}{c} \neq 0$$

In such a situation, layers cannot be defined. However, one sees very clearly in the preparation the stacked layers of equal thickness forming disclinations, focal curves associated with Dupin's cyclides and dislocations, very similar to those observed in smectics (Bouligand, 1972b). It seems that locally, the vector $\underset{\sim}{c}$ is not solenoidal, but that a number of defects, namely screw-dislocations, create a global twist of vector $\underset{\sim}{c}$ in the mesophase. It remains that one can find regions with a necessary twist of $\underset{\sim}{c}$, which however appears to be devoid of screw-dislocations. This is not abnormal, since the presence of a layering is not indispensable. A deeper geometrical and topological investigation is necessary in these domains of cholesteric liquid crystals, where the nucleation of screw-defects is favoured. These problems leads to the topological analysis of singularities of cholesteric liquids, a question which illustrates the fact that the cholesteric layering does not exist everywhere in the liquid.

4. TOPOLOGY IN NEMATICS AND IN CHOLESTERICS

In his contribution, M. Kléman (1983) shows the interest of topological methods in the study of ordered media (crystals, liquid crystals, ferromagnets, etc...) and he introduced himself, with Toulouse, a classification of defects using homotopy groups (1976). Liquid crystals are non orientable media in general and, along a closed circuit, the director $\underset{\sim}{n}$ can transform into $-\underset{\sim}{n}$, after one turn. One finds evident obstructions to continuity, when one explore the structure of the liquid along a surface limited by such circuits (Bouligand, 1974a). Non orientability is clearly demonstrated in fig. 21 for smectics. Non orientability also exists in nematics and cholesterics. Both $\underset{\sim}{n}$ and $\underset{\sim}{c}$ are non orientable. Several classical singular lines in cholesterics are represented in figs. 27-30 and one easily observes that along the circuits surrounding these lines, the envelope of $\underset{\sim}{n}$ and $\underset{\sim}{c}$ are either Mœbius strips or are orientable ribbons. The four possibilities are represented. In cholesterics showing a large helicoidal pitch, one often observes in phase microscopy ribbons which correspond to the locus of horizontal directors, say the molecules lying normally to the axis of

24

25

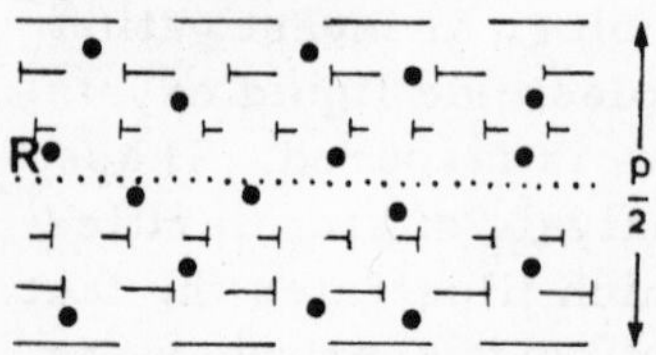

26

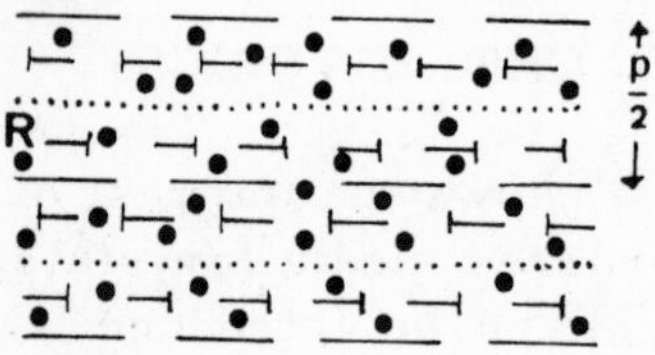

27

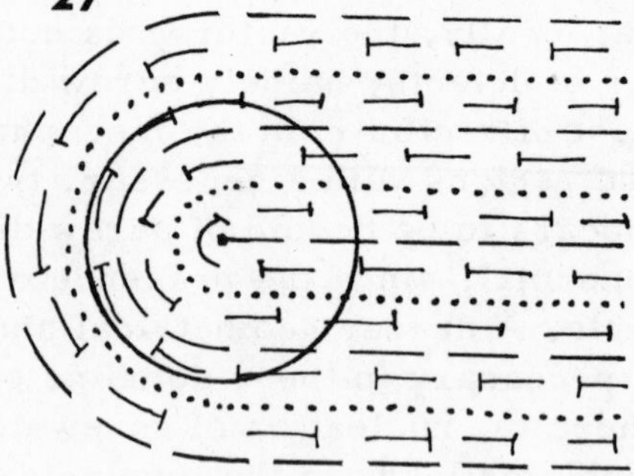

28

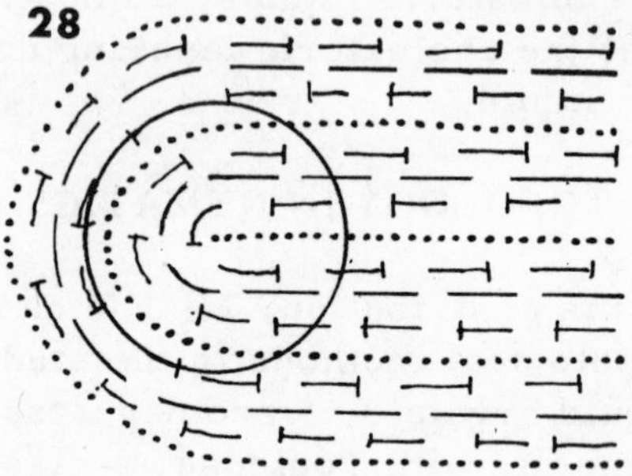

29

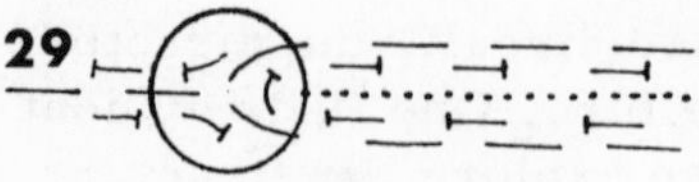

30

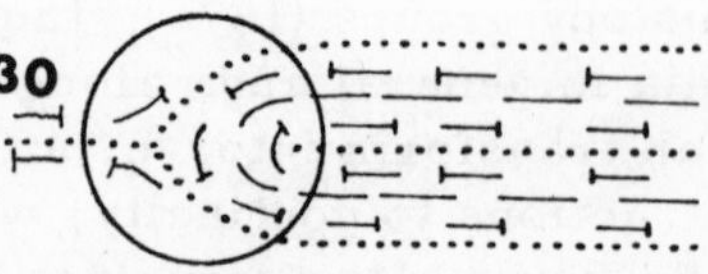

31

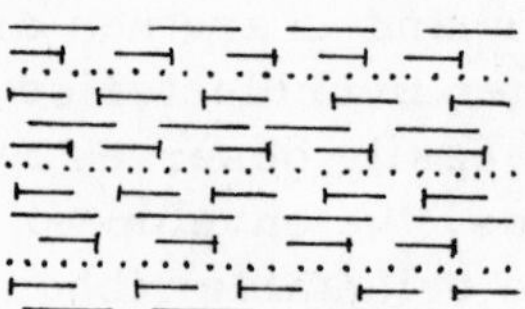

32

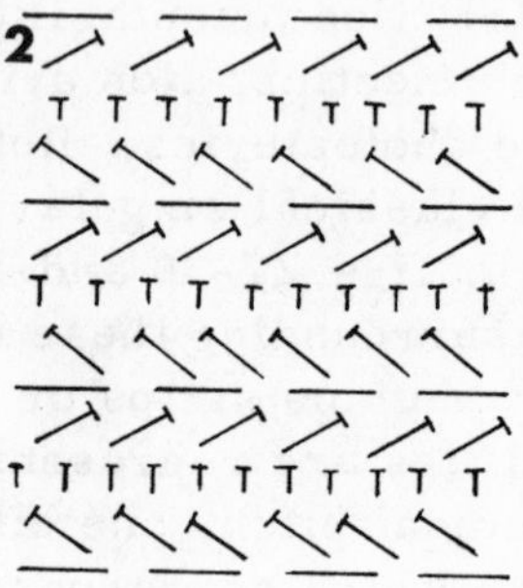

Figures 24 to 32 -

24 : nematic phase ; the elongated molecules are supposed to be aligned in the figure plane. 25 : influence of a soluble substance R, showing a molecular rotatory power upon a nematic liquid. The R molecules are represented by black spheres. The nematic molecules are indicated by segments, by nails or points according to their orientation to the drawing plane. A cholesteric twist appears in presence of R. 26 : the twist increases with the concentration of R. 27 : a $+\pi$ disclination in a cholesteric liquid ; such a system is not orientable; the sense of molecules and the sense of the cholesteric axis is changed along the closed circuit. 28 : a $+\pi$ disclination in a cholesteric liquid, with an orientable distribution of molecules ; however the cholesteric axis is not orientable. 29 : edge-dislocation ; the molecular field is not orientable, but the cholesteric axis is orientable along the circuit. 30 : both fields (molecules and twist axes) are orientable along the closed circuit. 31 : representation of a cholesteric structure in a plane containing the cholesteric axis. 32 : representation of a cholesteric liquid in a plane oblique with respect to the cholesteric axis.

the microscope. These ribbons form Mœbius strips, when they are closed around a defect line which associates the director $\underset{\sim}{n}$ with the opposite orientation $-\underset{\sim}{n}$. This is the only example we know of natural Mœbius strip and we published some micrographs of them (Bouligand, 1974a, 1981b).

According to our conventions described above, a perfect cholesteric liquid crystal observed in a plane normal to layers is represented in fig. 31. We have considered 12 successive planes to represent a 2π rotation. This does not mean that there are discrete steps of rotation. The structure is continuous and the director components write :

$$n_x = \cos \frac{2\pi}{p} z \ , \quad n_y = \sin \frac{2\pi}{p} z \quad \text{and} \quad n_z = 0.$$

In an oblique plane, the nails align along series of nested arcs (fig. 32) and, if an axis Os represents, in this section plane, the direction normal to layers, with a certain choice of units, these curves write, x_0 being an additive constant :

$$x = x_0 + \text{Log} \mid \sin s \mid .$$

Such curves were observed in oblique thin sections of numerous biological tissues and the cholesteric structure was demonstrated from this aspect (Bouligand, 1965, 1972c).

Edge- and screw-dislocations are numerous in cholesteric liquids and these defects present kinks or jogs, with remarkable topological configurations (Bouligand, 1974b). In the study of such systems, we described a lot of situations which were later interpreted as examples of sophisticated homotopy classes (Bouligand et al., 1978 ; Pœnaru, 1979). One knows, for instance, from Hopf (1931) that :

$$\pi_3 (P_2) = \mathbb{Z}$$

We find very frequently the situation corresponding to $h = -1$ (Hopf index), but we did not still verify the existence of other Hopf indexes as -2, -3, etc... but such situations are easily conceivable.

A neighbouring homotopy class could exist in ferromagnets. Directors must be changed into vectors, the magnetic moments and one has similarly :

$$\pi_3 (S_2) = \mathbb{Z}$$

In biological systems, we have verified the existence of the following situations :

$$\pi_0 (P_2) = 0 \quad \text{and} \quad \pi_1 (P_2) = \mathbb{Z}_2$$

Singular points in nematics and cholesterics which illustrates $\pi_2 (P_2) = \mathbb{Z}$ were not yet found in our biological examples. The reason is probably the difficulty to find such points exactly in the thickness of a section.

5. DISCUSSION AND CONCLUSION

The introduction of homotopy groups in my work is due to the collaboration of G. Toulouse and other colleagues (Bouligand et al., 1978) who reinterpreted my results in the classical terms of topology. As a zoologist, I adopted the mind of detailed descriptions and I applyed it to liquid crystals and their biological counterpart. I must confess that I find these concepts a little heavy, this being probably due for me, to a lack of experiment. However, I must indicate to topologists that my original descriptive paper (Bouligand, 1974b) contains a great number of situations, which still wait for a complete topological treatment. I also present in this paper my working methods, which are recipees to find topological equivalences. The purpose is to transform continuously a fibre bundle which appears very complicated at first, and to reach its simplest configuration at the end of a rational sequence of operations. Problems of hydrodynamics are extremely interesting (and complicated) in liquid crystals (see de Gennes, 1974 ; Dubois-Violette et al., 1978) and also present their topological counterpart which has not yet been studied.

ANNEX I.

Let be $\underset{\sim}{n}$ a vector field in $\mathbb{R}^3$, with

$$|\underset{\sim}{n}| = 1 \; , \; \underset{\sim}{n} \wedge \operatorname{curl} \underset{\sim}{n} = 0 \; , \; \operatorname{div} \underset{\sim}{n} = 0 \;\text{ and }\; \underset{\sim}{n} \cdot \operatorname{curl} \underset{\sim}{n} \neq 0 \quad (1\text{-}4)$$

The components of $\underset{\sim}{n}$ are n_1, n_2, n_3 in an orthonormal frame $0x_1\, x_2\, x_3$ and let us call n_{ij} the partial derivatives which are supposed to exist and to depend continuously on (x_1, x_2, x_3).

$$n_{ij} = \frac{\partial n_i}{\partial x_j}$$

One shows first that the rank of the square matrix $N = [n_{ij}]$ is 1. By derivation of $n_1^2 + n_2^2 + n_3^2 = 0$, one gets:

$$[n_{ji}] \cdot \underset{\sim}{n} = 2 \qquad (5)$$

From (2) and (5), one deduces

$$[n_{ij}] \cdot \underset{\sim}{n} = 0 \qquad (6)$$

The equation (2) $\underset{\sim}{n} \wedge \operatorname{curl} \underset{\sim}{n} = 0$ indicates that the integral lines are straight. Consider one point M running on each straight line, $\underset{\sim}{n}$ being the velocity. If x_1, x_2, x_3 are the coordinates of M at time zero, they will be $x_1 + t\, n_1\, (x_1, x_2, x_3)$, $x_2 + t\, n_2\, (x_1, x_2, x_3)$, $x_3 + t\, n_3\, (x_1, x_2, x_3)$, at time t. Since, by (3), this flow is volume preserving, one has :

$$\begin{vmatrix} 1+tn_{11} & tn_{12} & tn_{13} \\ tn_{21} & 1+tn_{22} & tn_{23} \\ tn_{31} & tn_{32} & 1+tn_{33} \end{vmatrix} = 1 \qquad (7)$$

Setting $\lambda = -\frac{1}{t}$, and denoting by I (resp. $N = N(x_1 x_2 x_3)$), the square unit matrix of rank 3, one sees that $\det |N - \lambda I| = -\lambda^3$, from which follows that the three eigen values of N are zero.

Consider now the matrix N corresponding to a given point (x_1, x_2, x_3) : thanks to (5) and (6), in a basis whose third vector is $\underset{\sim}{n}\, (x_1, x_2, x_3)$,

$$\text{N reduces to} \quad \begin{pmatrix} \alpha & \beta & 0 \\ \gamma & \delta & 0 \\ 0 & 0 & 0 \end{pmatrix}$$

The three eigen values being zero, one can even choose the basis so that it further reduces to

$$\begin{pmatrix} 0 & 0 & 0 \\ \gamma & 0 & 0 \\ 0 & 0 & 0 \end{pmatrix}$$

and the rank of N is $\leqslant 1$; however this rank is not zero, since $\underset{\sim}{n} \cdot \text{curl}\, \underset{\sim}{n} \neq 0$. The rank of N being 1, the constant rank theorem implies that there is a family of cylindrical surfaces whose generators are the integral lines of $\underset{\sim}{n}$. These cylinders are planar, since $\underset{\sim}{n} \cdot \text{curl}\, \underset{\sim}{n} \neq 0$. If they were curved cylinders, one would have unavoidable intersections of straight lines.

I thank greatly A. Chenciner for this demonstration. Some related problems are not yet solved. Consider now the case $|\underset{\sim}{n}| = 1$; div $\underset{\sim}{n} = 0$; $|\underset{\sim}{n} \wedge \text{curl}\, \underset{\sim}{n}|$ = cte. A solution, with a constant twist, is the following one : (a and p being constants)

$$n_x = a \cos \frac{2\pi}{p} z \ ; \ n_y = a \sin \frac{2\pi}{p} z \ ; \ n_z = \sqrt{1 - a^2}$$

What is the general solution ? Is it possible to find a similar vector $\underset{\sim}{P}$? This problem is obviously related to the structure of chiral smectics C.

Another problem arose from the study of thin sections of fibrous materials like crab carapace, after demineralization. One observes in all section planes, that the integral curves present a translation symmetry parallel to a vector $\underset{\sim}{t}$ lying in the section plane.

This means precisely that we have in $\mathbb{R}^3$ a continuous and derivable field $\underset{\sim}{n}$ (x, y, z) ; you suppose $|\underset{\sim}{n}| = 1$. Consider a plane P ; at each point M of P, the vector $\underset{\sim}{n}$ projects orthogonally onto P along the vector $\underset{\sim}{\nu}$. The only thing we know is that $\underset{\sim}{\nu}$ has a constant direction along straight lines of P, parallel to $\underset{\sim}{t}$ in P, but the length of $\underset{\sim}{\nu}$ is not necessarily constant. This is enough to give integral lines

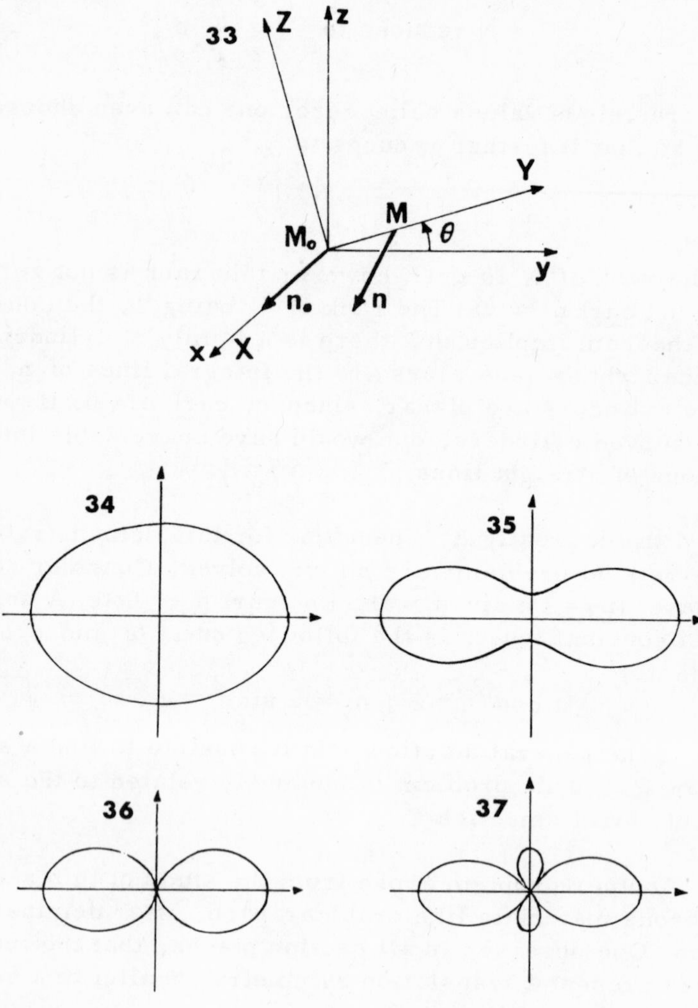

Figures 33 to 37 -

33 : relative position of two frames M_0xyz, a fixed one, and M_0XYZ, rotating about the Ox axis, in the study of the twist along all the possible axes in the plane M_0xy normal to molecules in M_0. 34 to 37 : four possible indicatrices of the twist measured along the axes normal to a given molecular direction.

showing a translation symmetry parallel to $\underset{\sim}{t}$. The problem is the following : if all the section planes show such translation symmetries, each one presenting its own direction $\underset{\sim}{t}$ (or, exceptionnally, an infinity of such directions $\underset{\sim}{t}$, this meaning that $\underset{\sim}{\nu}$ presents then a constant direction in each particular planes), prove that the vector field $\underset{\sim}{n}$ presents a translation symmetry in two directions. This is a planar translation symmetry ; if 0z is chosen normally to the two directions of translation symmetry, $\underset{\sim}{n}$ depends only on z.

ANNEX II

Let M_o x y z be a local frame with $\underset{\sim}{n} = \underset{\sim}{n}_o$ in M_o, $\underset{\sim}{n}_o$ being parallel to 0x (fig. 33). One has at the first order :

$$n_x = 1$$

$$n_y = n_{22}\, y + n_{23}\, z$$

$$n_z = n_{32}\, y + n_{33}\, z$$

Consider now a second frame M_oXYZ, with X = x and an angle $\theta = (y, Y) = (z, Z)$. Let be t_Y the twist along M_oY, expressed by

$$t_Y = -\frac{\partial n_z}{\partial Y}$$

One has $\quad n_Y = n_y \cos\theta + n_z \sin\theta \qquad (1)$

$$n_y = n_{22}\,(Y\cos\theta - Z\sin\theta) + n_{23}\,(Y\sin\theta + Z\cos\theta) \quad (2)$$

$$n_z = n_{32}\,(Y\cos\theta - Z\sin\theta) + n_{33}\,(Y\sin\theta + Z\cos\theta). \quad (3)$$

The term in Z obtained from (2) and (3), replaced in (1) gives

$$t_Y = n_{32}\sin^2\theta + (n_{22}-n_{33})\sin\theta.\cos\theta - n_{23}\cos^2\theta$$

Let us define ω and λ by the following equations

$$\tan 2\omega = \frac{n_{33} - n_{22}}{n_{23} + n_{32}} \quad \text{and} \quad -\lambda = \frac{n_{33} - n_{22}}{\sin 2\omega} = \frac{n_{23} - n_{32}}{\cos 2\omega}$$

One has then :

$$t_Y = \frac{1}{2}\left[\lambda\cos 2(\theta - \omega) + n_{32} - n_{23}\right]$$

and $$\frac{dt_Y}{d\theta} = -\lambda \sin 2\,(\theta - \omega)$$

The twist t_Y measured along M_0Y varies as a circular function of θ and there is a direction of maximum twist in absolute value. Indicatrices of the twist in the plane M_0yz are shown in figs. 34-37 for different fields $\tilde{n}$ around point M_0.

REFERENCES

Bouligand, Y. : 1965, C.R.Acad.Sci., 261, 3665, 4864.
Bouligand, Y. : 1972a, J.Physique, 33, 525.
Bouligand, Y. : 1972b, J.Physique, 33, 715.
Bouligand, Y. : 1972, Tissue & Cell, 4, 189.
Bouligand, Y. : 1974a, J.Physique, 35, 215.
Bouligand, Y. : 1974b, J.Physique, 35, 959.
Bouligand, Y. : 1978a, Solid State Physics, suppl. 14, 259.
Bouligand, Y. : 1978b, in 'Liq.Cryst.order in Polymers', Blumstein ed., Acad.Pr., 262.
Bouligand, Y. : 1981a, in Physics of defects, Ballian et al. ed., North Holl., 780.
Bouligand, Y. : 1981b, (ibid), 668.
Chandrasekhar, S., Sadashiva, B.K. and Suresh, K.A. (1977) Pramana, 471.
Dubois-Violette, E., Durand G., Guyon E., Manneville P. and Pieranski P. ; 1978, Solid State Physics, Suppl. 14, 147.
Frank F.C. : 1958, Disc. Faraday Soc., 25, 19.
Friedel G. and Grandjean F. : 1910, Bull.Soc.Fr.Minér., 33, 409.
Gennes, P.-G. (de): 1974, The Physics of Liquid Crystals, Clarendon, 1 vol., 333 p.
Hopf, H. : 1931, Math.Ann., 104, 637.
Kléman, M. : 1983, Contribution to this book.
Kléman, M. and Parodi, O.J. : 1975, J.Physique, 36, 671.
Kléman, M., Williams, C.E., Costello, J.M. and Gulik-Krzywicki, T. : 1977, Phil.Mag., 35, 33.
Lagerwall, S.T. and Stebler, B. : 1981, Physics of Defects, Balian et al. ed., 757-776. North Holl.
Pœnaru, V. : 1979, in Ill-condensed Matter, North Holl., 266.

Rosenblatt, C.S., Pindak, R., Clark, N.A., Meyer, R.B.: 1977, J. Physique, 38, 1105.

Sethna, J.P. and Kléman, M. : 1982, submitted.

Toulouse, G. and Kléman, M. : 1976, J. Physique, Lett., 37, 149.

INDEX

LIST OF PARTICIPANTS

A. ARAGNOL	Université d'Aix-Marseille II
H. BASART	Collège de France
H. BERESTYCKI	CNRS, Université Paris VI
Y. BOULIGAND	Ecole Pratique des Hautes Etude, Ivry
C.P. BRUTER	Université Paris XII
M. CAHEN	Université Libre de Bruxelles
A. CHENCINER	Université de Paris VII
A. COUETTE	Université de Caen
M. CRUMEYROLLE	Université de Toulouse
P. DOUSSON	Université de Saint-Etienne
J. DUPORT	Université de Chambéry
S. FAUVE	Ecole Normale Supérieure, Paris
J.P. FRANCOISE	IHES, Bures sur Yvette
M. GOLUBITSKY	Université de Californie, Berkeley
A. GRUNBAUM	Université de Californie, Berkeley
S. GUTT	Université Libre de Bruxelles
J. LERAY	Collège de France
A. LICHNEROWICZ	Collège de France

C. LOBRY	Université de Nice
Y. MADAY	Université de Paris XII
C. MARLE	Université de Paris VII
F. MAZAT	Université de Bordeaux
M. MENDES-FRANCE	Université de Bordeaux
F. MIGNOT	Université de PARIS VI
L. NIRENBERG	Courant Institute of Mathematical Science, New York
E. PARDOUX	Université de Provence
F. PELLETIER	Université de Dijon
J. RAPPAZ	Université de Lausanne
C. REDER	Université de Bordeaux
W. SHIH	CNRS, IHES
W. TULCZYJEW	Université de Calgary
J.E. WHITE	Université d'Atlanta